TECHNOLOGY TRANSFER AND THE NEW EU COMPETITION RULES

Intellectual Property Licensing after Modernisation

TECHNOLOGY TRANSFER AND THE NEW EU COMPETITION RULES

Intellectual Property Licensing after Modernisation

STEVEN D. ANDERMAN
JOHN KALLAUGHER

OXFORD
UNIVERSITY PRESS

OXFORD
UNIVERSITY PRESS

Great Clarendon Street, Oxford OX2 6DP

Oxford University Press is a department of the University of Oxford.
It furthers the University's objective of excellence in research, scholarship,
and education by publishing worldwide in

Oxford New York

Auckland Cape Town Dar es Salaam Hong Kong Karachi
Kuala Lumpur Madrid Melbourne Mexico City Nairobi
New Delhi Shanghai Taipei Toronto

With offices in

Argentina Austria Brazil Chile Czech Republic France Greece
Guatemala Hungary Italy Japan Poland Portugal Singapore
South Korea Switzerland Thailand Turkey Ukraine Vietnam

Oxford is a registered trade mark of Oxford University Press
in the UK and in certain other countries

Published in the United States
by Oxford University Press Inc., New York

First published 2006

British Library Cataloguing in Publication Data
Data available

Library of Congress Cataloging in Publication Data
Data available

Typeset by RefineCatch Limited, Bungay, Suffolk
Printed in Great Britain
on acid-free paper by
Antony Rowe Ltd., Chippenham, Wiltshire

ISBN 0–19–928214–5 978–0–19–928214–2

1 3 5 7 9 10 8 6 4 2

CONTENTS—SUMMARY

CONTENTS

TABLE OF CASES

A. Alphabetical—All Cases

B. Numerical—Judgments of European Court of Justice

C. Numerical—Judgments of Court of First Instance

D. Commission Decisions

TABLE OF LEGISLATION, TREATIES AND CONVENTIONS

*Paragraph numbers in **bold** indicate that the text is reproduced in full*

Part I

INTRODUCTION

1

INTRODUCTION

A. Introduction

1.01 The purpose of this book is to explore the implications for technology licensing of the fundamental changes in the application of EU competition law that have occurred over the last 10 years. Not too long ago, applying EC competition law to an intellectual property licence was usually a relatively simple matter of identifying possible 'restrictive' clauses in an agreement and checking them against a block exemption regulation issued by the Commission. To perform this exercise, it was not necessary to understand competition policy or to follow developments in the broader law under Article 81. If a licence conformed to the relevant block exemption, practitioners could be reasonably certain that competition law would not be an issue for the lifetime of the agreement, while outside the block exemption, 'restrictive' provisions were unenforceable and could void the entire agreement. Where no block exemption was available, practitioners simply needed to 'hope for the best', with a few indications from Commission policy pronouncements or decisions in individual cases. Article 82 cases (like the *Magill* case in the late 1980s and early 1990s) could be spectacular, but did not impact on everyday practice in the same way.

1.02 The application of EC competition law to intellectual property licensing has changed dramatically, however, as a result of the Commission's 'modernisation' initiatives. The regulatory structure has changed, so that there is no longer a need for comprehensive block exemptions to save the Commission the labour of assessing every agreement entered into in the EU on an individual basis. More importantly, the underlying policy has evolved. The application of EU competition rules and specifically of Article 81 EC Treaty is now meant to involve an economic analysis of consumer harm and economic benefits, which we describe in the text as the new Article 81 paradigm. The block exemptions, which were a key tool for application of competition law under the old regime, have been reduced to the status of safe harbours—providing 'legal certainty' for agreements where a detailed analysis would almost certainly find no significant competitive harm. As a result of these changes, application of competition rules to intellectual property licensing is no longer a matter of applying a self-contained body of special rules. It requires an understanding of policy and law under Articles 81 and 82, in the context of the special policy concerns raised by intellectual licensing.

1.03 For practitioners, the new rules raise problems in three contexts. The first involves review of new licence agreements for conformity with EU law. The 'bad news' here is that full legal certainty cannot be achieved at this stage, because under the new paradigm the enforceability of many provisions may depend on the evolution of the markets where the parties operate. The 'good news' is that the law is sufficiently uncertain that administrative fines are unlikely as long as the parties have made a good faith effort to ensure compliance. This will mean in particular that the parties need to avoid 'hard core' restrictions like absolute territorial protection for a licensee, unless they are very confident in the economic case for maintaining such provisions.

1.04 The second context in which the new paradigm and new procedural rules affect practitioners involves the defence against investigations by a competition enforcement authority or lawsuits from third parties, such as licensors who claim 'foreclosure' from a market. The complexity of the issues and the difficulty of obtaining evidence will mean that third party lawsuits will probably be rare, unless they supplement or follow an investigation by an enforcement authority. Past experience also suggests that enforcement authority investigations will be rare and are likely to focus more on Article 82 issues than on the enforceability of particular licence clauses outside Article 82. Where such investigations do occur, however, it will be imperative that practitioners (and enforcers) understand the full application of the new paradigm and frame their arguments and their economic assessments accordingly.

1.05 Probably the most important context where the new paradigm changes the rules of the game for practitioners is the all too common case where the parties to an agreement are no longer happy working together according to its terms.

Sometimes this may be the result of disappointment in the results of collaboration. Sometimes this may be the result of one party being acquired by a rival of the other party. Whatever the reason, at this point one party may seek to escape from its obligations under the agreement by invoking competition law concerns. Under the old system, the relevant block exemption regulation was usually determinative. Outside the block exemption the only issue was whether the provisions in question were 'restrictive' according to the formalistic standards then applicable under Article 81(1). Today, however, the space is open for the party seeking to escape a restriction to argue the economic case for invalidity and for the other party to defend the provision on the same basis. At this point both parties need to work with the new paradigm in all its complexity, as will any arbitral tribunal or court called upon to resolve their dispute.

The Commission has issued extensive Guidelines[1] that discuss the application of **1.06** the new paradigm to technology licensing, including the application of the new paradigm to common provisions in intellectual property licensing. The Guidelines represent a considerable achievement for the Commission staff. They have synthesised highly complex material in a relatively short space of time, in a field where the formalistic nature of the old system meant that the Commission had little practical experience. The Guidelines are therefore an essential first port of call for any practitioner or administrator needing to apply the new paradigm. The conclusions reached in this book may sometimes disagree with observations in the Guidelines, but the contribution of the Guidelines to understanding the issues must be acknowledged.

The Guidelines have their drawbacks, however. Due to the Commission's lack **1.07** of experience, the Guidelines cannot always take a definitive view on application of Article 81 in a context that could prejudice future investigative activity. The Guidelines are also difficult to follow in some places where it would be easy to misread their guidance if it is not understood from the perspective of the broader Article 81 paradigm. There are some issues (e.g. definition of 'technology markets' or treatment of block patent licences) where the Guidelines do not address the important issues. There are other areas, particularly in the context of standard-setting and patent pools, where the discussion in the Guidelines is not fully consistent with the analysis required under the new paradigm. Most importantly, the Guidelines do not always provide the kind of practical advice that practitioners require in assessing individual agreements.

This book expands upon and, where appropriate, comments on the approach **1.08** laid out in the Guidelines. It is intended to assist in the application of the new

[1] Guidelines on the application of Article 81 of the EC Treaty to technology transfer agreements (OJ C 101, 27.4.2004). Cited as TT Guidelines.

paradigm in practice by showing how specific types of agreements, contractual provisions, and commercial relationships that involve licensing fit within the general framework for analysis of restrictive agreements under Article 81. It is not possible to identify every provision that can raise issues or to predict in advance how Article 81 should be applied in all circumstances. Once the methodology and the policy concerns are understood, it should be possible to ask the right questions: Where is the competitive harm? Where is the economic benefit? This book may help in answering these questions, or at least in pointing to the criteria that may be relevant to finding the answers.

1.09 The book is divided roughly into four parts. In the first part (Chapters 1 and 2) we address the way in which the new approach to intellectual property in EC competition law fits with the overall policy considerations raised by the relationship between intellectual property and competition law, and the way in which the new system fits (somewhat uneasily) with the previous history of intellectual property regulation under Article 81. In the second part (Chapters 3 to 5) we describe the new Article 81 paradigm, with particular consideration of the central concepts of competitive harm and economic benefit, and explain how this paradigm applies in principle to intellectual property licensing. The third part (Chapters 6 to 9) describes the way in which the new paradigm needs to be applied to licensing in practice, starting with market definition and then covering, in turn, three categories of agreements: traditional 'vertical' licences between non-competitors; licences between competitors; and other forms of competitor collaboration, in particular joint licensing. The final part (Chapter 10) deals with Article 82. Although application of Article 82 has not (yet) been directly affected by the modernisation project, it is addressed here because of the increasing practical significance of Article 82 in the licensing context.

1.10 It should be emphasised that the various chapters are meant to work together to form a coherent whole. It will be difficult to follow the analysis of individual provisions in Chapters 7 to 9 without some feel for the framework of analysis set out earlier. In this regard the paradigm as set out in Chapter 3 is central to the analysis in subsequent chapters. Even readers who are generally familiar with Article 81 may wish to have a look at this discussion, since it breaks down the analysis further than is usual in works on this subject.

B. Background—the Relationship between Competition Law and IP

1.11 The introduction of the new EC paradigm comes at a time when advocates of competition policy and proponents of intellectual property have reached an accommodation that recognises that both competition law and intellectual property rights legislation constitute complementary components of a modern industrial policy.[2] Although both policies pursue the common aim of improving innovation and consumer welfare each does so using rather different means. Intellectual property rights legislation such as patents, copyright and design rights laws offer intellectual property right holders a period of exclusive rights to exploit their property right as both a reward to the individual and as an incentive to the wider process of innovation and R&D investment. Trademarks perform a different function. Their exclusivity is meant to protect the consumer as well as to reward the originator. Modern competition policy attempts to keep markets innovative and competitive by maintaining effective competition. The means it uses to pursue this aim include maintaining access to markets and preventing 'foreclosure' or monopolisation of markets.

At first sight there may seem to be a potential clash in the methods used by the **1.12** two systems of legal regulation to achieve their common aim. The concern to maintain access to markets appears to be implacably opposed to the concept of exclusive rights to make, use and sell a product. And indeed, there was a period when the misunderstanding of the economic effects of intellectual property rights led EC competition law and policy to attempt to place overly strict limits on the exercise of intellectual property rights, particularly in the field of patent licensing.[3] Today, however, the interrelationship between the two systems of law is characterised more by its accommodations than by its conflict. These accommodations tend to occur most often as an incidental result of the ordinary doctrines of each system. Thus, intellectual property laws make a contribution to effective competition and maintaining access to markets by devices within their own internal doctrines that strive to maintain a balance between 'initial' inventors and creators and 'follow on' invention and creation. Good examples are the 'fair use' doctrine in copyright law, the doctrine of 'non-obviousness' and the provision of compulsory licensing in patent law and interoperability imperatives and decompilation rights in the computer program directive. On rare occasions, as in the EC database directive, the accommodation will be explicitly spelt out in the intellectual property law itself.

Within EC competition law the accommodation also tends to occur more **1.13** owing to the incidental effect of the logic of the general doctrines of competition

[2] See e.g. US FTC-DoJ Hearings and Report.

[3] In the period from the 1950s to the 1970s there was a strong concern to limit the scope for licensing because of the influence of the patent misuse doctrines of the US judiciary and the 'Nine No-Nos of the US DoJ/FTC in respect of patent licensing. In the 1960s and 1970s there were internal EC worries about market partitioning heightened by the *Consten and Grundig* Case [1996] ECR 235. During this period little thought was given to the pro-competitive features of IPRs and IP licensing or the damaging economic effects of a restrictive competition policy upon investment in IPRs and IP licensing.

law rather than in the form of special treatment.[4] When one looks at EC competition law the most obvious example of special treatment is the 'exceptional circumstances' test embedded within the abuse of refusal to supply under Article 82.[5] The exceptional circumstances test represents an important acceptance by competition law that intellectual property rights are not the same as all other forms of property rights even while maintaining that the exercise of intellectual property rights must be subject to the regulatory limits of competition policy.[6]

1.14 This observation offers a good perspective for viewing the relationship between Article 81 and intellectual property licensing under the new post-modernised legal framework. The accommodation with intellectual property rights in the new Technology Transfer Block Exemption Regulation (TTBER) and Guidelines occurs almost entirely within the logic of the doctrines of competition law. Most of the accommodation takes place as an incidental benefit of the ordinary interpretation of Article 81 under the modernisation programme. In the Guidelines and Recitals there is evidence that the competition authorities have made a considerable effort to understand the nature of intellectual property rights and intellectual property rights licensing. Thus, they acknowledge that the creation of intellectual property rights often entails substantial investment and that it is often a risky endeavour. They state plainly that '[i]n order not to reduce dynamic competition and to maintain the incentive to innovate, the innovator must not be unduly restricted in the exploitation of the IPR that turn out to be valuable'. In particular, they must be able to seek compensation for successful projects that takes failed projects into account. The Commission also acknowledges that technology licensing may require the licensee to make considerable sunk investments in the licensed technology and production assets

[4] Some good examples of special treatment are offered by the competition legislation of Japan and Australia which creates a specific immunity, or partial immunity, from competition law for the exercise of IPRs.

[5] This doctrine includes the corollary proposition that the 'normal' exercise of IPRs will not abuse a dominant position. Only in extreme cases where IPRs are used unjustifiably by their highly dominant owners to exclude competitors from markets does EC competition policy reserve a right to intervene to limit the exercise of IPRs. It defines the special responsibilities of a dominant firm towards weakened competition so as to allow the dominant IPR owner to compete by preventing copying even if the exercise of this right denies access to primary markets to competitors. Even here, however, the test of 'exceptional circumstances' can be read as an application of general competition law based principles, since justifications based on the commercial imperatives of a market may also block any obligation to supply, even in the absence of IPRs. This question is discussed more fully in Chapter 10.

[6] IPR specialists express reservations about whether competition law has drawn the line at the right position. In terms of the current need to establish an industrial policy that promotes innovation both by IPRs and competition policy, it is certainly necessary to subject the 'exceptional circumstances' test to continuous review to ensure that it represents the best balance between EC competition policy and the exercise of IPRs.

necessary to exploit it.[7] Moreover, the Guidelines have accepted that the great majority of licensing agreements are pro-competitive and compatible with Article 81.[8]

This approach represents a conscious rejection of the argument that a special self-contained regime (like the old block exemptions) is necessary to satisfy the special requirements of intellectual property licensing. As the Guidelines confidently proclaim '[in] assessing licensing agreements under Article 81, the existing analytical framework is sufficiently flexible to take due account of the dynamic aspects of technology licensing'. The new framework clearly harmonises the treatment of licensing agreements with that of other commercial agreements under Article 81 and, on the whole, this results in a reasonable treatment of intellectual property licensing. However, as we shall see, the process of harmonisation has not been complete; the analytical framework has had to make certain adjustments to take due account of the special features of technology licensing. **1.15**

C. The New Legal Framework

The new TTBER and Guidelines are an integral part of the broader reform of practice and policy under Article 81. As part of its modernisation reform, the Commission has chosen a form of accommodation between competition law and intellectual property rights licensing which fits the assessment of licensing agreements predominantly within the new general legal framework of Article 81.[9] The modernisation project consists of a new procedural regulation (Regulation 1/2003), three framework BERs and at least four Commission Notices. The new regime is characterised by four features: **1.16**

1. The enforcement of Article 81 as a whole has been decentralised. The national authorities and courts now have jurisdiction to apply the balancing test of the four conditions in Article 81(3) as well as 81(1). In the interests of modernisation, the Commission has relinquished its longstanding monopoly over Article 81(3).
2. There is no longer any need for formal application for exemption under Article 81(3) to ensure provisional validity. The legal status of agreements will be judged at any time on the basis of the compatibility of the agreement

[7] Technology Transfer Guidelines para 8.

[8] Guidelines para 9. See too para 17 in which the Commission sets out the pro-competitive potential of licensing agreements.

[9] Indeed, the new TTBER and Guidelines have been brought into being two years early to fit within the overall project to modernise the application of Article 81 to agreements.

and its provisions with the prohibition of Article 81, in particular Article 81(3). The Guidelines to the application of Article 81(3) apply to all the modernised BERs.

3. It changes the application of Article 81(1) to agreements by requiring more of an *economic analysis* of the issue whether agreements have the object or *effect* of restricting competition.
4. The BERs now only offer a safe harbour and not the legal certainty of the old style BERs. Previous BERs in the IP licensing field conferred an exemption on the parties without market shares and subject only to a formal power of withdrawal by the Commission or a national competition authority. If there was a blacklisted provision in the agreement, the agreement was unexemptible and unenforceable. In the current group of BERs, including the TTBER, the BER takes a reduced role within the legal framework. Today, if an agreement is outside the scope of the safe harbour offered by the BER, the parties may nevertheless find a basis for exemption under Article 81(3) using the general 81(3) guidelines and the guidelines for the specific type of agreement.

1.17 This fourth feature of the new system is particularly important in the context of licensing. The most important thing that any practitioner familiar with the old system must recognise is that *the new block exemption is no longer useful as a source of legal protection on its own.* The block exemption is only applicable to standard licence arrangements where the licensor has less than a 30% share in the market for the licensed technology *and* where the licensee has less than a 30% market share in the downstream market where the technology is used. (Where the agreement is between competitors the relevant thresholds drop to 20%.) If the relevant market share of either the licensor or licensee exceeds this level, the TTBER ceases to be applicable. Yet, as the detailed discussion below will show, Article 81 will almost never be applicable to agreements where both parties are below these market share levels in any event. As a result, detailed study of the TTBER and the formal conditions for its application is not the key for the practical application of Article 81.[10]

1.18 The TTBER is important, however, because it identifies certain restrictions that are regarded as 'restrictive by object'. These restrictions, which are listed in Article 2 of the TTBER, are treated as 'hard core' restraints. They include fixing of resale prices, restrictions on sales to identified territories or customers (with some exceptions), and certain restrictions on research and development by licensees. Regardless of the market position of the licensor and licensee, these

[10] For readers that desire a more extensive summary of the terms of the TTBER, a detailed discussion is attached as Appendix 1.

provisions are almost certain to infringe Article 81. It is therefore imperative for practitioners reviewing an agreement for compliance with Article 81 to refer to the list in Article 2 TTBER.

The Guidelines are the real centrepiece of the reformed treatment of technology transfer under Article 81. The Guidelines commence with a summary of the general principles applicable under the new paradigm. The Guidelines continue with a detailed and useful summary of the TTBER and the conditions for its application. They then lay out a 'general framework for analysis'—applying the general principles set out in the first section to intellectual property licensing. They conclude by applying this analytical framework to a list of specific restrictions commonly found in IP licences. The central theme of the Guidelines is the need to identify competitive harm and to identify economic benefits that might outweigh those competitive harms in order to determine whether a licence agreement raises Article 81 issues. **1.19**

The system of assessment created by the Guidelines entails a fundamental change in the way that practitioners must address competition law issues in the intellectual property context. In place of a purely legalistic evaluation of the contents of licensing agreements, the new system demands a new type of legal and economic assessment of individual agreements. The old system offered legal certainty and this was important for technology transfer agreements because of the ever present risk that licensees might go into competition with their licensors once they mastered the problems of efficient manufacture of the new technology. However, it also created a tight corset or legal straitjacket. The parties had to adjust their commercial arrangements to fit the legal rules or forgo the benefits of legally certain enforceability of the commercial agreement throughout its duration. For those commercial agreements whose curves could fit within the corset of the BER, there were considerable benefits. However, for those commercial agreements whose contours were either too angular or too plump to fit within the golden corset, there was a form of legal limbo. **1.20**

The new reform entails some benefits to intellectual property owners. First, the Commission has finally recognised that 'vertical' licensing agreements, or agreements between non-competitors should systematically be treated more leniently than 'horizontal' licensing agreements, agreements between competitors. A lighter regulatory burden on purely vertical licensing agreements is entirely appropriate because such agreements present fewer risks to competition. Secondly, there are benefits for the parties to licensing in the reduced list of hard core restrictions in the new TTBER. These changes have meant that the new framework offers considerably greater flexibility to the parties to draft intellectual property licensing agreements to reflect their preferred underlying commercial bargain. No longer is it so necessary to distort commercial arrangements **1.21** to fit the strictly defined categories offered by the existing BERs. The Guidelines to the interpretation of the TTBER as well as to the more general application of Article 81 to licensing agreements outside the scope of the TTBER are also an improvement in the regulatory framework.

1.22 However, these improvements in the regulatory structure are accompanied by serious costs. First, and most obvious, has been the loss of legal certainty in the applicability of the safe harbour of the TTBER owing to its market share limits. Under the previous TTBER an agreement normally retained its exemption for its entire duration.[11] This allowed lawyers to enjoy the authority to determine the lawfulness of the licensing agreement with some finality. That type of legal certainty has disappeared to be replaced by a legal structure requiring a new methodology to ensure continued legal enforceability for intellectual property licensing agreements.

1.23 Second, the new legal methodology itself requires a considerable adjustment by lawyers to a new methodology for assessing licensing agreements for the purpose of exemption. Since the TTBER has practically no value other than as a reference for identifying hard core restraints, effective counselling requires lawyers to learn the new skills required by the new paradigm as applied in the Guidelines. This in turn requires lawyers to accept the need to combine legal and economic analysis in vetting licensing agreements under Article 81.

---

[11] The exemption could be withdrawn only by a formal withdrawal by the Commission or the competition authority of a Member State.

2

THE DEVELOPMENT OF THE REGULATORY FRAMEWORK FOR IP LICENSING IN THE EC

A. Introduction

The new modernised TTBER and Guidelines constitute a third stage in the development of EC competition policy towards intellectual property licensing. This chapter examines the two previous stages of regulatory policy and judicial decision to help to explain certain features of the current stage of law and policy. **2.01**

The first stage of Commission policy, the period before the introduction of the first Block Exemption Regulation in 1984, veered from benign neglect to excessive intervention. Initially, in the 1960s and early 1970s, licensing was treated leniently under EC Competition law as the Commission's Notice on Patent Licensing Agreements of 1962 (the so-called 'Christmas Message')[1] announced a policy of non-interference. By the 1970s, however, the mood had changed **2.02**

[1] It could also be seen in the earlier formulation of Article 4(2) of Regulation 17/62 which provided that notification was not required with respect to bilateral agreements which only imposed restrictions on the exercise of the rights of the assignee or user of intellectual property rights.

quite dramatically. The Commission's decision to monitor exclusive licensing agreements due to concerns about market partitioning after the *Grundig* case was reinforced by the 'Nine No-Nos' policy of the US Department of Justice Antitrust Division in respect of patent licensing. During this period little thought was given to the pro-competitive features of intellectual property rights and intellectual property licensing or the damaging economic effects of a restrictive competition policy upon investment in intellectual property rights. It was left to the European Court of Justice to lead the way to a more moderate view of the reach of Article 81(1) in a series of cases redefining the scope of restriction of competition as it applied to licensing agreements.

2.03 The second stage was marked by the introduction of the first block exemption regulation for patent licensing agreements. By the mid-1980s, a number of the Commission's concerns had been lessened by important changes in competition policy. There was an important change in approach within the USA as the severe criticisms of the Chicago School led to a more economically realistic assessment of the market power of patent holders by the courts and the legislators and the US Department of Justice quite elaborately dismantled the regime of the Nine No-Nos. This intellectual property friendlier policy in the USA coincided with the acceptance of vertical exclusive intellectual property licences as potentially pro-competitive in the *Nungesser* case in the ECJ and the adoption of the first Patent Licensing BER European Commission and Council in 1984. The first Know-how BER followed in 1989. During this period, too, EC concerns about market partitioning were partially eased by the growth of the doctrine of exhaustion and the build-up of more established pathways of interstate trade. The period culminated with the unified 1996 Technology Transfer Block Exemption Regulation. Shortly before, in 1995, the US Department of Justice and the Federal Trade Commission produced their Antitrust Guidelines to the Licensing of Intellectual Property, offering a rather different regulatory model, but this had a very limited effect on the 1996 exercise in the EC. It was to take until the 2004 TTBER before the lessons of the US Guidelines were more fully absorbed by the EC administrators.

B. Stage I: Before the BERs

(1) The pre-Grundig years

2.04 Intellectual property rights licensing initially enjoyed a special position under Article 81(1) in the period prior to the *Grundig*[2] case. The Commission

---

[2] [1966] ECR 299.

accepted that exclusive patent licensing agreements, unlike exclusive distribution agreements, could be viewed as not restrictive of competition as long as the contents of the licence remained within the 'scope of the patent'. Certain practices which extended the original grant of exclusivity, such as tie-ins of non-patented products with patented products; price and territorial restrictions on the sale of a patented product; post-expiration royalties; licensee veto of other licensees; and territorial restrictions on a sale of unpatented products made with a patented process were viewed as a misuse of the patent and illegal *per se*. Other licensing practices that amounted merely to a limited licence of the powers conferred by the patent were viewed as not restrictive of competition under Article 81(1). At this stage, the Commission was mainly concerned with provisions in the licensing agreement that attempted to extend the economic power of the licensor beyond its inherent scope.[3] As long as the licence conferred merely a subdivision of a licensor's rights granted by legislation, it deprived no third party of their freedom of action even if the contractual restriction conferred an exclusive right upon the licensee.

The influence of this doctrine could be seen in the Commission's Christmas Message Notice of 1962.[4] This Notice provided that a patent licence containing only limitations as to technical applications or fields of use, quantity of products to be manufactured, or restrictions on time, or persons to carry out the licensing role would not be caught by Article 81(1). These restrictions were viewed as essentially divisions of the patentee's grant of exclusive rights. Hence, it accepted that licensors involved in a single licensing relationship could give an 'exclusive' licence to a licensee for a particular territory, i.e. a licence which restricted the licensor not only from appointing any other licensee in that territory but also from making, using or selling itself in that territory, and not be caught by Article 81(1). The implication, at this stage, was that an agreement which divided the right under a patent to manufacture, use, and sell a protected product geographically came within its scope just as much as a division of technical application. Neither provision was thought to fall within the prohibition of Article 81(1). **2.05**

The Christmas Message also accepted that certain other obligations in licensing agreements were not caught by Article 81(1), namely: (a) the obligation to mark the products with an indication of the product, (b) the obligation to maintain quality standards or procure supplies of products indispensable to the technically **2.06**

[3] First Report on Competition Policy, 1972, at pp 65–74.

[4] It could also be seen in the earlier formulation of Article 4(2) of Regulation 17/62 which provided that notification was not required with respect to bilateral agreements which only imposed restrictions on the exercise of the rights of the assignee or user of intellectual property rights.

perfect exploitation of the patent, and (c) non-exclusive grant backs of know-how or licences of inventions. These were regarded as indispensable in view of the shared nature of the intellectual property right in a licensing relationship.

2.07 However, the Christmas Message Notice plainly indicated that other types of clauses had to be examined individually and did not provide a general clearance for more complex licensing arrangements such as joint ownership of patents, reciprocal licences, multiple parallel licences, or to any clauses attempting to extend beyond the period of validity of the patent. Indeed it implied that agreements with the latter clause might have difficulties in getting through the Article 81(3) barrier.

2.08 The Commission's reasons for clearing patent licensing agreements in this way were pragmatic. It thought, at the time, that patent licences were 'not likely to affect trade between Member States as things stand in the Community at present'. Moreover, it was preoccupied with the issues of formulating a block exemption for distribution agreements after *Grundig* and would not have welcomed a flood of notifications of patent licensing agreements.[5] In the event, the scope of the patent doctrine, as incorporated in the Notice, helped to shelter patent licensing from the effects of the Commission's 'freedom of action' doctrine as it applied to exclusive distribution agreements throughout the 1960s.

(2) The post-Grundig years

2.09 In the early 1970s, the attitude of the Commission to licensing of intellectual property rights dramatically changed. This was partly due to fears of the potential of intellectual property rights licences to seal off markets and limit interstate competition created by the *Grundig* decision.[6] Yet the Commission failed to draw a distinction between exclusive territorial manufacturing licences and territorial sales restrictions; instead, it chose to apply *per se* prohibitions to both alike.

2.10 In *Grundig*, the Court effectively widened the scope of Article 81(1) by defining it to apply to market partitioning agreements. Thus it stated that the infringement of Article 81 consisted of the attempt by the licensor and licensee to isolate 'the French market for Grundig products and maintain . . . artificially, for products of a very well known brand, separate national markets within the community'. This was prohibited under Article 81 as an attempt 'to distort competition in the common market'.[7]

2.11 In the course of the *Grundig* litigation, the case was made that the distribution

[5] Alexander, W, 'Patent Licensing Agreements in the EC' [1986] IIC 1.
[6] [1966] ECR 299. [7] Ibid.

agreement was not a restriction of competition because it added a new competitor to the French market to compete with other brands. Consten and Grundig argued that the Commission should have based its approach to Article 81(1) upon a rule of reason and considered the economic effects of the disputed contracts on competition between the different brands. More specifically, it should have taken into consideration the fact that a vertical sole dealer relationship by introducing a new competitor into the French market increased inter-brand competition in that market. The Court accepted that 'competition between producers is generally more noticeable than that between distributors of the same make'. However, it was particularly concerned to exclude this as a mitigating factor in a situation where the agreement inhibited interstate trade. As the Court stated:

> 'The principle of freedom of competition concerns the various stages and manifestations of competition. Although competition between producers is generally more noticeable than that between distributors of the same make, it does not . . . follow that an agreement tending to restrict the latter kind of competition should escape the prohibition in Article [81(1)] merely because it might increase the former.'[8]

This decision effectively grouped vertical agreements together with horizontal agreements as equal subjects of EC competition law regulation.

In the period after *Grundig*, and with the Exclusive Distribution Block Exemption Regulation 1968/68 approved, the Commission decided that it could turn to the need to monitor restrictions in networks of bilateral licensing agreements that affected interstate trade. The large number of licensing agreements which had been notified to it had made it plain that networks of multiple parallel licensing agreements created exclusive territories which coincided with the territory of Member States. Yet the exclusive territorial rights to manufacture did not in themselves contribute to the sealing-off effects. It was the accompanying territorial sales restrictions. **2.12**

Where an exclusive licensing agreement restricted to one territory is part of a licensing network, there is always the potential for competition from other licensees who have been allocated exclusive territories. The right to manufacture in and sell from a particular territory did not restrict the free movement of goods once the good was sold anywhere in the Common Market with the consent of its owner.[9] Consequently, licensees can engage in direct sales to other licensees' territories, whether actively by advertising or passively by responding **2.13**

---

[8] Ibid.

[9] See e.g. Beier, F-K, 'Industrial Property and the Free Movement of Goods in the Internal Market' [1990] IIC 131.

to unsolicited orders. Moreover, the customers of a licensee can engage in parallel trading, exporting and importing the licensed goods and services between the different territories of the licensees as well as that of the licensor. In such a situation, while territorial sales restrictions imposed on the licensees might reduce the flow of interstate trade, exclusive territorial restrictions on manufacture had no effect as such on direct or parallel sales. The only effect of an exclusive licence is to limit the number of licensees with a right to manufacture and sell in the protected area.

2.14 Nevertheless, the Commission decided to change its position and state that where the owner of a patent conferred an exclusive right to another undertaking in an assigned area, 'he loses the freedom to enter into agreements with other applicants. The exclusive character of such a licence may amount to a restriction of competition and thus fall within the category of prohibited agreements in so far as it has an appreciable effect on market conditions.'[10] From that point on, licensing agreements with exclusive territorial protection were viewed as restricting competition under Article 81(1) unless they had no appreciable effect.

2.15 In *Davidson Rubber*,[11] a licence network involving exclusive licensees in West Germany and France with market shares of 20% and 40% respectively, was held to be caught by Article 81(1). The Commission reasoned that the restriction on the licensor's ability to appoint other licensees limited the freedom of potential suppliers and exporters of the licensed product. As the Commission saw it, 'but for the exclusive rights Davidson might have granted a licence for the same area to someone else and, if it had, the licensee might have exported to other member states'. The Commission was prepared to grant an exemption because the exclusivity was necessary to induce the licensee to take on the risks in investing in the manufacturing facilities for a new product, but this meant that the agreement would have had to be notified when made.

2.16 With the Commission's Christmas Message Notice made suspect by these decisions, there was no longer an easy method of clearance for patent licences. The parties to licensing agreements now found themselves in a similar position to the parties to exclusive distribution agreements but without the benefit of the exclusive dealing block exemption. Industry was now faced with the need to notify patent licensing agreements to the Commission in order to obtain certainty as to their legal validity, and responded by engaging in intensive lobbying

---

10 See Commission's First Report on Competition Policy.
11 OJ EC L 143/31 (1972).

of national governments. There were also efforts made to introduce provisions in the drafting of the Community Patent Convention in 1975 to offset the Commission's policy.[12] Article 43(1) of the draft Convention provided that a Community patent may be licensed in whole or in part for the whole or parts of the territories in which it is effective and that a licence may be exclusive or non-exclusive. Article 43(2) then added that the rights conferred by the Community patent might be invoked against a licensee who contravenes any restriction in his licence which is covered by Article 43(1). These were meant to be a signal that such restrictions were not contrary to the Treaty.

The Commission responded to this concern by preparing a new patent licensing block exemption to deal with the expected large numbers of notifications. It also introduced a Notice on sub contracting agreements to help certain types of subcontracting agreements involving licences to be cleared by a group Notice.[13] It was not prepared to bow to the lobbying pressure in respect of territorial exclusivity under Article 81(1). That exclusivity provided a necessary incentive to the licensee to penetrate a new market which had not been, and could not be, exploited by the licensor was viewed as a basis only for exemption under Article 85(3) and not for clearance under Article 81(1). **2.17**

The Commission also decided that the obligation upon the licensor not to sell into the licensed territory and the obligation upon the licensee not to sell the licensed product to any country in which the licensor had licensed or assigned the product were normally prohibited under Article 81(1). The first restriction could be exempted provided it was limited to the first sale and otherwise met the test of the four conditions of Article 81(3). The second could be exempted provided that the prohibition applied to direct sales only, was limited in time and met the test of Article 81(3), in particular the indispensability requirement.[14] In its Fourth Report on Competition Policy in 1975, the Commission attempted to reconcile its decisions with the earlier pronouncements of the Court of Justice concerning intellectual property rights. **2.18**

On a legal plane, the Commission faced the problems exposed by the Court of Justice in its distinction between the existence of nationally protected industrial property rights, which is not to be affected by community law, and the exercise of these rights, which can be subject to the Treaty rules. The assessment of patent licensing agreements under the Treaty called upon a consideration of interests and issues that went beyond the field of competition **2.19**

[12] Alexander, W, 'Patent Licensing Agreements in the EC' [1986] IIC 1.
[13] Notice on Contractors and Subcontractors [1979] 1 CMLR 264.
[14] Patent Licensing Block Exemption Regulation 2349/84.

policy.[15] Export restrictions, for example required an exemption under Article 81(3) and thus were beyond the scope of the patent.[16] The Commission seemed to treat the sharing of exclusivity geographically as outside the scope of the patent without applying an appreciability test under Article 81(1).

2.20 During this period too the Commission adjusted the concept of the 'scope of the patent' in respect of non-territorial restrictions in order to meet competition concerns. In *AOIP* v *Beyrard*[16A] for example, the Commission found that certain non-territorial provisions were prohibited by Article 81(1), for example the no-challenge clause, non-competition clause, the obligation to pay royalties during the lifetime of the most recent original or improvement patent, whether or not the original patent was being exploited by the licensee. These were all viewed as going beyond the scope of the patent and not only caught by Article 81(1), but also inherently non-exemptible under Article 81(3).[17]

2.21 In its Fourth Report, the Commission indicated that subdivisions of the grant of the intellectual property right could be compatible with Article 81(1) subject to the absence of competition effects. For example, restrictions on the duration of the licence were allowed within the frame of the duration of the original grant. Restrictions on field of use were allowed but only as long as they were not disguised market sharing arrangements. Quantitative output restrictions were caught by Article 81(1).

2.22 This restrictive policy towards both territorial and non-territorial restraints in licensing agreements was reflected in the draft block exemption for patent licensing officially published in 1979. It allowed territorial exclusivity to all licensees, but restricted all territorial sales restraints and export bans to SMEs only, defining these as undertakings with a turnover of less than 100 million ECD. It incorporated a 'blacklist' of 14 clauses, the presence of any of which would preclude the application of the block exemption. It also offered a 'whitelist' of nine restraints which would not prevent exemption which were more based on the test of clearance under Article 85(1), stating only that they may or may not be caught under Article 85(1).

---

[15] Annex to *Seventh General Report on the Activities of the Communities*, 1975, at points 19 and 20.

[16] In *Kabelmetal-Luchaire* [1975] OJ EC L 222/34 the Commission held that a contractual prohibition on direct sales by the licensee outside its territory was prohibited by Art 85(1).

[16A] [1976] 1 CMLR D14.

[17] Similarly, in *Vaessen/Morris* [1979] OJ L 19/32 the Commission found that a tie-in of certain products to a patented product and process was prohibited by Art 85(1) because it was not essential to the proper exploitation of the patent. Other products were perfectly adequate and the effect of the tie-in was to deprive the licensee of its business freedom to obtain its supplies elsewhere, if it chose. We shall explore the development of the law of non-territorial restraints in Chapter 8.

A factor that appears to have reinforced the Commission's hard line towards non-territorial restrictions in patent licensing agreements was the knowledge that the Antitrust Division of the US Justice Department, under the influence of the 'patent misuse' doctrine developed by the US Supreme Court,[18] had developed a set of administrative guidelines to antitrust enforcement in the form of Nine No-Nos for patent licensing.[19] The following restrictions were viewed as *per se* unlawful: **2.23**

1. Tie-ins
2. Grant-backs
3. Resale restraints
4. Tie-outs
5. Licensee vetoes
6. Mandatory package licences
7. Royalties not reasonably related to the licensees' sales of the patented product
8. Restrictions on the sale of an unpatented product manufactured with a patented process
9. Price restrictions on sales of a licensed product[20]

These restraints were not subject to a rule of reason. At this stage, the Justice Department appeared to be unconcerned about the effects of strict antitrust rules upon the process of innovation.[21]

The influence of the Nine No-Nos reached into the European Commission in the preparation of the draft block exemption. As Hartmut Johannes, the administrator within DGIV (the Competition Directorate) with responsibility for intellectual property rights, put it in 1978, 'In the art of antitrust, the Americans are the teachers and the Europeans are the pupils.' He indicated that US thinking was not accepted uncritically; in particular its *per se* rules and the horizontal/vertical distinction were not transferable. However, he also acknowledged that when the Commission prepared its draft patent licensing block **2.24**

[18] Under the impetus of Mr Justice Douglas. See e.g. *Morton Salt Co* v *G S Suppiger Co* 314 US 488 (1941); *United States* v *United States Gypsum Co.* 333 US 364 (1947).

[19] 28 Wilson, Dept of Justice Luncheon Speech, 'Law on Licensing Practice—Myth or Reality Straight Talk from Alice in Wonderland'.

[20] Oppenheim, Weston and McCarthy, *Federal Antitrust Laws Tenth Commentary*, 4th ed, 1981, pp 885–887.

[21] In 1967, the Assistant Attorney General announced, 'I do not believe that the impact of antitrust on patent licensing restrictions has any effect on innovative activity whatsoever . . . to me it seems probably so even if antitrust law were to go so far as to prohibit not only price fixing but also field of use restrictions, quantity restrictions, territorial restrictions, and any other restriction on the complete freedom of the licensee.' See Turner, 'Antitrust and Innovation' (1967) 12 *Antitrust Bulletin* 277 at p 281.

exemption. 'It knew the no nos but of course not the American critics of these no nos; the critics dated from 1981.'[22]

2.25 For intellectual property right holders in the EC, however, the main worries created by the Commission's policy surrounded its refusal to accept the logic of the need for protection of new licensees against competition from other licensees as well as the licensor during the period the licensee was tooling up and familiarising itself with the new technology. Industry was uneasy because it was faced with a pervasive notification requirement for intellectual property rights licences until an appropriate block exemption could be prepared. The Commission's policy at this stage appeared to be doubly unfriendly to innovation. It insisted upon an automatic prohibition of exclusive territorial licences under Article 81(1), in spite of the costs and delays of notification. Yet in its draft block exemption it appeared to be unwilling to translate Article 81(3) to allow protection for territorial sales restrictions to any but the smallest firms.

2.26 EC competition policy at that stage was so focused on monitoring exclusivity in licensing agreements that it risked deterring investment in innovation and technology transfer. It appeared to ignore the need under the Treaty to strike a balance between the protection of interstate trade and the promotion of the diffusion of the manufacture of new technologies throughout the common market.

(3) The Court and the concept of restriction of competition

2.27 In the early 1980s, the Court of Justice took the lead in easing the competition law framework for intellectual property rights licensing. The Court had already indicated in *STM*[23] that territorial exclusivity in vertical agreements should be dealt with by a test of qualitative appreciability. In the landmark decision of *Nungesser* v *Commission*,[24] the Court of Justice introduced the concept that exclusivity could be an indispensable inducement to the licensee to invest in new technology in a new territory. The case concerned an assignment of

---

[22] This was a reference to the groundswell in antitrust thinking that had been introduced by a combination of the writings of Chicago School adherents such as Bork and Posner and the experience of the administrators within the Justice Department. See Fox, E, 'The New American Competition Policy—from Antitrust to Pro-Efficiency' [1981] 2 ECLR 439; see also e.g. Bark, R, *An Antitrust Paradox: A Policy at War with Itself*, 2nd ed, 1978 reissued 1993, Basic Books, New York; Posner, R, *Antitrust Law: An Economic Perspective*, 1976, University of Chicago Press. By 1987 the Department's views had begun to change quite radically. It signalled the change by suggesting that there was a need to differentiate between horizontal and vertical restraints and to move away from a formalistic approach to an economic approach consisting of an evaluation of licensing restraints. See Lipsky, Abbot, 'Current Antitrust Division Views on Patent Licensing Practices', Antitrust Section of the ABA, 15 Nov. 1981, 1981 *Trade Regulation Reporter* (CCH) 55, 985.

[23] *STM* v *Maschinenbau Ulm* [1966] ECR 235.

[24] [1982] ECR 2015.

breeders' rights to a maize seed variety by INRA, a French research institution, to Eisele who registered the right in Germany. Under the agreement, Eisele was given an exclusive right to manufacture and sell INRA's maize seed variety in Germany. In addition, INRA agreed itself to refrain, and to prevent others, from importing its maize seed variety in Germany. Eisele had relied upon its intellectual property right to prevent a parallel importer from importing into Germany from another French source and had obtained a court approved settlement. Receiving a complaint, the Commission decided, following its policy of the 1970s, that all provisions of the agreement were caught by Article 81(1) and could not be exempted under Article 81(3). This meant that the Commission was prepared to treat exclusive territorial licences to manufacture and sell as equally anticompetitive as territorial sales restrictions under Article 81(1). Both required exemption under Article 81(3). The Commission's view was that the exclusive territorial licences were restrictive of the freedom of action of the parties to the agreement and had an adverse effect on the freedom of action of third parties to import and export the maize seed between France and Germany.

2.28 The Court objected to the Commission's insistence that an exclusivity clause in a vertical agreement such as an intellectual property right licence was by its nature caught by Article 81(1) as a restriction on competition,[25] because this gave insufficient weight to the fact that the new seed technology could only move to new Member States if the exclusive territorial restriction was part of the contract and was the only means to promote competition.

2.29 It thought that there was a need to distinguish between exclusive licences to manufacture and sell in a particular territory on the one hand, and territorial sales restrictions that consist of a contractual obligation on other licensees not to sell directly into a protected territory on the other. The first case, which it called an 'open exclusive licence' was one where 'the exclusivity of the licence relates solely to the contractual relationship between the owner of the right and the licensee, whereby the owner merely undertakes not to grant other licences in respect of the same territory and not to compete himself with the licensee on that territory'. In the second case, which it called a 'closed exclusive licence', the exclusive licence is joined with territorial sales restrictions and intellectual property rights to create an absolute territorial protection under which the parties to the contract propose, as regards the products and the territory in question, to eliminate all competition from third parties, such as parallel importers or licensees for other territories.[26]

2.30 The Court held that an open exclusive licence, i.e. an exclusive licence to

[25] Ibid. at para 2 (c). [26] Ibid. at para 53.

manufacture and sell in a particular territory, was 'not in itself incompatible with Article 81(1)',[27] reasoning as follows:

> 'in case of a licence of breeders' rights over hybrid seeds newly developed in one Member State, an undertaking established in another Member State which was not certain that it would not encounter competition from other licensees for the territory granted to it, or from the owner of the right himself, might be deterred from accepting the risk of cultivating and marketing that product, such a result would be damaging to the dissemination of a new technology and would prejudice competition in the Community between the new product and similar existing products.'

2.31 One problem with the exception for open exclusive licences as defined by the Court was that it only allowed for a limited form of protection from the other manufacturers within the licence network.[28] Any[29] attempt to place limits on other licensees, as well as on the more obvious category of parallel importers, not to sell into the territory would take the licence outside the scope of an open licence.[30] Moreover, although the Court defined 'open exclusive' licences as a category that is not automatically caught by Article 81(1), it did not hold that all such licences were cleared. The Court signalled quite clearly that the exception was to be carefully circumscribed to cases where the technology was new, the commercial risks would have deterred the licensee in the particular case, and the protection offered consisted only in territorial exclusivity for manufacturing and selling purposes and did not include territorial sales restrictions. In taking this decision, the Court was not concerned to draw a line between intellectual property rights and competition policy as such, but rather between the need for adequate incentives for the integration of manufacturing processes and the need to protect interstate trade. No mention was made of the existence/exercise distinction or the specific subject-matter issue. The case turned on the specific nature of the technology and the extent of protection against risks that was necessary to achieve its dissemination.[31]

---

[27] [1982] ECR 2015 at para 53. The Court stated that where an exclusive licence of plant seed varieties 'does not affect the position of third parties such as parallel importers and licensees for other territories, the exclusive licence' it is not in itself incompatible with Article 81(1) of the EEC Treaty.

[28] Thus, the licensee could have protection under the contract against either the licensor or other licensees locating manufacturing operations in the territory. The Court reasoned that this limited form of security to the licensee was necessary to induce risky investment and it offered no contractual protection against technical products sold in other territories and imported into the territory.

[29] See Siragusa, 'EEC Technology Transfers—A Private View' [1982] *Fordham Cor Law Ins* 116–118.

[30] [1982] ECR 2015 at para 58.

[31] See e.g. Peeters, Jan, 'The Rule of Reason Revisited: Prohibition on Restraints of Competition in the Sherman Act and the EEC Treaty', 37 *The American Journal of Comparative Law* 521 at p 566.

In *Coditel II*,[32] the Court went a step further and held that an agreement conferring an exclusive right to exhibit a film for a specified period in the territory of the Member States with absolute territorial protection is not necessarily caught by Article 85(1). In the *Coditel* cases, a Belgian company, Ciné-Vog Films, had obtained exclusive exhibition rights to a French film, *Le Boucher*, for a five year period. It discovered that Coditel, a group of Belgian cable companies, had obtained the film from a transmission broadcast by the German licensee with exclusive rights in Germany. In *Coditel I*,[33] Ciné-Vog was able to obtain an injunction to prohibit Coditel from transmitting the film on Belgian television without offending Article 59 (free movement of services). Coditel had argued that such exclusive rights limited to Member States would result in a partitioning of the common market as regards the undertaking of economic action in the film industry. The Court, however, took note of the specific characteristics of the product, i.e. that it was a literary or artistic work available to the public that may be infinitely repeated and presented problems of copyright different from other works such as books which are circulated in a material form. It also took note of the contract which stipulated that the right to show the film on Belgian television could not be exercised until a fixed number of months after the first showing of the film in cinemas in Belgium. The Court considered that this highlighted the fact that the right of a copyright owner and his assigns to require fees for any showing is part of the essential function of copyright in this type of literary and artistic work. The Court reasoned that the effect of these factors was that: while copyright entails the right to demand fees for any showing or performance, the rules in the Treaty cannot in principle constitute an obstacle to the geographical limits which the parties to a contract of assignment have agreed upon in order to protect the author and his assigns in this regard. The mere fact that these geographical limits may coincide with national frontiers does not point to a different solution in situations where television is organised in the Member States largely on the basis of legal broadcasting monopolies.[34] The Court held that the assignee could rely on his performing right to prohibit an unauthorised showing by cable television without offending Article 59. **2.32**

In *Coditel II* the cable company group Coditel appealed from the decision of a Belgian court awarding damages to Ciné-Vog for the unauthorised transmission. The theory of Coditel on this occasion was that Article 85 could be applied to the agreement and that its prohibition applied to the exclusive licence or assignment of the copyright as an improper exercise as the means of a restrictive agreement. The Court accepted that in principle the exercise of an intellectual **2.33**

[32] [1982] ECR 3381. [33] [1980] ECR 881. [34] Ibid. at para 16.

property right in the form of an agreement may be incompatible with Article 85 but it held that:

> 'the mere fact that the owner of a copyright in a firm has granted to a sole licensee the exclusive right to exhibit that film in the territory of a Member State and consequently to prohibit, during a specified period, its showing by others is not sufficient to justify the finding that a contract must be regarded as the purpose, the means or the result of an agreement, decision or concerted practice prohibited by the Treaty.'[35]

2.34 The Court was willing to allow the licensor to bestow absolute territorial protection upon the licensee because the industry was engaged in an integrative activity and investment in licensing the product protected by the intellectual property right could only be secured if the full territorial protection was allowed. The Court reiterated the distinction it drew in *Coditel I*, i.e. between licences of works which are performing rights which can be infinitely repeated without anything physical being put on the market and other types of literary and artistic works such as books or records which cannot be circulated in the market apart from their material form.[36] In the Article 85(1) context, this was not so much an attentiveness to the nature of the intellectual property right; rather, the Court was willing to justify the absolute territorial protection on the basis that it was necessary to prevent the purpose of the licensing agreement from being entirely frustrated.[37]

2.35 The Court's concern for the protection of the intellectual property right and a need to ensure a just return on the investment made in developing the industrial property should also be seen as prompted more by the need to ensure that the film industry continued to penetrate national markets than by an insistence upon allowing inventors to appropriate their just return. The Court indicated that its indispensability test for territorial protection was related to the special 'characteristics of the [film] industry and its markets in the Community, especially those relating to dubbing and subtitling for the benefit of different language groups, to the possibilities of television broadcasts, and to the system of financing [film] production in Europe'.[38] In other words, where the financing of the integration of film production through the medium of television throughout Europe would be jeopardised by inadequate intellectual property rights protection, and the only adequate protection was absolute territorial protection, the Court was prepared to allow such protection to ensure that the process of interstate film distribution would actually take place.

---

[35] [1982] ECR 3381. [36] Ibid.

[37] See Joliet, R, 'Territorial and Exclusive Trade Mark Licensing under the EEC Law of Competition' [1984] IIC 21; Joliet, R, 'Trademark Licensing Agreements under the EEC Law of Competition' [1983–5] *Northwest Jnl of Int' Law and Business* 755.

[38] [1982] ECR 3381.

The Advocate General had suggested that a just return for intellectual property rights in the film industry related to the specific object of the intellectual property right and was therefore not caught by Article 81(1) but this was not referred to in the Court's judgment. Instead, the Court referred only to a fair return upon the investment made. It was the task of the national court to determine whether the exclusive agreement created barriers that were artificial or unjustifiable in terms of the needs of the industry, of undue duration or based on fees which exceeded a fair return on investment.[39] Moreover, the Court was careful to set limits to the width of the exception it was creating for Community-wide film. It would be curtailed where the object or effect was to prevent or restrict the distribution of films or where competition within the film market was distorted.[40] Nevertheless, the Court had created the basis for a further exception in the definition of restraint of competition under Article 81(1). **2.36**

In *Erauw-Jacquery*,[41] the ECJ showed once again that the width of the exception in Article 81(1) would vary with the type of product protected by an intellectual property right and the industry in which it is used. In this case, the Court was prepared to find an export ban indispensable but the circumstances were unusual. The licence was for basic seed (i.e. seeds used to produce the seeds used by growers for the production of cereals) from the licensor to the propagator and the licensor insisted on an export ban for the purpose of quality control. The export ban and customer restrictions on licensees were held to be compatible with Article 81(1) because they were objectively justified by the need to ensure the proper handling of seeds by the growers who enjoyed a licence only for propagation. The Court stated that 'a person who has made considerable efforts to develop varieties of basic seed which may be the subject matter of plant breeders' rights must be allowed to protect himself against any improper handling of those varieties of seed'. The protection against improper handling of seed could only be assured by a restriction of propagation to selected licensees. To that extent the clause prohibiting the licensee from selling and exporting basic seeds does not come within the prohibition laid down by Article 81(1) of the Treaty (para 10). **2.37**

The Advocate General compared the risk of the technology in the seed finding its way into the hands of rivals to the risk faced by a franchisor that its franchised know-how might benefit competitors. The Court was also strongly influenced by the substantial financial commitment made in investing in the development of the seed.[42] Although the Court referred to *Nungesser*, this reference was to the **2.38**

[39] [1982] ECR 3381 at para 19. [40] Ibid. at para 20. [41] [1988] ECR 1919.

[42] Ibid. at para 10. It has been argued that the Court based its findings not on the existence/exercise distinction but rather on the value of the IPR. Derbyshire, J, 'Computer Programs and Competition Policy: A Block Exemption for Software Licensing' (1994) 9 EIPR at 370.

financial commitment of the licensor in developing the product and the urgency of the need for protection for the licensor's investment rather than to the need for an adequate incentive of territorial protection for the economic risks undertaken by the licensee.[43]

2.39 It would be a mistake to conclude from the indispensability analysis of these exceptions that a general doctrine of ancillary restraints at that point was more widely available to clear territorial intellectual property rights restraints, given the Court's concerns with market partitioning under Article 85(1). Instead, the test was twofold: is the degree of territorial protection essential to ensure that the licensing agreement will result in a market opening for the product? Are its effects on intrabrand interstate outweighed by the benefits of the extension of manufacture of goods or provision of service? In the case of *Erauw-Jacquery*, for example, the Court placed great weight on the need for protection for quality control purposes. The costs of the ban on exports were outweighed by the penetration benefits of the basic licence. In *Pronuptia*,[44] the Court made it clear in the context of franchise agreements where brands were well-known, territorial exclusivity incentives for franchising were not to be regarded as genuine ancillary restraints for the purposes of Article 81(1) unless the franchise brand *were relatively unknown*.[45] However, for established franchise operations, territorial exclusivity as such would be caught by Article 81(1).[46]

2.40 Nevertheless, the sequence of cases since *Nungesser* made the point that the test of 'qualitative' appreciability applies to territorial exclusivity in intellectual property rights licensing agreements. The Commission acknowledged the existence of these as exceptions to their general rule, but applied them sparingly. The new technology exception in *Nungesser was* strictly construed so that the test of 'newness' can be difficult to meet. For example, in *Velcro/Aplix*, the Commission refused to clear an exclusivity clause because the technology was not new.[47] In *Rich Products/Jus Rol*, the fact that other firms in the UK and in other Member States had developed processes for freezing yeast, even though they were not easily accessible, meant that the technology was not 'new' under Article 85(1).[48]

---

43 These cases may be applied by analogy to related products protected by IPRs. A number of authors have drawn attention to their possible application to computer software. See e.g. Derbyshire, n 42 above; Forrester, I, 'Software Licensing in Light of Current EC Competition Law' [1992] *ECLR* 5; Moritz, HW, 'EC Competition Law Aspects and Software Licensing Agreements—A German Perspective' [1994] *nc* 357 and 515.

44 [1986] ECR 353.

45 See e.g. Peeters, Jan, 'The Rule of Reason Revisited: Prohibition on Restraints of Competition in the Sherman Act and the EEC Treaty', 37 *The American Journal of Comparative Law* 521 at p 566.

46 This in fact was one major reason for the Commission's preparing a block exemption for franchising agreements.

47 [1989] 4 CMLR 157.

48 [1988] 4 CMLR 527.

In *Delta Chemie*[49] the fact that the licensee had previously acted as the distributor meant that the technology was not new. Moreover, the Commission stated that the exception in *Erauw-Jacquery* is to be limited to licences for the propagation of basic seed.[50]

(4) Non-territorial restraints and restrictions on competition

There was more of a consensus between Commission and Court over the concepts applicable to clear the non-territorial restraints in licensing agreements. The Court's endorsement of the concept of 'ancillary restraints' led to important basis for finding clauses in licensing contracts to be not appreciably restrictive of competition.[51] During the 1970s, the Commission began to acknowledge that in the case of certain agreements that were pro-competitive, some of their contractual restraints which were objectively indispensable to the nature and purpose of the agreement could be regarded as not being restrictive of competition. In *Pronuptia*,[52] the Court had occasion to apply ancillary restraints reasoning under Article 81(1) to the licensing of know-how together with a trademark in a distribution franchising agreement. The Court held that the compatibility of franschising agreements with Article 81(1) could be assessed only by reference to the provisions in the agreement in their economic context. The Court identified two conditions that were indispensable to the proper functioning of a franchising system. The first was that: **2.41**

> 'the franchisor must be able to communicate his know-how to the franchisees and provide them with the necessary assistance in order to enable them to apply his methods without running the risk that his know-how and assistance might benefit his competitors even indirectly'.

The second condition the Court identified as essential was that 'the franchisor must be able to take the measures necessary for maintaining the identity and the reputation of the network bearing his business name or symbol'. It followed that provisions which establish the means of control necessary for that purpose were also not restrictions on competition for the purpose of Article 85(1). The Court accepted that there were eight provisions justified as ancillary restraints under this criterion. One example was not to transfer the shop to another party without prior approval. However, the ECJ was unwilling to extend the logic of the concept of ancillary restraints to territorial exclusivity incentives for franchising where brands were well-known. It stated that territorial restrictions in the

[49] [1989] 4 CMLR 535. [50] [1989] 4 CMLR 535.

[51] See discussion in Gonzalez Diaz, E, 'Some Reflections on the Notion of Ancillary Restraints' [1995] *Fordham Cor Law Ins* 325. See *Reuter/BASF* 2 CMLR D44; *Remia BV & Verenidge Bedrijven Nutricia* v *Commission* [1987] 1 CMLR 1. 74; [1986] ECR 353.

[52] [1986] ECR 353.

franchising agreement were not genuine ancillary restraints or indispensable for the purposes of Article 81(1).

C. Stage II: The Development of Block Exemption Regulations for Technology Transfer Agreements

2.42 During the first stage, although the Court decisions introduced a series of exceptions to a wide interpretation of Article 81(1), they narrowly defined those exceptions. Many undertakings were still required to notify many normal categories of commercial agreements. To meet the need for expeditious exemption, the Commission decided that it was necessary to expand the categories of block exemption regulations, particularly in the IP licensing field.

(1) The Patent Licensing Block Exemption Regulation (2349/84)

2.43 The process of formulating block exemptions required a translation of the balancing test in Article 81(3) into specific lists of acceptable and unacceptable clauses for particular categories of agreements. In principle, the Commission was supposed to build the block exemption on the basis of experience with those categories of agreement. The first patent licensing block exemption approved by Council in 1984.[53] was far more favourable to patent licensing than the 1979 draft. In Recital 11, the Commission announced:

> 'Exclusive licensing agreements . . . are not in themselves incompatible with Article [81(1)] where they are concerned with the introduction and protection of a new technology in the licensed territory, by reason of the scale of the research which has been undertaken and the risk that is involved in manufacturing and marketing a product which is unfamiliar to users in the licensed territory at the time the agreement is made.'

Since the exemption related to agreements that on balance met the criteria of Article 81(3), the Commission felt that it could widen the entitlement to the degree of exclusivity it exempted to agreements 'concerned with the introduction and protection of a process for manufacturing a product which is already known'.

2.44 In fact the policy of the block exemption was to limit the category of technology

[53] Alexander, 'Block Exemption for Patent Licensing Agreement EC Regulation No 2349/84' [1986] IIC 1; Pichard, S, 'The Commission's Patent Licensing Regulation—A Guide' [1984] ECLR 384; Korah, V, *Patent Licensing and the EEC Competition Rules—Regulation 2349/84*, 1985, ESC Publishing, Oxford.

licences to pure patent licences[54] or mixed patent know-how technology licences (Article 1(1)).[55] Other intellectual property rights could be included in the licensed technology package but these had to be 'ancillary', i.e. less important, than the patent. The Regulation did not apply to pure know-how licences or mixed licences once the patent had expired.

If a licence agreement fitted into a category of agreement covered by the block exemption, the licensor could offer exclusive territorial rights to manufacture and sell the licensed product in any territory in the common market including the common market as a whole. It could also offer protection to licensees from other licensees as well as the licensor, consisting of a protection against active sales for a period extending for the duration of the patent protection in the relevant territory. The abandonment of the turnover restrictions for eligibility for territorial sales restrictions meant that the blacklist was viewed by the Commission as its 'front line' of defence against anticompetitive licences slipping through the net. There was a reserve power to withdraw a block exemption in the event of anticompetitive conduct but this was only available after the agreement had been made and exempted. Article 82 was known to provide a further reserve. Nevertheless, the Commission viewed its series of *per se* prohibitions as an effective filter to eligibility in the block exemption itself. **2.45**

Although the block exemption made a major concession to the economic arguments for protections to new licensees against sales from established licensees and the licensor during the tooling-up period, it was inflexible in two important respects. First, the blacklist had to be strictly complied with. A deviation could result in the agreement's not being exempted. A non-opposition procedure provided a chance for the parties with doubtful clauses to submit these to the Commission for approval with a six month time limit—fixed for the Commission to oppose it.[56] However, the non-opposition procedure could not test whether the agreement could be exempted despite the presence of a blacklisted clause. The blacklisted clauses were regarded as making the licensing agreement non-exemptible. **2.46**

Secondly, the patent licence block exemption was strict in limiting the category of agreements to which it applied. Its exemption of restrictions applied only to patent licences or mixed patent know-how licences, if they involved the *manufacture* of the licensed product (Article 1(2)). It did not apply to licences for **2.47**

[54] 'Patents' were defined to include *inter alia* patents, European patents, patent applications, utility models, *certificats d'utilité* and *certificats d'addition* under French law.

[55] 'Know-how', even if non-ancillary, could be included as long as the patent was necessary to achieve the objects of the technology and the know-how was secret and permitted a better exploitation of the licensee's patents.

[56] A non-response could be taken as approval.

sales only. It did not apply to patent pools, reciprocal licences or joint ventures (Article 5(1)). It did not apply to mixed technology licensing involving trademark licences or copyright licences if they were so important that the patents were not necessary. Moreover, the licensee could be restricted from applying the know-how in the licensor's and other licensees' territories and fields of application only as long as the licensed patents accompanying it were necessary to achieve the objects of the licensed technology and at least one of the parallel patents registered in the licensed territory remained in force (Recital 9).

(2) The Know-how Block Exemption Regulation (556/89)

2.48 By the mid-1980s, the there was a clear need to complement the patent licensing BER with one for know-how licensing.[57] The Commission's decision in *Boussois/Interpane*[58] indicated that the patent licensing block exemption was too inflexible to apply to certain types of mixed technology licences in which the know-how predominated and the licensed patents were not necessary to carry out the licensed process. The Commission drafted a second block exemption for pure know-how licensing agreements and mixed patent know-how technology licensing agreements which could not qualify under the patent licensing block exemption. In Recital 1 it stated that the communication of know-how is often irreversible and greater contractual certainty is necessary so that each party can expect to appropriate to itself the benefits of its investment. The Commission defined protectable know-how as 'a body of technical information that is secret, substantial and identified in any appropriate form' and carefully defined these concepts defined in the Regulation to provide some guarantee to the Commission that there was know-how of innovative value as the basis for the exempted licence agreement. The Commission could not allow the parties to determine for themselves whether or not the technical information was sufficiently valuable to make the licensing agreement worthwhile. This filter was viewed as a key protection against anticompetitive agreements disguised as know-how licences.

2.49 The Commission adopted a structure for the know-how block exemption that was similar to the patent licensing block exemption. Exclusive territorial

[57] There was evidence that particularly US firms were hesitant to grant licences of their valuable technical information in the EC when know-how licensing agreements were of doubtful legal validity. See e.g. Hoyng and Biesheuvel, 'The Know-how Group Exemption' [1989] CMLR 219 at 220. See too Frazer, T, 'Vorsprung durch Technik: The Commission's Policy on Know-How Agreements' [1989] *Yearbook of Eur Law* 1; Winn, D, 'Commission Know-how Regulation 556/89: Innovation and Territorial Exclusivity Improvements and the Quid Pro Quo' [1990] ECLR 135.

[58] [1988] 4 CMLR 124.

protection was given to the licensees of know-how agreements against active sales by other licensees[59] and the licensor[60] for 10 years from the date of the making of the first licence as long as the know-how remained secret and substantial. Protection against passive sales was allowed for the first five years after the date of the first licence entered into by the licensor for the same technology in the EEC. The whitelist again contained permitted clauses, all justified by reference to ancillary restraints. Again it was matched by an extensive blacklist which contained major safeguards against anticompetitive clauses using a *per se* prohibition approach partly because of the absence of a market share limit.

There were signs that the Commission was beginning to recognise the need to relax the policy of the block exemptions in certain respects in the interest of encouraging investment in innovation.[61] One example of this was the change in the Commission's policy to royalties stated in Recital 15: **2.50**

> 'as a rule, parties do not need to be protected against the foreseeable financial consequences of an agreement freely entered into and should therefore not be restricted in their choice of appropriate means of financing the technology transfer'.[62]

Under the patent regulation, the royalties clause had only allowed the parties to agree to spread the upfront payment while the patent was still in force. The Commission allowed its fear of extending the royalty payments for non-patented information to interfere with the need to give the parties the flexibility to adjust payments to avoid a prohibitive upfront payment. By the time of the know-how block exemption, the Commission appeared to have accepted the need for a change of approach. The Regulation provided[63] that royalties could be paid 'in the amounts and according to the methods freely determined by the parties'. There were still difficulties with compliance with the technical rules of the know-how block exemption,[64] but at least the Commission had begun to reduce the inflexibility in the block exemption policy to royalty payments.

The intellectual property rights licensing block exemption policy had improved **2.51**

[59] In the case of a licensee, the starting date for the time period is the date of the first licence entered into by the licensor for the same technology *in the EEC*.

[60] The starting date for the period of protection against active sales by the licensor was the date of the first licence entered into by the licensor in the territory for the same technology.

[61] At this time too, the Commission began to relax its approach to the assessment of licensing in joint ventures under Art. 85(1), proclaiming a policy of new realism. See *XIIIth Report on Competition Policy*. See too Anderman, S, 'Co-operation, International Competitiveness and Competition Policy' in Deakin, S and Michie, J (eds), *Contracts, Cooperation and Competitiveness: Studies in Economics Management and Law*, 1997, Oxford University Press.

[62] Other examples could be found in the more relaxed approach to tie-ins and non-severable improvements.

[63] 12 Art 2.1(7).

[64] Whaite, 'Licensing in Europe' (1990) 3 EIPR 88 at 91.

from that in 1984. However, it continued to present certain difficulties to the parties to technology licensing agreements:

(a) The excessive inflexibility and technicality of its conditions. The blacklisted clauses could now be exempted by a process of individual exemption if the market conditions were right, but the need for the contents of the licence to conform with exactitude to the requirements of the block exemption remained.
(b) Mixed technology licensing agreements had to be drafted to fit either within the Patent Licensing Regulation (2349/84) or the Know-how Regulation (556/89) and there were some difficulties in choosing which of the two Regulations applied.
(c) Important gaps remained in the coverage of the block exemptions in relation to major categories of intellectual property right licences. Trademark and copyright licences could not qualify under a block exemption unless these forms of intellectual property right were 'ancillary' to the know-how or patents. If they were equally important or crucial to the licensing agreement, they did not fit into the category of agreements qualifying for the block exemption.[65]

(3) The Unified Technology Transfer Block Exemption Regulation (240/96)

2.52 The renewal of the patent licensing block exemption was taken as an occasion to reshape the Commission's policy towards block exemptions for technology licensing. The two Regulations were merged into a new, unifying Technology Transfer Block Exemption (240/96)[66] which applied to pure patent licences, pure know-how licences, and mixed patent know-how licences, the most common form of technology licensing agreements.[67] This new Regulation took its place alongside the regulatory regimes for specialisation agreements, research and development agreements, franchise agreements, cooperative joint ventures,

[65] See e.g. the type of production franchise or trade mark licence in *Campari* [1978] 2 CMLR 397, in which the trade mark was too important to the overall agreement to be able to be viewed as ancillary to the know-how. See too *Moosehead/Whitbread* [1991] 4 CMLR 391.

[66] The concept of patents is deemed to include a wide variety of IPRs, e.g. patent applications, utility models and applications, topographies of semiconductor products, and plant breeder rights (see Art 8). The new Regulation provides that it can apply to agreements which include licensing of other intellectual property rights such as trademarks, copyright, and design rights or as long as such rights are of assistance in achieving the object of the licensed technology (Art. 1.5). In other words they must be 'ancillary' to the patent or know-how (Recital 6). This means that industrial or production franchises were excluded from the block exemption (see e.g. *Moosehead/Whitbread* [1991] 4 CMLR 391). Moreover, computer software was not covered.

[67] See Korah, V, *Technology Transfer Agreements and the EC Competition Rules*, 1996, Oxford University Press.

and mergers[68] which together provided a virtual 'second tier' of regulation of intellectual property licensing by competition law in addition to the underlying intellectual property legislation.

This occasion offered the Commission an opportunity to make a substantial adjustment to the regulatory regime so that it had less of an inhibiting effect upon intellectual property right licensing. It could, for example, have chosen to design the block exemption to treat vertical licensing relationships differently from horizontal relationships but that route was not taken. It continued to evade such a choice, leaving the individual rules to apply to a mixed and varying model of whether licensing is predominantly horizontal or vertical.[69] **2.53**

The Commission clearly acknowledged the need to encourage technology licensing in the new Regulation. In Recital 3 it states that the purpose of the reform is 'to encourage the dissemination of technical knowledge in the Community and to promote the manufacture of technically sophisticated goods'. Yet the block exemption itself did not attempt a radical reform of competition policy. Instead, it made five types of adjustments to the existing structure. First, it increased the time periods for territorial exclusivity in most cases, thus marginally increasing the incentives for investment in technology transfer. Secondly, it lengthened the 'whitelist' adding to the categories of clauses which were either not caught by Article 81(1) or were exempt under Article 81(3). Thirdly, it reduced the blacklist to a core of seven clauses, signalling a greater willingness to accept that vertical licensing arrangements could be pro-competitive even if they contained restrictions such as tie-ins or minimum-quality specifications technically necessary for the satisfactory exploitation of the technology. **2.54**

The blacklist consisted of a modern list of *per se* prohibitions in technology licensing agreements: the restrictions on price fixing, customer sharing, and quantity limits for licensees were designed to deter cartel and market sharing operations. The BER also blacklisted provisions which impeded parallel trade, non-competition clauses, grant-back provisions which required the licensee **2.55**

[68] See respectively Reg 417/85, amended by Reg 151/93; [1993] 4 CMLR 155; Reg 418/85, as amended by Reg 151/93; [1993] 4 CMLR 163; Reg 4087/88; Commission's Notice on Co-operative Joint Ventures [1993] OJ C 43/2; EC Merger Reg, Council Reg 4064/89, [1990] OJ L 257/14.

[69] On the one hand, the territorial restraints were designed as if the block exemption applied exclusively to vertical relationships. On the other hand, the block exemption treated all licensing agreements which were initially vertical as potentially horizontal so that they may not contain restraints, such as exclusive grant-backs of improvements, which could prevent them blossoming into subsequent competitive relationships. Yet, by providing that its benefit could be withdrawn if technology licensing occurs in a horizontal relationship which shows signs of an anticompetitive tendency, such as cross-licensing or patent pooling, or if the parties had market shares approaching dominance, it implied that all other horizontal licensing relationships were eligible provided they otherwise qualified for the block exemption.

to assign in whole or in part its rights to improvements to, or new applications of, the licensed technology and any clause which improperly prolonged the duration of the licensing agreement owing to new improvements.

2.56 As in the two previous block exemptions for intellectual property rights licensing, the 2.56 Commission retained a power to withdraw the benefit of the Regulation where it found that an exempted agreement had effects which were incompatible with Article 81(3).[70] The Commission could exercise its power of withdrawal of the BER by making a formal decision and this made unenforceable the restrictive provisions in the licensing agreement and make the parties liable to fines from that date onwards. The decision, and the grounds on which it was based, could be appealed to the CFI and ECJ. Not surprisingly, the use of the withdrawal power was rare. Nevertheless, the Commission did use the threat of its exercise to obtain changes in conduct by the parties to licensing agreements. One famous example was that of *Tetra Pak*[71] in which the Commission responded to an acquisition of a competitor's exclusive licences by a dominant firm by indicating that it would have withdrawn the exemption until Tetra Pak modified the licences from an exclusive to a non-exclusive licence.

2.57 At one stage, the Commission attempted to introduce an additional market share test as a precondition to exclusive territoriality because of its concern that dominant undertakings with exclusive licences 'might prevent access to the market of the technology by third parties and eliminate competition in respect of a substantial part of the products in question'. Initially, these market share tests were wide in scope, applying to all licensing agreements whether horizontal or vertical. In the penultimate version of the Regulation, they were limited to agreements between competing manufacturers. In the final version, however, they were taken out altogether as a precondition for exemption and instead made a basis for withdrawal.

2.58 The Commission changed the 1996 block exemption to be more inclusive.[72]

---

[70] Under Article 7, there were four examples: (1) where the effect of the agreement was to prevent the licensed products from being exposed to effective competition in the licensed territory from identical or substitutable goods or services, 'which may in particular occur where the licensee's market share exceeds 40%'; (2) where the licensee refused, without any objectively justified reason, to meet unsolicited orders from users or resellers in the territory of other licensees; (3) where the parties without any objectively justified reason either refused to meet orders from users or resellers in their own territories who would market outside that territory, or make it difficult for users or resellers to obtain the products from other resellers in, the Common Market; (4) where the parties were competing manufacturers at the date of the grant of the licence and the effect of a minimum quantity or best endeavours clause would have the effect of preventing the licensee from using competing technologies.

[71] See e.g. Case T-51/89 *Tetra Pak Rausing* v *Commission* [1991] 4 CMLR 334.

[72] Under the patent and know-how regulations, the concept of an ancillary provision relating to other IPRs was translated into a requirement that the other type of IPR must not be as

Nevertheless, it left strict limits to its coverage of intellectual property rights licences.[73] Moreover, it also retained the requirement that the parties apply the Articles of the block exemption to the clauses of the licensing agreement with a fine-tooth comb. If the parties miscalculated the requirements of the whitelist or blacklist and failed to use the opposition procedure, the block exemption was non-applicable.[74] Consequently, the block exemption did little to change the formalistic approach to evaluating the contents of licensing agreements and compared unfavourably with its major competitor system, the US, which had formulated new Guidelines for intellectual property licensing in 1995.

The legal framework for technology transfer created by TTBER 1996 was moreover form-based and legalistic in the sense that certain clauses were either exemptible or non-exemptible depending almost entirely on their form. To qualify for exemption the parties were required to draft an agreement that which was consistent with the Commission's threefold categorisation of numerous restraints in licensing agreements: its 'whitelist' of 26 clauses, the clearly exemptible clauses under the block exemption;[75] its 'blacklist' of seven clauses which **2.59**

important or more important to the agreement than the patent or know-how. For example, in *Boussois/Interpane* the know-how was held to have dominated the patented technology and did not have the effect of merely contributing to the better exploitation of the licensed patents. In *Moosehead/Whitbread* the lager production licence did not qualify under the know-how block exemption because the trade mark was not ancillary to the licensed know-how as required by that exemption. The Commission's view was that the principal interest of the parties was in the exploitation of the trademark rather than the know-how. In the new Technology Transfer Regulation, however, the concept of ancillary restraint was defined in rather different terms. Article 10(15) defined ancillary provisions as: 'provisions relating to the obligation of intellectual property rights other than patents which contain no obligations restrictive of competition other than those also attached tothe licensed know how or patents and exempted under this Regulation'. Under the Technology Transfer Regulation there was greater possibility for other IPRs to be part of a mixed technology licence and nevertheless qualify for exemption provided that they met the following two tests: (i) the provisions relating to other IPRs had to be ancillary, i.e. contain no obligations restrictive of competition other than those also attached to the licensed know-how or patents used and exempted under the regulation (Article 10(5)); and (ii) the IPRs had to contribute to the objects of the licensed technology (Recital 6).

[73] In Article 5, the BER excluded from the block exemption the types of agreements which would require a careful examination of the legal and economic context in order to determine their net effect on competition, such as, patent or know-how pools (Article 5(1)(1)) and cross-licensing arrangements (Article 5(1)(3)) between competing firms. These had to qualify for exemption, if at all, by individual exemption. Article 5 further excluded licensing agreements between competing firms with interests in a joint venture, such agreements could be exempted under the R&D block exemption, and either cleared or exempted under the Commission's guidelines to cooperative joint ventures. Similarly, franchise agreements were excluded from the Technology Transfer Regulation but exempted by the franchising block exemption (4087/88). Licences solely for the purpose of sales were are not covered by the Technology Transfer Regulation, but could be exempted under the exclusive distribution block exemption (1983/83).

[74] See e.g. *Delimitis* [1991] ECR I-935.

[75] In fact they consisted of a mix of cleared clauses, i.e. those not caught by Article 81(1) and exemptible clauses, i.e. those caught by Article 81(1) but automatically exempted under Article 81(3).

consisted of clearly non-exemptible clauses If a clause was blacklisted, the *agreement* as a whole was non-exemptible and unenforceable. There was no severance policy for blacklisted clauses. Instead, there was a third category of 'grey,' clauses (all those which are not whitelisted or blacklisted). These clauses could be approved, opposed or approved by being non-opposed under a 'quick look' procedure subject to a deadline of four months for the Commission.

2.60 This Commission's threefold characterisation encompassed a wide variety of typical clauses in licensing agreements: territorial restraints, customer allocation, field of use, output restrictions, price restrictions, non-compete obligations; tying grant-backs and no-challenge clauses. It somewhat controversially subjected the contractual restrictions on territorial restriction to micro regulation more for political concerns to promote market integration than for conventional competition reasons.

2.61 Consequently, although the 1996 Regulation allowed industry to transfer the more conventional technology, on the whole, sufficiently freely and with legal certainty, within the EU, it also left a regulatory structure that did not differentiate sufficiently between genuinely vertical and horizontal agreements. Concern with the political-economic problems of maintaining intrabrand competition and integration of the Member States into a single market led to less than optimum incentives to technology licensing agreements from a purely economic competition point of view. In particular, strict rules for territorial exclusivity, even where the economic costs and benefits, or 'efficiencies', suggested little competition risk, meant that the needed incentives for second and subsequent waves of licensees were sometimes greatly reduced. The Commission itself has openly acknowledged that its territorial restrictions have been imposed 'because of the added market integration objective which EC competition policy has'.[76]

2.62 On the other hand, TTBER 1996 facilitated licensing by not setting market share limits to the safe haven of the block exemption leaving high market share agreements to be regulated by Article 82 or the possibilities of a 'withdrawal procedure' whereby the benefits of the block exemption could be formally withdrawn by the Commission or the competition authority of a Member State in carefully defined cases in the TTBER. This approach was adopted in the final draft of the Regulation after a strong campaign by industry throughout Europe was waged against a Commission proposal to introduce market share limits to the 1996 block exemption.

---

[76] See Evaluation report para 55; see too Guidelines on Vertical Restraints point 7.

D. Stage III: The Revision of TTBER in a Modernised Setting

By 2001, the Commission had decided that there was a clear need to change its policy towards technology transfer agreements even before Regulation 240/96 was due to expire in April 2006. This was caused partly by the change in the Commission approach to the regulation of vertical agreements more generally under Article 81. It had begun to move away from a legalistic and form-based approach to a more economic and effects-based approach to the regulation of other types of agreements, taking into greater account the economic analysis of possible costs and benefits, or 'efficiencies,' of certain restrictions and recognising the different economic effects of vertical and horizontal agreements respectively. This could be seen in the main characteristics of the Commission's reforms of vertical distribution agreements, such as exclusive and selective distribution and franchising (the Vertical Agreements Block Exemption (EC 2790/1999) and horizontal agreements, such as research and development (EC 2659/2000) and specialisation agreements (EC 2658/2000). The new Vertical Agreements Block Exemption, introduced a more flexible regulatory approach by widening its scope to include a variety of categories of agreement. It also abandoned white and grey lists of clauses retaining only an outright prohibition upon a limited number of 'hard core' restrictions. This easing of the regulatory straitjacket allowed the parties greater freedom to draft their agreements but the benefits of the safe haven offered by the block exemption were restricted by market shares limit of 30% built into the block exemption as part of a more economic approach to regulation. The new regulatory framework placed a premium upon self-certification by the parties in cases where their market shares exceeded the limits. To offer assistance in this endeavour, the Vertical BER was supplemented by the Vertical Guidelines as were the later R&D BER and Specialisation BER supplemented by the Horizontal Guidelines. **2.63**

At a more technical level, the reform of the Commission resulting in the replacement of Regulation 17 by the new Regulation 1/2003 gave the courts and competition authorities of the Member States the competence to apply Article 81(3) directly. This not only ended the need for precautionary notification of agreements; the Commission decided that it also meant that the opposition procedure or quick look facility for grey clauses would have to be abandoned and clauses would 'either be covered by the block exemption or treated as hardcore'.[77] **2.64**

[77] Commission Evaluation Report on the Transfer of Technology Block Exemption Regulation No 240/96, 'Technology Transfer Agreements under Article 81' at p 6.

2.65 A second important factor was the publication of the Commission's Evaluation Report on the Technology Transfer Regulation[78] that stressed that the 1996 Regulation had a number of shortcomings and needed reform. The Commission identified four types of shortcomings in the Regulation. First, because of its legal formalism and narrow definitions of scope, the Regulation was described as creating a 'legal straitjacket' in the sense that companies often had to redraft their commercial agreements to fit within its confines. Secondly, Regulation 240/96 was too narrow in scope covering only a limited number of exclusive licensing arrangements, mainly to pure and mixed patent and know-how licences. Thirdly, the regulatory framework lacked economic realism. The Regulation was too restrictive in the sense that the blacklist covered items such as non-compete clauses and tie-ins that were not always anticompetitive and could have efficiency-enhancing effects where the licensees were non-dominant. The Regulation made no distinction between horizontal and vertical licensing agreements despite their different legal effects. On occasion, restrictions with different legal form but with similar economic effects on markets were given different treatment.[79] Finally, some exempted clauses raised competition concerns particularly where licensees had strong market power.

2.66 These criticisms were taken on board by the Commission in its decision to reform the regulatory framework for technology transfer. In choosing the form of a new technology transfer regulation, the Commission used the Vertical Agreements Block Exemption Regulation as a role model and gave priority to a harmonising TTBER with its other block exemption regulations as part of its preparation for competition policy in the post-modernisation world of 25 Member States. By 2001, the Commission had decided that there was a clear need to change its policy towards technology agreements even before Regulation 240/96 was due to expire in April 2006. This was caused partly by the change in the Commission's approach to the regulation of vertical agreements more generally under Article 81. It had begun to move away from a legalistic and form-based approach to a more economic and effects-based approach to the regulation of other types of agreements, taking into greater account the economic analysis of possible costs and benefits, or 'efficiencies,' of certain restrictions and recognising the different economic effects of 'vertical' and 'horizontal' agreements respectively. This could be seen in the main characteristics of the Commission's reforms of vertical distribution agreements, such as exclusive and selective distribution and franchising (the Vertical Agreements Block Exemption (EC 2790/1999) and horizontal agreements, such as research and development

[78] COM(2001) 786 final, 20.12.2001.

[79] For example territorial and customer restrictions were treated differently.

(EC 2659/2000) and specialisation agreements (EC 2658/2000). These new regulations all had in common the following four features:

- their greater flexibility of scope;
- their abandonment of white and grey clauses concentrating their form-based regulations upon a limited number of hard core restrictions;
- their introduction of market shares limits as part of a more economic approach to regulation; and
- their resort to extensive guidelines both to explain the existing regulations and to offer ancillary guidance to other related policies of the Commission.

At a more technical level, the reform of the Commission resulting in the replacement of Regulation 17 by the new Regulation 1/2003 gave the courts and competition authorities of the Member States the power to apply Article 81(3) directly. This not only ended the need for precautionary notification of agreements; the Commission decided that it also meant that the opposition procedure or 'quick look facility' for grey clauses would have to be abandoned and 'either be covered by the block exemption or treated as hardcore'.[80] **2.67**

The significance of this new blueprint for directed self-regulation under Regulation 1/2003 is that there is no longer a need to submit an agreement for precautionary notification to the Commission to ensure provisional validity to a licensing agreement. Its validity stands or falls depending on its contents at the time it is challenged. If an agreement is caught by Article 81(1) and falls outside the block exemption, it is no longer automatically void under Article 81(2). If the agreement is outside the block exemption, it can be analysed by analogy with the contents of the Regulation and the framework for analysis in the Guidelines to determine whether it is exemptible under Article 81(3). **2.68**

[80] Commission Evaluation Report on the Transfer of Technology Block Exemption Regulation No 240/96, 'Technology Transfer Agreements under Article 81' at p 6.

Part II

THE NEW FRAMEWORK FOR ANALYSIS OF IP ISSUES

3

THE 'MODERNISATION' OF EU COMPETITION LAW ENFORCEMENT—THE NEW ANALYTICAL PARADIGM FOR APPLYING ARTICLE 81

A. Introduction

The entry of Regulation 1/2003 into force on 1 May 2004, concluded a revolution in the application of Article 81 EC Treaty. Regulation 1 marks the culmination of a series of regulatory initiatives and policy pronouncements, dismantling the elaborate structure for regulatory control over contracts that had been in place since the 1960s.[1] In its place, the Commission has created a structure for **3.01**

[1] In addition to Regulation 1/2003 itself, the principal instruments for affecting this change have included: *Commission Regulation (EC) No 2790/1999 of 22 December 1999 on the application of Article 81(3) of the Treaty to categories of vertical agreements and concerted practices*, [1999] O.J. L 336/21, ("Vertical Restraints Block Exemption"); *Commission notice – Guidelines on Vertical Restraints*, [2000] O.J. C 291/1 ("Vertical Guidelines"); *Commission Regulation (EC) No 2659/2000 of 29 November 2000 on the application of Article 81(3) of the Treaty to categories of research and development agreements;* [2000] O.J. L 30/7 ("R&D Block Exemption"); *Commission Notice – Guidelines on the applicability of Article 81 to horizontal co-operation agreements* [2001] O.J. C 3/2 ("Horizontal Guidelines"); and *Commission Notice – Guidelines on the application of Article 81(3) of the Treaty* [2004] O.J. C 101/97 ("81(3) Guidelines").

control of agreements that is based on an economic assessment of both the benefits and the anticompetitive effects of agreements. The new 'modernised' structure is meant to make Article 81 a more effective tool for implementing competition policy and to allow the Commission and Member State authorities to focus enforcement resources on issues that have a significant impact on companies and consumers.

3.02 The 'modernisation' revolution is particularly important for agreements involving intellectual property, because these agreements were subject to detailed prescriptive regulation under the old regime. The old regulatory system imposed a model on licensing arrangements that, in the view of some commentators, artifically limited the ability of parties to construct commercially sensible arrangements. The scrapping of the old regime means, however, that the application of competition law to licensing arrangements can never be certain—some arrangements that were always legally valid under the old rules may raise issues if the parties have a high level of market power. The purpose of this chapter is to outline the way that Article 81 is meant to be applied in the post 'modernised' world. We start by describing the nature of competition regulation under the old Article 81 regime, as background for understanding the policy basis for the modernisation project. We then outline the structure of analysis under the 'post-modern' approach to Article 81, which we refer to as the 'new analytical paradigm'.

B. Article 81 in the Old Regime—Ordoliberalism, Legal Formalism, and Prescriptive Block Exemptions

(1) The Ordoliberal ideal of competition law

3.03 The competition rules of the EC Treaty were originally conceived as the centrepiece of an Ordoliberal structure for the new 'common market'. Reacting against both socialist forms of market organisation and the cartel-dominated economics of Germany prior to the Second World War, the German economists and jurists of the Freiburg school had developed a particular understanding of the interplay between market organisation and political power. These Ordoliberal theorists regarded a controlled market economy as the form of economic organisation that would best achieve the objectives of economic growth and fair distribution of its benefits without excessive accumulation of political power. Competition policy played a central role in this Ordoliberal conception of economic organisation. By creating a 'constitution' for the economy that set clear limits for the behaviour of firms, competition rules would ensure that the market generated the intended benefits while controlling 'economic power' (just as a national constitution distributes and controls political power).[2]

An Ordoliberal competition policy required the creation of a strong central competition authority. The theory of the 'economic constitution' required the creation of clear rules backed by credible sanctions. While derogation from those rules might sometimes make economic sense, such derogations could only be granted on the basis of the objectives of competition policy. The structure of Article 81 EC Treaty as applied in the original enforcement regulation, Regulation 17/62, reflected the Ordoliberal conception of the role of the competition authority. Exceptions from the Article 81(1) prohibition could only be granted by the Commission. National authorities and courts were not empowered to apply the exception rules set out in Article 81(3). **3.04**

(2) Legal formalism in Article 81(1)

The perceived need to establish clear rules led to a formalistic application of Article 81(1). The Commission, with some support from the Court of Justice, interpreted the reference in Article 81(1) to 'an agreement . . . that has the object or effect of restricting competition' to mean that some agreements or provisions in agreements constituted 'restrictions' of competition. A restriction of competition was any contractual provision that limited the ability of a party to an agreement to conduct its own business in any way that was competitively relevant. Although the Court required that this restriction be 'appreciable' in its effects, the standard for appreciability was based on the size of the firms involved and could apply where the arrangements involved 5% of the affected market or even less. The effect of this approach was to make almost any agreement that included a restriction on competitive conduct subject to the Commission's jurisdiction; case-by-case consideration of competition policy was reserved for the Commission in the exercise of its discretion to apply Article 81(3). **3.05**

The lack of economic rigour in the interpretation of Article 81(1) in the old regime should not, however, be blamed on the peculiarities of a German Ordoliberal approach.[3] While competition law enforcement regimes have always **3.06**

---

[2] See Gerber, D, *Law and Competition in Twentieth Century Europe: Protecting Prometheus*, 1998, Oxford University Press; Kallaugher and Sher, 'Rebates Revisited: Anti-competitive Effects and Exclusionary Abuse under Art. 82' [2004] ECLR 255; Möschel, W, 'Competition Law from an Ordo Point of View' in Peacock, A and Willgerodt, H (eds), *German Neo-Liberals and the Social Market Economy*, 1989, Macmillan, p. 142–59.

[3] In fact, German competition law has always given an important role to economics as a basis for developing competition policy and the German enforcement agencies have always included a high proportion of staff trained in economics. Indeed, in respect of non-price vertical restraints, German law has always had a more market-oriented approach than EU law. This approach has also been reflected in a relatively liberal approach to IP licensing issues. The Ordoliberal contribution was the emphasis on prescriptive rules with a possibility of legal exceptions in

recognised that 'competition' leads to lower prices and higher output, there was no strong analytical model for assessing how conduct subject to competition rules (mergers, agreements, or unilateral conduct) affects this result in individual cases. As a result, many jurisdictions, including the United States, relied on form-based assessment and *per se* rules to foster something that 'looked like' competition. Article 81 was not at all unique.[4]

(3) The prescriptive block exemption

3.07 In contrast with the position under the Ordoliberal-inspired German competition law, the formalistic definition of a restriction of competition under Article 81 also applied to agreements between firms and their customers (other than individual consumers). This meant that any distribution agreement could be subject to Article 81 if it contained any provisions that limited the competitive autonomy of either the manufacturer or a distributor. The application of Article 81 to the great mass of distribution agreements prompted the great innovation of Community competition law: the exemption by category, usually referred to as a 'block exemption'. In order to deal with the backlog of applications for individual exemption under Article 81(3), the Council of Ministers authorised the Commission to grant exemptions to categories of agreement by regulation. The Council enabling regulation required that the Commission define in each block exemption regulation, *inter alia*, the provisions in agreements to which the exemption would apply and the provisions that would render the exemption inapplicable. This prescriptive approach to the drafting of block exemptions meant that parties drafted their agreements to fit the model set out in the regulation. This tendency was accentuated by the reluctance of the Commission to entertain applications for individual exemption in respect of agreements of a type covered by an existing block exemption.

(4) The limits of the old regime and the drive for change

3.08 The Ordoliberal enforcement structure for Article 81 made sense in 1958 (when the EC Treaty came into effect) or 1962 (when the enforcement regulation was finally adopted). Outside Germany understanding of competition law concepts was limited at best. National courts in particular could not be expected to apply competition rules rigorously, particularly where to do so would require

---

horizontal cases (which EU law applied to vertical restraints as well) and an emphasis on structure, particularly in merger and dominance cases.

[4] The Ordoliberal idea that competition rules provide the framework for an economic 'order' with constitutional implications may, however, have encouraged development of these rules to their logical conclusion. The emphasis in the German legal system on legal certainty may also have encouraged the German courts to distil application of competition law in specific cases into formal rules of general application.

a dramatic change in locally accepted patterns of dealing. In the absence of any developed body of precedent, firms and their advisors could not apply competition law without active intervention from the body charged with developing competition policy—the Commission. By the 1990s, however, this position had changed. Throughout the EU an increasingly sophisticated body of academics and practitioners could understand and apply competition law principles. Courts and national governments no longer instinctively viewed competition law as an alien intrusion on commercial freedom. Competition law was accepted, even where it was not fully assimilated.

The world outside had changed as well. Other jurisdictions that had once applied variations of formalism to assessment of allegedly anticompetitive agreements had embraced economic analysis as the basis for applying competition policy to such agreements. In the United States, commentators, including but not limited to the 'Chicago School', had criticised the lack of strong economic support for many well-established antitrust rules. Richard Posner suggested that consumer welfare should be the guiding criterion for assessing allegedly anticompetitive arrangements. He also suggested that price theory could provide the basis for making this assessment—i.e. an arrangement is anticompetitive if it leads to an increase in price or a decrease in output. This insight provided the tool for a new analytical paradigm, a break from *per se* rules and structural assessment.[5] **3.09**

When the EU adopted rules requiring the notification of mergers to the Commission for clearance in 1988, the Commission quickly developed an approach to merger assessment squarely based on economic efficiency and consumer welfare. The development of an economics-based policy for mergers was hard to reconcile with the formalistic approach to Article 81. To an economist, the concepts that drove application of competition policy in a merger case are equally applicable to a case involving supposedly restrictive agreements. At the same time, judgments from the European courts, particularly the ruling by the Court of First Instance in *European Night Services*,[6] suggested that the formalistic reliance on restrictions on competitive conduct for application of Article 81(1) was not sufficient; at least in cases that not involve a restriction 'by object', some market analysis was required. At the same time, the proposed enlargement of the European Union promised to put the system of individual notification of agreements under even greater strain. These factors led the Commission to undertake a fundamental rethinking of the nature of competition policy under Article 81, and ultimately to jettison the old Ordoliberal-inspired regime. **3.10**

[5] See Posner, R. *The Social Cost of Monopoly* 83 Journal of Political Economy, para 807. Posner, R *Antitrust Law* (Chicago, 1975) for a good modern summary see Pera, A and Auricchio, V, *Consumer Welfare, Standard Proof and the objectives of Competition Policy*, 1, European Competition Journal (2005).

[6] *European Night Services* v *Commission* [1998] ECR II-3141.

C. The 'Modernised' Approach to Article 81

3.11 Under the new approach to Article 81:

- The application of Article 81(1) is based on an economic analysis of whether agreements have the object or effect of restricting competition.
- National authorities and courts now have jurisdiction to apply Article 81(3) —the grant of an 'exemption' from application of Article 81(1) is no longer a preserve of the Commission.
- Formal 'notification' by the parties to an agreement is no longer a prerequisite for application of Article 81(3).
- Block exemption regulations now serve as a 'safe harbour' for application of Article 81(3)—they do not impose a template for permissible agreements.

3.12 The effect of these changes is to make Article 81 function 'as a whole'; Article 81(1) and Article 81(3) complement each other as an integrated set of rules for review and control of potentially anticompetitive arrangements. Now that national authorities and courts have jurisdiction to apply Article 81(3), the broad discretion claimed by the Commission in application of Article 81(3) under the old regime is no longer sustainable; the principles for application of Article 81(3) need to be applied on a legally consistent basis. Equally important, now that application of Article 81(3) is decentralised, it is no longer necessary to maintain a maximalist interpretation of Article 81(1) in order to retain jurisdiction over policy issues to the Commission under Article 81(3). This, in turn, allows the most important change of all—the adoption of an economically based consumer welfare standard as the policy basis for applying both Article 81(1) and Article 81(3).

3.13 In considering the substantive application of the new paradigm, it is useful to understand how the Commission could engineer a fundamental change in the law without a change in the wording of Article 81 itself. A combination of factors has made this possible. First, as regards the interpretation of Article 81(1), the rulings of the Court of Justice have left some ambiguity regarding the nature of the market assessment required to find a restriction of competition, particularly where an agreement is alleged to be restrictive by effect. As already noted, more recent cases have tended to read economic content into the requirement that effects be assessed in the context of actual market conditions. Secondly, as regards interpretation of Article 81(3), the courts have traditionally accorded the Commission a substantial margin of discretion—appellate review has focused on formal legality and not on the nature of the substantive assessment. It is therefore open to the Commission to characterise the exercise of discretion under the old regime as informed by economics-based considerations, even if the decisions did not always articulate an explicit economic assessment. Thirdly, the Commission has an acknowledged role in the Community legal structure as the guardian of 'competition policy'. Where the application of competition law raises new issues, the Commission's policy determinations are entitled to a certain deference. This role becomes particularly important in the context of the shift in enforcement powers to Member State authorities and courts, where the provision of guidance from the Commission on the appropriate standards to apply is arguably both appropriate and necessary. Finally, to the extent that the old regime was anchored in the structure and language of the old procedural regulation, the Commission measures applying that regulation,[7] and the old block exemptions, changes in the regulations can affect the substantive legal assessment.

A particularly important feature of this exercise has been the use of interpretative **3.14** 'notices' as part of the Commission's exercise of its policy-setting role. Four Commission notices lay out the modern approach to Article 81 enforcement.[8] The Vertical Restraints Notice, published in 1999, sets out the revised policy regarding agreements between manufacturers and their customers (wholesalers and retailers). The Horizontal Restraints Notice, published in 2001, sets out the Commission policy regarding agreements between competitors and potential competitors. The Notice on Application of Article 81(3), published in 2004, elaborates on the analytical structure for application of Article 81(3) set out in the first two notices to create a generally applicable framework for assessing both competitive harm and consumer benefits. The Commission 'Guidelines on the application of Article 81 of the EC Treaty to technology transfer agreements'[9] applies these concepts in the context of IP licensing and is therefore particularly important for consideration of the issues considered in this work. In the discussion below we refer extensively to these notices, with reference where appropriate to the older case law.

[7] For example, the form for notification of agreements, Form A/B, institutionalised a formalistic approach.

[8] These notices are officially published documents setting out the Commission's view of the appropriate interpretation of Article 81 in a specific context. The notices establish the Commission's enforcement policy. More importantly in the post-modernisation environment, the notices are intended to provide guidance to Member State enforcement authorities and Member State courts that are now, for the first time, entrusted with application of Article 81 in its entirety.

[9] OJ [2004] C 101/2–42.

D. Applying the New Paradigm—Structure of Analysis under Article 81

(1) Article 81(1)

3.15 Article 81(1) prohibits:

> 'All agreements between undertakings, decisions by associations of undertakings, or concerted practices which may affect trade between Member States and which have as their object or effect, the prevention, restriction or distortion of competition within the common market'.

On its face, this language sets out four requirements for the application of Article 81(1):

(i) the existence of an agreement, decision of an association, or concerted practice;
(ii) participation by more than one 'undertaking';
(iii) an effect on trade between Member States; and
(iv) the object *or* effect of preventing, restricting or distorting competition.

The first three requirements do not represent a change from the position under the old regime. The real issue in applying economic analysis to Article 81(1) involves the fourth requirement—when does an agreement have as its object or effect the prevention, restriction or distortion of competition?

3.16 Under the old regime, this question was simple. 'Competition' was viewed as the rivalry between firms in a market. Thus the concept 'restriction of competition' covered any limitation of a competitor's ability to engage in 'competitive' activity—i.e. in conduct advancing their rivalry. The competitive autonomy of the firms—their freedom from restrictions on rivalry—was the key reference point for applying Article 81(1). From the economic perspective, in contrast, an agreement that limits rivalry is 'anticompetitive' only to the extent that it could lead to anticompetitive harm, a coordinated effect or a unilateral effect. Arguably at least, a purist economic interpretation of Article 81 would focus simply on whether an agreement leads to anticompetitive harm.[10] In practical terms, this distinction may not appear terribly important, since most agreement that could lead to anticompetitive harm involves a limitation on competitive autonomy. The competitive autonomy concept is nonetheless important because it retains a role in application of Article 81(1). This role is shown by analysis of when an agreement has the *effect* of restricting competition under the new paradigm.

[10] This would be consistent with the wording of Article 81, which refers to prevention and distortion of competition—not simply to restrictions.

(a) When does an agreement have the 'effect' of restricting competition?

Under the old regime, an agreement (or provision in an agreement) had the effect of restricting competition if it restricted the competitive autonomy of either party and had an 'appreciable' effect—usually determined by reference to the share of the market affected by the agreement. The new paradigm still appears to contemplate that a restrictive provision must involve a restriction on rivalry. This is reflected in the discussion in both the Article 81(3) Guidelines and the Intellectual Property Guidelines regarding effects on actual or potential competition.[11] The discussion in the Guidelines suggest that it remains necessary to show that a 'restraint' limits actual or potential competition that would have existed either without the agreement or without the allegedly restrictive provisions. The Guidelines leave little doubt that the 'competition' for this purpose involves the conduct—actual or potential rivalry—of firms in the market.[12] **3.17**

The Guidelines introduce economic analysis in the separate discussion of whether a restriction has anticompetitive effects. The Guidelines indicate that for an agreement to have the effect of restricting competition for the purposes of Article 81(1) it must affect actual and potential competition to such an extent that, on the relevant market, negative effects on prices, output, innovation or the variety or quality of goods and services can be expected with a reasonable degree of probability.[13] Such negative effects must be *appreciable:* the prohibition rule of Article 81(1) does not apply when the identified anti-competitive effects are *insignificant.*[14] Thus the prohibition rule of Article 81(1) only applies where, on the basis of proper market analysis, it can be concluded that the agreement has likely appreciable anti-competitive effects on the market.[15] The burden of proof for showing that such effects exist normally rests with the Commission.[16] **3.18**

The effect of these pronouncements is to create a two-tier requirement for **3.19**

---

[11] See Article 81(3) Guidelines paras 17–19.

[12] Consider the following passage from the Article 81(3) Guidelines, para 18: 'For instance, where two undertakings established in different Member States undertake not to sell products in each other's home markets, (potential) competition that existed prior to the agreement is restricted. Similarly, where a supplier imposes obligations on his distributors not to sell competing products and these obligations foreclose third party access to the market, actual or potential competition that would have existed in the absence of the agreement is restricted.' In each case it is the conduct of the producers, distributors or third parties that is 'restricted'.

[13] Article 81(3) Guidelines para 24. [14] Ibid.

[15] This approach takes into account the economic approach on which the Guidelines are based and 'gives Article 81(1) a fairly narrow scope of application'. Gauer, C, Dalheimer, D, Kjølbye, L, and Desmijter, E, *'Regulation 1/2003, a modernized application of EC Competition Rules'*, I Competition Policy Newsletter 3.8 (2003).

[16] Article 2, Council Regulation (EC) No 1/2003 on the implementation of the rules of competition laid down in Articles 81 and 82 of the Treaty, [2003] OJ L 1/1. See Schwarze, Jürgen and Weitbrecht, Andreas, *Grundzüge des europäischen Kartellverfahrensrechts,* § 11, 29 et seq. Nomos, 2004.

showing that an agreement is subject to Article 81(1) because it is restrictive by effect. First it is necessary to show that the agreement or provision constitutes a 'restriction of competition' according to the competitive autonomy approach of the old regime. Second, it is necessary to show that the agreement or provision has an appreciable anticompetitive effect, applying the modern economics-based approach. Put another way, it is necessary to show both a 'restriction' [of competition] and a 'restrictive effect' [on competition].[17]

(b) Restrictions by 'object'—the modern approach

3.20 The same double threshold for applying Article 81(1) applies where agreements are potentially restrictive by 'object'. Under the old regime, restrictions by object included those provisions that had as their sole or principal purpose limiting of one of the important parameters of rivalry; setting of prices, allocation of customers, division of territories. If a contractual provision fell into one of these categories, it was 'automatically' subject to Article 81(1), unless it could be shown to be 'ancillary' to an otherwise lawful agreement.

3.21 Under the new paradigm, in contrast, 'restrictions of competition by object are those that are presumed in light of the objectives pursued by the Community competition rules, to have such a high potential of negative effects on competition that it is unnecessary for the purposes of applying Article 81(1) to demonstrate any actual effects on the market'.[18] This presumption is based on experience showing that they are likely to produce negative effects on the market and to jeopardise the objectives pursued by the Community competition rules.[19] The assessment of whether or not an agreement has as its object the restriction of competition is based on a number of factors. These factors include, in particular, the content of the agreement and the objective aims pursued by it. It may also be necessary to consider the context in which it is (to be) applied and the actual conduct and behaviour of the parties on the market. In other words, an examination of the facts underlying the agreement and the specific circumstances in which it operates may be required before it can be concluded whether a particular restriction constitutes a restriction of competition by object.[20]

---

[17] It is worth asking why this two-stage approach is necessary. It would after all be possible to assess any agreement or concerted practice under Article 81 in order to determine whether it created competitive harm, without the intermediate step of considering the preliminary question of whether there is a restriction in a formal sense. One explanation is that this approach helps to maintain consistency with the language, if not the spirit, of the old case law. This approach also has an advantage, however, in filtering out cases where competitive harm is very unlikely without the need for even a preliminary economic analysis.

[18] Article 81(3) Guidelines para 21. [19] Ibid.

[20] Article 81(3) Guidelines para 22.

Thus even a restriction by 'object' under the new paradigm must satisfy the two-tier test. First, it must be 'restrictive' on its face. Secondly, it must fall within the category of restrictions that are inherently likely to cause competitive harm under the economic definition. If in the context where the agreement is implemented, the application of the presumption of anticompetitive harm is unjustified, then the restriction should not be regarded as anticompetitive by object.[21] **3.22**

(c) Implications of the two-tier test—non-restrictive agreements and 'ancillarity'

The double threshold for application of Article 81(1) means that two different kinds of arguments are relevant for determining whether Article 81(1) is applicable. The first type of argument concerns the question of whether the agreement constitutes a restriction of competition. If there is no limitation on some significant form of competitive conduct on the part of either a party to the agreement or some third party, then the agreement or practice falls outside Article 81(1) because there is no *restriction*. Likewise, if the parties to the agreement are not actual or potential competitors, a restriction on the parties may not be caught by Article 81(1) because it is not a restriction of *competition*. The extensive case law developed under the old regime dealing with the question of when a restriction is necessary to enter a new market or provide a new product continues to be relevant in this context. **3.23**

The second type of argument concerns whether the agreement has the *effect* of restricting competition, i.e. whether it causes competitive harm. This is usually an empirical question, to be considered on the basis of the economic factors discussed later in this work. A special question arises where a 'restriction' occurs in the context of a broader transaction or agreement. Where the broader agreement itself is not restrictive of competition and where the 'restriction is directly related to the main agreement, . . . is subordinate to the implementation of that transaction and is inseparably linked to it', then the subordinate restriction cannot cause competitive harm.[22] (If the restriction *did* cause competitive harm it would mean that the larger agreement caused competitive harm.) Since restrictions of this nature, which are often referred to as 'ancillary restrictions', cannot have competitive harm on their own they cannot be subject on their own to Article 81(1).[23] **3.24**

[21] This analysis is supported by the treatment of restrictions by object in the new TTBER. Where a restriction on price, customers or territories is regarded as exemptible under the Regulation, it is deemed to be restrictive by effect, and no longer by object.

[22] Article 81(3) Guidelines, para 29.

[23] It may be noted that the concept of 'ancillarity' here differs from the concept of ancillarity as it is understood under US antitrust law. In US law, an ancillary restriction is entitled to consideration under a 'cult of reason' analysis, even though it might otherwise be treated as a 'hard core' restriction – but it may still be found unlawful.

(2) Article 81(3)

3.25 According to Article 81(3), the prohibition of agreements caught by Article 81(1) may be declared inapplicable to:

> '— any agreement or category of agreements between undertakings;
> — any decision or category of decisions by associations of undertakings;
> — any concerted practice or category of concerted practices,
>
> which contributes to improving the production or distribution of goods or to promoting technical or economic progress, while allowing consumers a fair share of the resulting benefit, and which does not:
>
> (a) impose on the undertakings concerned restrictions which are not indispensable to the attainment of these objectives;
> (b) afford such undertakings the possibility of eliminating competition in respect of a substantial part of the products in question.'

(a) General principles applicable to Article 81(3) analysis

3.26 Article 81(3) explicitly applies to: an 'agreement or category of agreements between undertakings'; a 'decision or category of decisions by associations of undertakings' or to a 'concerted practice or category of concerted practices'. Article 81(3) cannot therefore be applied to an individual restriction isolated from its context: an enforcement authority or court applying Article 81(3) must consider the restrictive effects and economic benefits of the agreement or practice as a whole, in the form presented by the parties.

3.27 Article 81(3) does not exclude *a priori* certain types of agreements from its scope. As a matter of principle all restrictive agreements that fulfil the four conditions of Article 81(3) are covered by the exception rule.[24] Indeed Article 81(3) on its face expressly acknowledges that restrictive agreements may generate objective economic benefits so as to outweigh the negative effects of the restriction of competition.[25] When the pro-competitive effects of an agreement outweigh its anticompetitive effects the agreement is on balance pro-competitive and compatible with the objectives of the Community competition rules.

3.28 The application of Article 81(3) is subject to four cumulative conditions, two positive and two negative:

(a) The agreement must contribute to improving the production or distribution of goods or contribute to promoting technical or economic progress.
(b) Consumers must receive a fair share of the resulting benefits.

---

[24] Case T-17/93 *Matra* [1994] ECR II-595 para 85. See also Whish, Richard, *Competition Law*, 5th ed, LexisNexis Butterworths, at p 150 with further examples.

[25] Article 81(3) Guidelines para 33.

(c) The restrictions must be indispensable to the attainment of these objectives.
(d) The agreement must not afford the parties the possibility of eliminating competition in respect of a substantial part of the products in question.

The application of each of these requirements is discussed in turn below.

(b) Promoting 'economic benefits'

3.29 The first requirement of Article 81(3)—'improving the production or distribution of goods or contribute to promoting technical or economic progress'—is the key to application of Article 81(3) under the new paradigm. This provision requires assessment of the 'economic benefits' of a agreement or practice in terms of static or dynamic efficiencies.[26] The proponents of the agreement need to demonstrate tangible benefits. In keeping with the overall economic analysis, these must be welfare benefits—increases in product quality, increases in production or decreases in prices. Demonstrating economic benefits does not, however, mean that the first requirement of Article 81(3) is satisfied. For an agreement, decision or practice to qualify for Article 81(3) exemption, its *net effect* must be beneficial, i.e. the benefits of the agreement as a whole must be shown to outweigh the restrictive effects identified in an Article 81(1) analysis.[27] This analysis requires a quantitative economic assessment of the agreement, decision or practice in order to carrying out the balancing of restrictive effects against benefits.[28] Even if an agreement was ruled anticompetitive 'by object' it

---

[26] See Horizontal Guidelines para 32: 'The first condition [of Article 81(3)] requires that the agreement contributes to improving the production or distribution of products or to promoting technical or economic progress. As these benefits relate to static or dynamic efficiencies, they can be referred to as "economic benefits" '.

[27] Horizontal Guidelines para 32; Article 81(3) Guidelines para 33. See Joined Cases 56/64 and 58/66, *Consten and Grundig* [1966] ECR 429: 'This improvement must in particular show appreciable objective advantages of such a character as to compensate for the disadvantages which they cause in the field of competition'. See in addition Joined Cases 209–215/78 and 218/78 *Van Landewyck* v *Commission* [1980] ECR 3125; Joined Cases 25 and 26/84 *Ford* v *Commission* ECR [1985] 2725; Case 75/84 *Metro II* [1986] ECR 3021; *Vichy* [1991] OJ L 75/57; *Schoeller* [1993] OJ L 183/1; *SCK/FNK* [1995] OJ L 312/79 at para 2.121. It will be noted that the Article 81(3) Guidelines take a somewhat different approach, assigning the balance of restrictive effects and benefits to the 'fair share for consumers' requirement of Article 81(3). (See Article 81(3) Guidelines at paras 85–86.)

[28] Article 81(3) Guidelines paras 55 and 101. See also Case T-206/99 *Métropole Télévision SA* v *Commission*, Judgment of 20 March 2001 at para 57:

> '. . . according to consistent case-law, where the Commission has a power of appraisal in order to carry out its duties, respect for the rights guaranteed by the Community legal order in administrative procedures is all the more fundamental. Those guarantees include, in particular, the duty of the competent institution to examine carefully and impartially all the relevant aspects of the individual case (Case C-269/90 *Technische Universität München* [1991] ECR I-5469, paragraph 14; Case T-44/90 *La Cinq* v *Commission* [1992] ECR II-1, paragraph 86).'

remains necessary to identify and quantify the real restrictive effects in order to carry out the Article 81(3) assessment.

3.30 As regards the benefits of the agreement, it follows from the application of a balancing test that where the results of this assessment find no empirical evidence of restrictive effects, *any* appreciable benefit that results from the agreement, decision or practice will fulfill the requirements of Article 81(3). Furthermore, it follows that where appreciable net benefits can be shown, it is irrelevant whether such benefits are 'low'; the significance of the agreement, decision or practice in the context of the markets affected is not relevant to the application of the first condition of Article 81(3).[29]

3.31 One issue on which the Commission approach appears to be unnecessarily conservative involves identification of the market or markets where the pro-competitive benefit is generated. The Commission has taken the position that these benefits must be shown either in the *same* market as the one where the anticompetitive harm is identified or in a closely related market[30] (see below). The Commission position is curious because Article 81(3) does not explicitly limit the context in which a benefit need be generated.[31] The narrow interpretation offered by the Commission could have the result of prohibiting agreements where there are major consumer welfare benefits.

(c) Consumers receive a fair share of the resulting benefit

3.32 The new Article 81 standard is a 'consumer welfare' standard. Arrangements that arguably improve allocative efficiency may still be blocked if the benefits of that improvement are retained entirely by the firms involved in the arrange-

[29] See also Article 81(3) Guidelines para 85: 'In line with the overall objective of Article 81 to prevent anti-competitive agreements, the net effect of the agreement must be at least neutral from the point of view of those consumers directly or likely affected by the agreement.'

[30] Article 81(3) Guidelines para 43.

[31] The Commission bases this conclusion on the judgment of the Court of First Instance in Case T-131/99, *Shaw* [2002] ECR II-2023, para 163, and on the requirement that consumers share the benefit (Article 81(3) Guidelines at note 56). The Commission reliance on the *Shaw* judgment to support this rule appears to be misplaced. In that case, operators of public houses subject to a tied sales agreement imposed by Whitbread applied for annulment of a Commission decision exempting the agreement under Article 81(3). The applicants contended that the Commission had erred in application of Article 81(3) because it had considered the restrictive effects and benefits of the Whitbread agreements as a class rather than by assessing each agreement individually. The Court rejected this argument, observing that it was the agreements taken together that led to the restriction, not the individual agreement, so it was appropriate to look at the benefits of the agreements as a whole. While the Court then refers to the benefits to the tied lessees taken as a whole, this simply follow the analysis of benefits set out in the Commission decision, it cannot be read as requiring that benefits are realised by the same customer group that suffers harm. The key requirement is that the agreements causing the restriction and the benefit are the same.

ments. The availability of new and improved products constitutes an important source of consumer welfare. As long as the increase in value stemming from such improvements exceeds any harm from maintenance or an increase in price caused by the restrictive agreement, consumers are better off than without the agreement and the consumer pass-on requirement of Article 81(3) is normally fulfilled.[32] Similarly, where it is shown that consumers will benefit from more favourable prices as a result of the agreement, decision or practice, the second condition is likely to be fulfilled.[33] The 'consumer benefit' condition is also likely to be fulfilled where the underlying markets are competitive as the benefits of the agreement, decision or practice will be passed on to the consumer as the various undertakings compete for business.[34]

(d) Indispensability

According to the third condition of Article 81(3) an agreement must not impose restrictions, which are not indispensable to the attainment of the economic benefits created by the agreement in question (*i.e.*, the benefits identified in response to the first condition). In the context of this condition, the decisive factor is whether economic benefits are produced as a result of the agreement that would not be fully realised in the absence of the agreement.[35] **3.33**

This condition requires that the efficiencies be specific to the agreement in question in the sense that there are no other economically practicable and less restrictive means for the parties to achieve these efficiencies.[36] In making this assessment the market conditions and business realities facing the parties to the agreement must be taken into account. Firms invoking the benefit of Article 81(3) are not required to consider hypothetical or theoretical alternatives. The Commission recognises in its Guidelines that it cannot second guess the business judgment of the parties. Intervention is only appropriate where it is reasonably clear that there are realistic and attainable alternatives. The parties must only explain and demonstrate why such seemingly realistic and significantly less restrictive alternatives to the agreement would be significantly less efficient.[37] **3.34**

It follows that there is thus no scope under the third condition of Article 81(3) for an assessment of whether theoretical alternative arrangements that would **3.35**

---

32 Article 81(3) Guidelines para 104.

33 See e.g. *BT/MCI* [1994] OJ L 223/36, rec. 55.

34 See e.g. Horizontal Guidelines para 34.

35 Article 81(3) Guidelines para 74.

36 Article 81(3) Guidelines para 75. See also *FN/CF* OJ [1971] L134/6; *GEC/Weir Sodium Circulators* OJ [1977] L327/26; *Amersham/Buchler* OJ [1982] L314/34; *Ivoclar* OJ [1985] L369/1; *X/Open Group* OJ [1987] L35/36; *ABI* OJ [1987] L43/51; *Iveco/Ford* OJ [1988] L230/39; *BBC Brown Boveri* OJ [1988] L301/68; *Delta Chemie/DDD* OJ [1988] L309/34; *Alcatel/Espace/ANT* OJ [1990] L32/19; *Concordato Incendio* OJ [1990] L15/25; *Grundig II* OJ [1994] L20/15; *Jahrhundertvertrag*, OJ [1993] L50/14; *Fujitsu/AMD* OJ [1994] L341/66; *Asahi/St Gobain* OJ [1994] L354/87; *Atlas* OJ [1996] L239/23; *Phoenix/GlobalOne* OJ [1996] L239/57.

37 Article 81(3) Guidelines para 75.

result in lesser benefits would be somehow 'preferable'. Indeed, the wording of Article 81(3) itself[38] makes it clear that it not sufficient that such an alternative arrangement reproduce the 'bulk of existing benefit'; any alternative must reproduce the *entire benefit* of the existing system.[39] There is also no scope for an assessment of whether *other parties* might create alternative arrangements that could duplicate or improve upon the benefits of the agreements, decisions or practices in question: Article 81(3) clearly refers only to the parties to the agreement, decision or practice that is under investigation.

3.36 Once it has been established that the agreement, decision or practice results in an appreciable net benefit under the first condition of Article 81(3), there can be no additional requirement to show proportionality between benefits and restrictive effects imposed under the third condition of Article 81(3). The Community law principle of proportionality has not been developed to determine whether a particular agreement, decision or practice promotes the competitive process, namely to win customers by offering better products or better prices than those offered by rivals. That is the rule fulfilled by the balancing process under the first condition of Article 81(3).

(e) The agreement should not 'eliminate' competition

3.37 The fourth condition of Article 81(3) bars application of Article 81(3) where an agreement leads to the *elimination* of competition in a substantial part of the EU. This provision is intended to ensure that the degree of competition necessary to ensure the observance of the basic requirements and the attainment of the objectives of the Treaty is preserved.[40] It is important to note that this condition requires that the elimination of competition be a result of the agreement under consideration. Even if there is no competition at all in the affected market, this condition is not applicable where the agreement does not affect that position (e.g., by creating a barrier to entry).

3.38 Whether competition is being 'eliminated' within the meaning of the fourth

---

[38] Article 81(3) refers to 'resulting benefit' (2nd condition) and 'these objectives' (3rd condition). It clearly does not refer to 'the bulk' of the resulting benefit or 'the bulk' of these objectives.

[39] See Case 27/76 *Metro I* [1977] ECR 1875; *SOPELEM/Vickers* OJ [1978] L70/47; *Rockwell/Iveco* OJ [1983] L224/19; *VW/MAN* OJ [1983] L376/11; *Optical Fibres* OJ [1986] L236/30; *ENI/Montedison* OJ [1986] L5/13; *Enichem/ICI* OJ [1988] L50/18; *IATA Passenger Agency Programme* [1991] L 258/18; *IATA Cargo Agency Programme* [1991] L 258/29; *Assurpol* [1992] OJ L 37/16; *Ford/Volkswagen* [1993] OJ L 20/14; *Pasteur/Mérieux* [1994] OJ L 30/1; *Philips/Osram* [1994] OJ L 378/37; *Lufthansa/SAS* [1996] OJ L 54/28; *Scottish and Newcastle* [1999] OJ L 186/28, para 162; *Whitbread* [1999] OJ L 88/26, para 175; *Bass* [1999] OJ L 186/1, para 193. See also Faull, J. and Nikpay, A, *The EC Law of Competition*, 2nd ed, 2006, Oxford University Press, at para 2.166; Sauter in Immenga/Mestmäcker, EG-Wettbewerbsrecht, Art 85 EGV, 25.

[40] See e.g. Case 26/76 *Metro* v *Commission* [1977] ECR 1875; Case T-7/93 *Langnese-Iglo* v *Commission* [1995] ECR II-1533.

condition of Article 81(3) will depend on the degree of competition existing prior to the agreement and on the impact of the restrictive agreement on competition, i.e. the reduction in competition that the agreement brings about.[41] It is evident that the standard required to show that competition is 'eliminated' is higher than that required to show that competition is 'restricted'.[42] The European Court has suggested that this point may be reached where competition is no longer 'effective' in a significant market as a result of the agreement,[43] i.e. that as a result of the agreement the parties to the agreement are no longer under effective competitive constraint from each other or from third parties.

There is no real precedent regarding what would constitute a 'substantial part' of the EU for purposes of this provision. Most commentators have equated this requirement with the similar language in Article 82, which is usually interpreted as referring to the geographic scope of the market that is affected. It is at least arguable, however, that this requirement should be interpreted to refer not just to the geographical extent,[44] but also to the overall importance of the affected market.[45] **3.39**

As a practical matter, the 'no elimination of competition' condition is unlikely to be significant in many cases. It will be unusual that an agreement could eliminate competition in a significant market and still meet the balancing test of benefits versus restrictive effects. This is most likely to occur where the benefits of an agreement are attributable to a different market then the one where the competitive harm 'occurs'. If the 'benefited market' is substantially larger than the 'harmed market' and competition is eliminated in the harmed market, the arrangements could fail the elimination of competition condition. This would mean that the agreements would be prohibited even though they generate a net benefit in consumer welfare or allocative efficiency. Since this result is difficult to justify from the perspective of an economics-based competition policy, special care will be required both in defining markets and in identifying the 'elimination of competition' to ensure that this condition can only be invoked where there is a serious economic cost in a significant market. **3.40**

[41] Article 81(3) Guidelines para 107.

[42] Otherwise Article 81(3) would be redundant since it only becomes applicable where there is a restriction of competition.

[43] See Case 26/76 *Metro* v *Commission* [1977] ECR 1875.

[44] In the Article 82 context, the territory of a Member State is generally regarded as meeting the substantial part requirement. In Article 81(3), however, the specific reference to the 'products in question' would appear to require a different approach.

[45] Note AG in *Ahmed Saeed*.

E. Conclusion—A Structure for Analysis under the New Paradigm

3.41 The analysis required under the new paradigm can be summarised as a 10 step process:

Step One Is there an agreement, concerted practice, or decision of an association of undertakings?

Step Two If so, does that agreement 'restrict' competition—is it an agreement between actual or potential competitors or do either the agreement itself or provisions in the agreement restrict a firm in respect of a significant parameter of competition? (Note that to answer this question it is necessary to define the relevant market.)

Step Three Is the 'restrictive' agreement or provision 'ancillary' to some arguably legitimate arrangement or agreement (i.e. necessary for that activity to proceed)? If so the arrangement or agreement itself will be the subject of Article 81 assessment.

Step Four If a restriction identified in Step Two is not ancillary—does the restriction have the 'object' of restricting competition? If the answer is 'yes', go to Step Six.

Step Five If the restriction is not ancillary and does not have the object of restricting competition, does it have the effect of restricting competition? This requires an economic analysis in the context of the market defined in Step Two.

Step Six If the restriction has either the object or effect of restricting competition, does it generate significant economic benefits? This requires an economic assessment of the efficiency gains that will be generated by the agreement or agreements under review. (If the 'restriction' identified in Step Two concerns a provision in an agreement rather than the agreement itself, it is still necessary to consider the benefits of the agreement; the restriction is dealt with in Step Eight).

Step Seven Do those benefits outweigh the restrictive effects identified in Step Five? (If the provision was classified as a restriction by object under Step Four, it is necessary to assess the actual competitive harm caused by the restriction to perform this analysis.)

Step Eight Is the restriction identified in Step Two 'indispensable' to achieve the benefits identified in Step Six?

Step Nine Do consumers share in the benefits identified in Step Six—i.e. does the agreement or practice contribute to *consumer* welfare?

Step Ten Does the agreement give the participating firms the possibility of eliminating competition in an affected market?

Steps One to Five involve application of Article 81(1). Where Article 81 is applied by a regulatory authority or court, the burden will be on the regulator or plaintiff to show that the questions set out at Steps One, Two, and either Four or Five are affirmative, and to show that the agreement or provision is not ancillary under Step Three. Steps Six to Ten involve application of Article 81(3). Where Article 81 is applied by a regulatory authority or court, the burden will be on the defendant to show that the answers to the questions set out at Steps Six to Nine are affirmative and that the answer at Step Ten is negative. Since the requirements of Article 81(3) are cumulative, a court or regulator may conclude that Article 81(3) is inapplicable by finding that the defendant cannot satisfy its burden on any of these points (in other words, it is not necessary to take Steps Six through Ten in order). **3.42**

From a comparative law perspective, the new paradigm bears a very close resemblance to the 'rule of reason' under US antitrust law. This is not coincidental. The architects of the new paradigm were clearly influenced by the development of a structured rule of reason analysis in the US. Furthermore, the role of guidelines and administrative safe harbours in application of the new paradigm mirrors the use of such tools by the US agencies. This is particularly the case in the context of applying Article 81 to intellectual property licensing. On the other hand, in contrast to the position under US law, there are no *per se* rules under the new paradigm. Furthermore there remain potentially significant differences in the approach to defining markets and in the assessment of market power. **3.43**

Veterans of the old competition law regime may ask, where do the block exemptions fit into this analytical structure? From a formal perspective, the block exemptions come in between Steps Five and Six, because where a block exemption is applicable, further assessment under Article 81(3) is not necessary. From a practical perspective, however, in a self-assement exercise the block exemptions are applied between Steps Two and Three. If a 'restrictive' provision clearly qualifies for exemption under a block exemption there is no need to consider whether it is ancillary or whether it has a restrictive effect.[46] Where the application of the block exemption is uncertain, however, it is necessary to perform the full analysis. The potential applicability of the block exemption then becomes relevant to a subsequent risk assessment. **3.44**

[46] Since application of the block exemptions drafted under the new paradigm require application of a market share filter it is necessary to identify the restriction and define the market under Step Two before the block exemption can be applied. Even where the block exemptions are not formally applicable the structure of assessment under the block exemptions can still be relevant to the general Article 81 assessment, particularly as regards classifying provisions as restrictive by object or by effect.

4

APPLYING THE ECONOMICS-BASED APPROACH TO ARTICLE 81: ASSESSING COMPETITIVE HARM AND ECONOMIC BENEFIT

A. Introduction

It will be evident from the previous chapter that the focus of Article 81 analysis under the new paradigm is on economic analysis. The application of Article 81(1) depends on whether an agreement or practice has a restrictive effect on competition, which depends in turn on whether the agreement leads to economic harm 4.01

in terms of higher prices, limited output, or reductions in product quality. The application of Article 81(3) depends on whether the agreement generates economic benefits, which are in turn based on an economic assessment of efficiency gains that may be associated with those agreements. This reliance on economic assessment to determine the legality of agreements creates a challenge for courts, regulators and legal advisors. This chapter outlines the nature of this assessment and identifies some of the principal concepts necessary for making this assessment.[1] This discussion provides the basis for applying the new paradigm to specific issues in intellectual property licensing in subsequent chapters.

B. Policy Goals and Key Concepts

(1) The goals of competition policy—economic efficiency, consumer welfare and the nature of 'competition'

4.02 The modernisation package is based on the premise that the goal of competition policy is to foster economic *efficiency* and increase *consumer welfare.* Economic efficiency describes how well a market functions compared to the ideal 'efficient' result.[2] Consumer welfare describes a result of an increase in efficiency; raising the aggregate benefit of economic activity to purchasers in a market (and ultimately to the customers of those purchasers as well).[3] 'In simple terms, "competition" leads to an increase in consumer welfare because it leads to lower prices and better quality products and services and to an optimal distribution of scarce resources.'

[1] The purpose of this discussion is to make these concepts accessible to non-specialists. Readers who require a proper introduction to the economics of competition law should consult one of the specialised works available on this subject: Posner, R, *Antitrust Law*, 2nd ed, 2001, University of Chicago Press; Bishop, S and Walker, M, *The Economics of EC Competition Law*, 2nd ed, 2002, Sweet & Maxwell; Hovenkamp, H, *Federal Antitrust Policy: The Law of Competition and its Practice*, (2nd edn, 1999, West Publishing Co.

[2] The ideal efficient result is referred to as 'Pareto efficiency' and describes the position where no change in the market that improved the position of one person could be achieved without making someone else worse off. Because Pareto efficiency is seldom achieved (and almost impossible to test) the more utilitarian standard of efficiency adopted is 'Kaldor–Hicks efficiency' which defines efficiency as where the benefits to one person or group of a change in market conditions exceed the losses that are realised by another person or group as a result of the change.

[3] It will be noted that a 'consumer' for this purpose is contrasted with a 'supplier' and does not necessarily represent an individual purchasing goods for final consumption. (Where a purchaser or purchasers in a market have market power, however, the consumer will usually be identified with *their* customers.) Although purchaser benefits will normally filter down to the final consumer level, it is not necessary and not always possible to trace the benefits directly to the final consumer. The consumer welfare standard is contrasted with a 'total welfare standard' whereby the efficiency gains are valued even where they are fully retained on the supplier side.

But what is 'competition'? From the perspective of classical economics, competition is the opposite of 'monopoly'. Thus where there are many sellers and buyers in a market, that market can be described as competitive. This is a 'structural' view of competition. From the perspective of the non-economist, 'competition' describes the conduct of firms in a market. Where firms introduce new products, increase capacity, lower prices, or invest heavily in advertising, the market is described as 'competitive'. Where firms build cartels or where a few large firms charge high prices for low quality products, the market is perceived as 'uncompetititive'. Confusingly, economists refer to conduct in the market as 'rivalry'. This can be misleading to the extent that it suggests that competitive conduct involves constant interaction between competitors, like runners in a race, to divert sales or customers from each other. Yet in many 'competitive' markets, sellers will have little or no idea of the precise position of their rivals or the identity of their customers. A final way of describing 'competition' refers to the state of the market itself; where the market generates substantial consumer benefit (i.e., where the market is close to a result that is Pareto efficient) that market is competitive, while a market that is less efficient is, in its result, less competitive. This perspective focuses on the 'performance' of the market. **4.03**

The three perspectives on competition identified above are all clearly related.[4] The 'structure' of a market influences the way in which firms act, and the way that firms act determines (together with external factors such as the costs of raw materials and the performance of the overall economy) how the market will perform. The focus in modern competition analysis on consumer welfare, however, with its roots in price theory, focuses the analysis squarely on the end result. Neither structure nor conduct can be considered in isolation. The key question is how either may lead to an improvement or deterioration in the way the market performs—an increase or reduction in efficiency. The goal of competition policy therefore becomes one of controlling conduct or structural change that could reduce efficiency, while allowing and indeed encouraging conduct or structural change that could increase efficiency. **4.04**

The central analytical tool used to implement modern competition policy is the concept of 'market power'. Only where there is market power is conduct or structural change likely to reduce efficiency. This reduction of efficiency may **4.05**

[4] A fourth perspective on 'competition' would focus on the 'process' by which markets change or adapt to new technologies and new forms of organisation. The so-called 'Austrian' view describes competition in this sense as 'a process of creative destruction.' (Schumpeter, J, '*Capitalism, Socialism and Democracy*, 1976, George Allen & Unwin) Proponents of this dynamic view of competition may regard reliance on regulatory control of market power as misguided, since it will slow the dynamic development of markets, leading to consumer harm in the medium and long term.

occur through *unilateral effects* (where a single firm protects or increases its market power, e.g. by abusive leveraging) or through *joint effects* (where firms can assert collective market power through their interaction on a market, e.g. by coordinating price increases). The existence of market power does not, however, mean that conduct or structural change will necessarily lead to a reduction in efficiency; it could in fact lead to an increase in efficiency. Somewhat confusingly, increases in efficiency are described as 'efficiencies'. These are in turn classified as static efficiencies (improvements in the current performance of the market) and dynamic efficiencies (improvements in the performance of the market over time). Each of these concepts is described in greater detail below.

(a) The key analytical tool—'market power'

4.06 Economic theory suggests that, under fully competitive conditions, prices will tend towards the point where they cover the marginal costs of the most efficient producer, including a competitive return on the capital invested by that firm. Where a firm can raise its prices above this 'competitive level' it is said to possess market power. Where market power reaches the point where a firm can unilaterally price at the point where it maximises its own profitability—the firm has 'monopoly power'—the extreme form of market power. Where market power is absent, there is no basis for applying competition rules because the market itself should produce an efficient allocation of resources and a maximisation of consumer welfare. While limited market power is commonplace (because firms seek to raise their profitability by *differentiating* their products), this process makes products more directly responsive to consumer needs and thus itself leads to optimal resource allocation and full consumer welfare. Thus, in a competition law system based on economic principles, intervention becomes necessary only where there is either already some significant degree of market power, or where an agreement or practice could allow a firm (or firms) to obtain a significant degree of market power. This makes the identification of actual or potential market power the key to initial assessment of any agreement or practice.

(b) Identifying market power—the need to define relevant markets

4.07 Although it is sometimes possible to identify the existence of market power by direct evidence,[5] the existence of market power is usually implied from the position of the firm or firms in a relevant market. This practice is based on the premise, largely supported by empirical observation, that the ability of firms to raise prices above a competitive level is related to their share of the market where they operate. The shares held by rivals and the extent to which rivals not

[5] *FTC* v *Staples Inc*, 970 F. Supp. 1066 (DDC 1997).

currently competing in the market might enter the market are also relevant to the determination of whether the firms under consideration have significant market power. The analysis of relative or absolute market shares, the identification and assessment of barriers to entry (including the importance of proprietary technology), and the consideration of other factors related to the nature of the competitors, their suppliers and their customers, together describe the 'structure' of the market. Where the number of competitors is small, a market is said to be 'concentrated'. Anticompetitive effects are usually only likely in concentrated markets where high barriers to entry are likely.

(c) The nature of competitive harm—unilateral effects and joint effects

Competitive harm—raising of prices above competitive levels or reduction in output or quality below competitive levels—can come about in two ways. Where a single firm is able to raise prices or reduce output to a significant degree, the result is a 'unilateral effect'. Where interaction between competitors leads to higher prices or a reduction in output or product quality, the result is a 'joint effect' or 'coordinated' effect. **4.08**

Typically a firm will be able to generate unilateral effects where it is significantly larger than its rivals and enjoys a substantial share of the total market. (Under these circumstances, a firm may be able to exercise market power directly (by raising prices or limiting output). These market conditions may also make it easier for a firm to preserve its existing level of market power by discouraging new competitors or making competition by existing rivals more costly. Unilateral effects may also be possible in markets where products are 'differentiated'; even a firm which is not the leader in the broader 'market' may be able to obtain a competitive advantage if it can raise the costs of its closest rivals and thus give itself latitude to impose price increases. **4.09**

The classic form of coordinated effect is an increase in prices due to a price fixing agreement between competitors. Coordinated effects can also occur, however, in the absence of an agreement between competitors, because 'coordination' between competitors becomes more likely where markets are concentrated and market conditions are transparent. Where a limited number of market participants are fully aware of each other's actions, the market price will be higher, because no competitor can expect to obtain an increase in sales by a price reduction (since all rivals will presumably follow suit).[6] The existence of **4.10**

[6] This kind of competitive harm is not necessarily unlawful under competition law, since economic theory teaches that it is an inevitable result of the market structure. Actions by firms to facilitate this kind of activity or mergers that change the market structure to make this kind of harm more likely can lead to action under competition law, unless there are benefits that outweigh these harms.

coordinated effects does not require substantial market power on the part of individual participants, in effect, the participants obtain market power that they otherwise would not have through their agreement or coordination.

(d) A special form of economic harm—frustrating geographic market integration

4.11 The central goal of the EC Treaty is to create an integrated geographic market. Removing barriers to trade within the EU across the full range of products and services is meant to improve allocative efficiency in the entire European economy, contributing directly to consumer welfare and making European firms more effective competitors on a global basis. Competition law is traditionally viewed as a tool for achieving this objective. An agreement or practice that divides markets within the EU could impede integration even where the market power of the firms involved is limited, because of the aggegrate impact of such agreements in perpetuating established trading patterns. This means that, for purposes of assessing economic harm under Article 81, an agreement that limits a party to a territory that is less than the EU as a whole will often be presumed to cause competitive harm.

C. The Benefits of Conduct or Structural Change—Static and Dynamic Efficiency

4.12 The second half of an economics-based competition policy involves the assessment of economic benefits that flow from conduct or structural change. It is not uncommon for commercial activity that leads to unilateral or coordinated effects also to lead to significant consumer welfare benefits. Thus a pharmaceutical company that develops a medicine for a previously untreatable condition may obtain substantial market power, but will at the same time contribute to a direct increase in consumer welfare. More generally any firm (or group of firms) that develops a product that meets a need that was previously either not met at all or met with products of a lower quality may gain market power while contributing to consumer welfare. Thus a competition policy that simply prohibited any conduct or structural change that led to an increase in market power will block conduct or transactions that actually benefit consumers. The challenge in formulating competition policy is to identify the situations where the harm (unilateral or coordinated effects) are greater then the benefits. This requires an economic analysis of those benefits. From an economic perspective, conduct or structural change can lead to welfare benefits in two ways: by contributing to *static efficiency* or by contributing to *dynamic efficiency.*

(1) Static efficiency—allocative and productive efficiencies

Static efficiency measures the efficiency of a market at a particular point in time. Static efficiency involves both 'allocative efficiency' and 'productive efficiency'. Allocative efficiency refers to the efficiency of the market as a whole; full allocative efficiency is achieved when the market is 'cleared' by the price mechanism matching the value consumers place on a product to its costs of production. Thus an agreement that leads to an increase in the efficiency of the market as a whole (e.g. by lowering transaction costs for all participants or by decreasing the cost of a common input) can lead directly to a better result in terms of allocative efficiency.[7] Productive efficiency refers to the more efficient use of the resources of an individual firm (as in achieving economies of scale) and lower costs. Productive efficiency is the most important criterion for assessment in many competition cases because it is capable, to some extent at least, of direct verification. Even where it necessary to project the future impact of an agreement or transaction on productive efficiency, current operating costs and patterns provide a basis for informed speculation. An increase in productive efficiency by a significant firm in a market by lowering costs should usually have an impact on allocative efficiency. The existence and extent of market power may, however, affect the extent to which an increase in productive efficiency leads to an increase in allocative efficiency. An increase in productive efficiency for a firm in a monopoly position can actually lead to a reduction in allocative efficiency because the production efficiencies are not passed on via price reductions. **4.13**

(2) Dynamic efficiency

Dynamic efficiency, refers to the efficiency of the market over time. It focuses on the medium- or long-term gain to future consumer welfare that may be generated by new products, processes, or ways of doing business. The importance of dynamic efficiency is universally acknowledged, particularly in the context of intellectual property rights. There is nonetheless a tension between considerations of dynamic efficiency and the modern model of competition analysis, which is based on static efficiency (both unilateral and coordinated effects describe reductions in allocative efficiency that lead to a loss of consumer welfare). It is sometimes suggested that the emphasis of competition policy on static efficiency actually impedes dynamic efficiency, because only firms that possess substantial market power (and want to keep it) can invest in research and development.[8] This tension is resolved by accepting that some conduct that **4.14**

---

[7] Agreements can facilitate practices that increase allocative efficiency by reflecting differences in demand. Thus charging a higher price to customers who put a higher value on a product ('Ramsay pricing') may lead to higher output without an increase in average prices.

[8] See Bishop, S and Walker, M *The Economics of EC Competition Law: Concepts, Application and Measurement*, 2nd edn, 2002.

creates market power is allowed where it could promote dynamic efficiency. Furthermore, dynamic efficiency considerations are also introduced, to a limited extent, to consideration of competitive harm. Conduct or structural change that leads to a reduction in dynamic efficiency may be condemned as a competitive harm, alongside reductions in static efficiency.

4.15 The difficulty in assessing dynamic efficiencies is that the benefits of present day conduct or structural change for dynamic efficiency are inevitably speculative. Even on an historical basis is it not always possible to determine whether agreements or structural changes have actually led to improvements in dynamic efficiency.[9] Looking forward, it is only possible to establish that practices or structural change create necessary conditions for improvements in dynamic efficiency. For the same reasons, the benefits of dynamic efficiencies can rarely be fully quantified. At most it may be possible to identify the range of potential benefits and the number of potential consumer beneficiaries.[10] To allow consideration of dynamic efficiencies it is not possible to apply the same standards of proof applicable to static efficiencies.[11]

D. Controlling 'Anticompetitive' Agreements—the Role of Article 81

(1) How 'agreements' lead to competitive harm

4.16 Agreements can lead both to coordinated effects and to unilateral effects. The classic form of anticompetitive agreement is the price-fixing cartel which leads to the classic form of coordinated effect—the direct reduction of competition between direct competitors. But even in the absence of a formal price-fixing cartel, agreements between competitors dividing the territories where they operate or the customers that they serve can reduce or eliminate the competitive constraints that they apply to each other. An agreement between competitors to exchange confidential information can also lead to coordinated effects because it facilitates 'collusion'; by reducing uncertainty about each competitor's activities

---

[9] It is striking in this regard that there is no empirical evidence that shows a link between innovation and the existence of a strong patent system except in certain sectors.

[10] In pharmaceutical markets it is sometimes possible to calculate the level of benefit for a possible new product by reference to the number of persons suffering or likely to suffer from the condition treated by that product.

[11] In principle this observation must also be applicable in cases where a reduction in dynamic efficiencies results in competitive harm. In this scenario, however, speculation cannot be a sufficient basis to invoke legal powers to prohibit arrangements that do not lead to significant unilateral or coordinated effects. There should be a showing that there is a real likelihood that dynamic efficiency will be negatively affected. We will return to this point.

it is easier to plan and implement price rises. Finally, an agreement can lead a loss in dynamic efficiency by eliminating firms that might produce new products or processes in the future or by reducing incentives for firms to compete on that basis.

(2) How 'agreements' lead to coordinated effects

Coordinated effects are regarded as particularly serious because they involve a direct disruption of the competitive process. Furthermore, while some coordinated effects may result from changes in market structure (markets with a limited number of players will have higher prices), agreements can lead to coordinated effects even where the market power of individual participants is limited. Coordinated effects by their nature only become an issue where the parties to an agreement are actual or potential competitors (because otherwise their 'coordination' could not increase their market power). This makes consideration of whether the parties to an agreement are actual or potential competitors an essential preliminary step in economic analysis under Article 81. **4.17**

(3) How 'agreements' lead to unilateral effects

Agreements can lead to unilateral effects where they either create market power for a firm or make it easier for a firm with market power to exercise that power. Unilateral effects can result from agreements between competitors (e.g. where competing firms allocate customers or territories). Unilateral effects can, however, also result from understandings with either suppliers or customers who are not competitors of the firm with market power. Where a customer with market power requires that its best suppliers not deal with its rivals, that agreement may have unilateral effects by making it easier for the firm imposing the condition to raise prices; if rivals are restricted to higher price inputs those rivals will not be able to constrain the price rise. Where a firm requires that its customers not resell its product below a set minimum price, that agreement may make it easier for that firm to increase its own prices, since the customers will be more willing to accept the higher price if they can make a higher per-unit margin of their own. Furthermore, the knowledge of each customer that other customers are under the same obligation to maintain higher prices means that each customer can adhere to the obligation without concern that it will be undercut by rivals selling the same product. In consequence, each customer may gain a degree of market power. **4.18**

(4) How 'agreements' can lead to a reduction in dynamic efficiency

Agreements lead to a reduction in dynamic efficiency where they slow the rate at which future efficiency improvements may occur. One example, which will be **4.19** addressed later in this work, would involve a joint venture for research and development among firms that are currently pursuing different research paths. When the joint venture agrees to focus on one research path, the potential benefits of two alternative technologies will be lost. If the path pursued by the joint venture proves abortive, then the agreement will mean that there will be no new technology where otherwise the rejected research path could have generated a successful product. Just as the evidence for consumer benefits based on improvements in dynamic efficiency is often speculative, so too the impact of an agreement in reducing dynamic efficiency is inherently uncertain. This should counsel restraint in basing regulatory intervention on the likely existence of such effects.

(5) How 'restrictive' agreements lead to consumer benefits

4.20 As a general matter, the benefits of agreements between firms for the efficient functioning of a market must be self-evident. The more limited question that must be addressed is how 'restrictive agreements'—agreements between competitors or agreements that include provisions limiting the autonomy of a party to make its own decisions regarding important forms of competitive conduct—can lead to consumer benefits. This can occur in three ways: (i) by making more efficient use of productive facilities; (ii) by making a productive activity possible or less costly; or (iii) by duplicating the benefits of integration.

(a) Consumer benefits of agreements—more efficient use of productive facilities—economies of scale and synergies

4.21 Agreements between competitors can create benefits where they allow more efficient use of existing facilities (economies of scale). Where competitors do not currently produce identical products or use identical facilities, obtaining access to a competitor's assets may allow a firm to produce a better product. Where the improved product is the result of a combination of the facilities and/or methods of the beneficiary firm and of the competitor firm, this improved product is the result of a 'synergy' between the assets of the two firms.

(b) Consumer benefits of agreements—making productive activity possible

4.22 Restrictive provisions may make a productive activity possible or less costly by limiting the risks of a productive transaction for a party. If your neighbour needs to do extra shopping on a Saturday, it may create a welfare benefit for you to lend your neighbour your car for the day. You are unlikely to allow your neighbour to use your car, however, if as a result your neighbour is either not required to return the car at all or may take your car on a 1,000 kilometre journey. The conditions that you impose on the use of the car (which may be implied in your discussion with your neighbour) allow you to make the car available and thus make your own small contribution to consumer welfare.

(c) Consumer benefits of agreements—duplicating the benefits of integration

If a firm is fully integrated (i.e., if it controls its entire production and distribution system from obtaining raw materials to selling to final consumers), it needs no agreements with outside firms at all. Such total integration is rare. Most firms need to choose between performing functions themselves or relying on third parties to perform those functions. Instead of producing all necessary inputs, a firm can contract with existing producers of those inputs. Instead of creating the distribution network necessary to sell products direct to all consumers, a firm can sell to intermediaries who will sell to final consumers. Agreements with customers and suppliers allow a firm to capture the benefits of integration in more flexible and less costly ways. Working with other independent firms in this way, however, has a price compared with a strategy based on integration. An integrated firm can plan the supply of inputs to match its own demand and can exert direct control on the costs incurred in creating those inputs. An integrated firm can create appropriate incentives for the salesforce in retail outlets to promote sales of its products. The purpose of 'restrictions' in agreements with non-competitors is therefore often to re-create the incentives and controls that would apply in a more integrated operation and thus to capture the productive efficiencies of an integrated firm. **4.23**

E. 'Horizontal' and 'Vertical' Agreements

Each of the varieties of anticompetitive harm identified above (i.e., unilateral effects, coordinated effects, or reductions in dynamic efficiency), requires a different analytical approach. In the same way, the different categories of consumer benefit identified above raise different issues. These differences lead to one of the fundamental distinctions in competition law—the difference between horizontal agreements (agreements between competitors) and vertical agreements (agreements between a firm and its customers and suppliers). Horizontal agreements often lead to coordinated effects (which are viewed as the most serious kind of anticompetitive harm), while their benefits are usually limited to economies of scale and synergies. The anticompetitive harm of a vertical agreement, in contrast, usually involves unilateral effects, while the benefits tend to involve either facilitating an efficiency, enhancing transaction or duplicating the benefits of integration. In light of these differences between horizontal and vertical agreements, it is not surprising that rules-based systems of competition regulation (like Article 81 under the old regime) tend to develop distinct rules **4.24** for each category of agreement. In a system founded on economic effects (like Article 81 under the new paradigm), it remains convenient to examine the issues raised by each category of agreement separately. This is reflected in the Commission Guidelines and in the modern block exemptions that draw a clear distinction between analysis of horizontal and vertical arrangements. Accordingly, we consider in the following sections the varieties of competitive harm that may result from agreements in each category, followed by a summary of the efficiency benefits identified with these categories.

(1) Horizontal agreements and consumer harm

(a) Competitive harm in horizontal agreements—direct agreements to raise prices or restrict output

4.25 The classic cartel involves an agreement between competitors to raise prices or restrict output. Since this accords directly to our definition of competitive harm, the potential economic harm from this agreement is apparent. The actual harm, however, depends on the identity of the participants and the structure of the market. Taking an often-used example, the agreement by two farmers in the American Midwest to raise the price of wheat will not actually affect the world price of wheat. The farmers have coordinated their market position, but there is no coordinated effect, because even acting together they do not have market power. Some minimum number of farmers must join together before their action can have an appreciable impact.[12] In principle, the impact of the action increases with the number of sellers participating. Other features of the market structure will, however, also affect the level of the potential price rise and thus the level of harm: the ability of buyers to find sources of supply outside the traditional market; the extent to which products are differentiated within the broader market (a cartel among producers of close substitutes will have more power); the ability of suppliers not participating in the cartel to increase production; and the likelihood that parties to the cartel agreement will 'cheat'—selling at a price below the cartel price. Thus, while joint market share is the key indicator of market power for the classic cartel, other factors may affect the actual level of economic harm in an individual case.

4.26 Formal agreements between direct competitors to raise prices or limit output are relatively rare in the EU today. More informal arrangements or 'understandings' do, however, continue take place. Under Article 81, these arrangements are sometimes analysed as 'concerted practices' rather than as 'agreements' and

[12] The farmers might also agree to plant only half their acreage with wheat that year. This agreement to restrict output would have the effect of raising or maintaining price levels if adhered to by a sufficient number of participants.

discussions or communication between firms that leads to such understandings is sometimes described on a general basis as 'collusion'.[13] From the perspective of assessing competitive harm, the same factors are relevant in assessing an informal arrangement as would inform the analysis in a case involving a real agreement on the same terms. The nature of informal understandings regarding market behaviour means that it will be easier for participants to cheat by selling at a lower price or exceeding an agreed production level.

(b) Competitive harm in horizontal agreements—collusion

'Collusion' also describes a very different pattern of conduct. In markets where (i) there is a relatively small number of sellers and (ii) prices and quantities produced are transparent (i.e., readily accessible to other market participants), prices will tend to be uniform—any price differences between products will tend to reflect a perceived quality difference. Price uniformity in this kind of market results from the fact that unilateral price reductions cannot create a benefit in the form of a higher market share for the price-cutting firm. Other firms will need to reduce their own prices to maintain sales, leaving all participants worse off. Similarly, where a firm in the market initiates a price increase, the other firms have an incentive to follow if they believe that customers will accept the increase (i.e. that total sales in the market will not go down to a degree that makes overall sales at the new price less profitable). Because the nature of the market gives each participant an insight into the strategy of its rivals, the level at which the oligopoly price stabilizes—the equilibrium price—will tend to be higher in this kind of market—the structure of the market gives all the participants a degree of market power. The necessary interaction of the firms on the market creates a form of coordinated effect. This interaction is often described as 'collusion'. **4.27**

The degree of concentration is clearly the most important factor in making this kind of collusion likely.[14] The likelihood of collusion resulting from a concentrated market structure depends, on a number of other factors as well. The first concerns the real level of transparency regarding pricing or output. Even where published prices are known, the availability and extent of discounts may be uncertain. Likewise, while current or past output may be known, future output may change rapidly. The level of product differentiation on the market can also affect price transparency. Where product specifications differ and particularly **4.28**

[13] Under US antitrust law this type of behaviour is described as an anticompetitive 'agreement' because the language of Section One Sherman Act does not cover 'concerted practices' separately.

[14] The Herfendahl–Hirschman Index (HHI) has been used in the US since 1982 to identify the degree of concentration in a market. Now used in EC merger assessments along with market shares. See the Horizontal Merger Guidelines [2004] OJ L 31/5.

where they are subject to rapid change, the inability to compare prices will make price alignment more difficult. The second concerns the speed with which rivals can react to a change in prices or capacity. If rivals can only react after a substantial period, then it will be possible for a competitor to grab a short term advantage. A third relevant factor may be differences in the cost structures of the competing firms. Where firms share a common cost structure then a price increase or decrease will affect them all equally. Major differences in cost structure, in contrast, may mean that the optimal price level for the firms can vary—the lower cost firm may not follow a price rise.[15]

4.29 The coordinated effects that can result from an oligopoly market structure are not themselves unlawful. Agreements between competitors can, however, make such effects more likely or increase their magnitude. Where this occurs the arrangement or agreement is said to facilitate collusion. This can occur where an agreement affects one or more of the non-concentration factors identified above. Thus, an agreement by which competitors exchange detailed information on pricing or exchange plans on capacity can increase transparency and lead to a coordinated effect, even though there is no agreement on the level of prices or capacity. Likewise, cooperation by competitors in setting standards for products in the market may facilitate collusion by making price comparison easier and making cost structures more similar. Cooperation by competitors in production or purchase of common inputs may make cost structures more similar, where the inputs involved constitute a substantial proportion of the total costs of the product and where a substantial proportion of competitors in the market participate in the cooperative venture.

4.30 Thus to show that an agreement leads to competitive harm based on facilitating collusion, it is necessary to show that: (i) the market structure is of a sort to make collusion possible, and (ii) the agreement or practice will have the effect of making collusion more likely or more effective. The combined market share of the firms involved in the agreement and the degree of concentration in the market are necessary factors in this analysis, but only a full economic analysis of the market will allow an affirmative finding of real economic harm.

(c) Competitive harm in horizontal agreements—market allocation

4.31 'Market allocation' has traditionally been treated along with cartels and collusive behaviour as a classic horizontal restriction. The two principal forms of market allocation are allocation by territory and allocation by customer. Allocation by territory may involve assigning a particular area to exploitation by each

[15] See Bishop, S and Walker, M, *The Economies of EC Competition Law: Concepts, Application and Measurement*, 2nd ed, 2002, Sweet & Maxwell.

party or it may involve assigning particular locations where the parties will operate, with the result that customers in the area near the assigned location will turn to that location for their supplies. Allocation by customer may involve assigning categories of customer to a party or it may involve allocating individual identified customers between the parties.

The economic analysis of an agreement that allocates markets between firms that would otherwise be actual or potential competitors differs from that applicable to the horizontal arrangements discussed above. Where price-fixing and collusion involve joint action to raise prices or reduce output, market allocation involves agreement by the firms involved to leave each other alone. Thus, where market allocation leads to an increase in market power this is primarily a unilateral effect.[16] Each of the parties gains market power in its assigned sphere. This means that to assess the effects of a market allocation arrangement, the position of each party to the arrangement needs to be considered separately. **4.32**

The most serious competitive effects of customer allocation result from 'bid-rigging'—agreements among competitors where one competitor is designated to 'win' a supposedly competitive bid, in return for mutual agreement that each other competitor will have its turn winning other bids.[17] This practice is not price-fixing, since there need not be agreement on the price charged by the designated winner. The overall effect is nonetheless the same; prices in the market will tend to be higher because the constraints of the bidding process are eliminated. **4.33**

In the absence of bid-rigging, the effects of an agreement to allocate customers will depend on: (i) the market power of each beneficiary firm, and (ii) the extent to which the other firms involved would otherwise provide a significant competitive constraint on that beneficiary firm. The market share of the beneficiary will usually provide a useful indicator of market power and the market share of the other parties will usually provide an indicator of likely competitive constraints.[18] Two other factors, however, need to be taken into account: (i) the **4.34**

---

[16] Where market allocation leads to a significant reduction in the number of sellers for a territory or customer group, collusion may become more likely, leading to coordinated effects as well. This effect would need to be analysed separately, since it will depend on whether the other requirements for successful collusion apply.

[17] The other 'bidders' could also be compensated by the promise of subcontracting work from the successful bidder. This practice is also sometimes described as collusive tendering. From a policy perspective, bid-rigging may raise additional issues because it involves a degree of deception that may be fraudulent. In terms of economic harm, however, the deceptive element is only important insofar as the reduction in transparency regarding price options for the customer impedes the efficient function of the market.

[18] In each case the assessment of market power also depends on other factors, such as the extent of barriers to entry and the strength of competitors not participating in the arrangements. The

extent to which rivals involved in customer allocation produce close substitutes, and (ii) the possibility of arbitrage.

(i) Close substitutes

4.35 The need to examine the closeness of substitutes becomes important where differentiated products are involved.[19] Where the level of substitutability between the two products is relatively low (but they are still in the same market) the increase in the beneficiary's market power will probably bear a closer resemblance to its increase in market share. Where the products are the closest substitutes, on the other hand, their relative market shares may substantially understate the effect that customer allocation has on their respective market power.

(ii) Arbitrage

4.36 Arbitrage can occur if products sold to one customer group can readily be resold to a second customer group. If the beneficiary of a market allocation agreement is selling to the second customer group, the price for the first customer group constrains that firm's pricing for the second group, since a significant price increase will lead the customers in the second group to obtain supplies originally sold to the first group. The extent to which arbitrage can control the beneficiary's market power will depend on the transaction costs incurred in connection with the purchase in the lower price market and the resale. The effectiveness of arbitrage may also vary depending on the beneficiary. Where two firms enter into a market allocation agreement, one firm may be more exposed to arbitrage and thus may benefit less from the arrangement.[20] As in the case of close substitutes, the relative impact of arbitrage changes the inferences that can be drawn from market shares.

4.37 Where a territorial restriction is involved, the effect of the arrangement will depend on the competitive constraints faced by each party in its protected territory. This does not mean that each territory needs to be assessed as a separate geographic 'market'. The definition of the geographic market does, however, play a significant role in the analysis. Where competitors divide territories within the same market, the external criteria (barriers to entry, importance

---

purchaser power of customers in each allocated group may also vary significantly and thus affect the extent to which each beneficiary gains from the transaction.

[19] See TT Guidelines para 21. cf. Commission Notice on the Definition of the Relevant Market for the Purposes of Community Competition Law [1997] OJ L 372/5.

[20] For example, if two firms making medical supplies agree that firm A will supply retirement homes and the firm B will supply pharmacies, the constraint of arbitrage on firm A will probably be higher than on firm B, since a significant increase in price for firm A will lead customers to purchase from the pharmacies. An increase in price by firm B, in contrast, may not be subject to the same constraint because there will be additional transaction costs incurred for pharmacies purchasing the products *from* retirement homes.

of third party rivals, strength of customers) are the same for assessing the position of both parties. Most importantly, where territories are divided within the same market, the parties will often impose a significant competitive constraint on each other, absent the agreement. Where, in contrast, market allocation involves separate geographic markets, the external constraints on each party's market power may differ, and the constraint that would be exercised by the parties on each other as actual or potential competitors in the absence of the agreement may also differ as well. The extent of arbitrage may play an important role in this assessment, since the customers of a party may exercise a competitive constraint, even where that party itself is not active in the market.

Where market allocation involves separate geographic markets and the parties were not previously significant competitors for the others in their respective markets, the competitive harm from the arrangements may be minimal. The general policy disfavouring agreements that partition the EU, will nonetheless mean that any arrangements that allocate territories within the EU will be presumed to cause some competitive harm, unless the position of the parties are insubstantial. The extent to which the parties take steps to limit arbitrage, e.g. by imposing vertical restraints on their respective customers (as discussed below) or refusing to deal with customers that wish to resell in another market, will play a key role in this assessment.[21] **4.38**

(d) Competitive harm in horizontal agreements not related to price or output—consumer choice and dynamic efficiency

Since the new Article 81 paradigm is principally grounded in price theory, the assessment of competitive harm tends to focus on short-term effects on price or output. Some agreements between competitors may, however, lead to a welfare loss even though there is no direct short-term impact on prices or output. The principal ways in which this might occur are where agreements have an impact on the variety of products available to consumers or on the activities of competitors in developing new or improved products. **4.39**

Where competitors agree on common standards for products, there is often a welfare loss for consumers. Just as an increase in product differentiation can create consumer benefits by creating products that meet the needs of specific consumers more exactly, the reduction of product differentiation can create a welfare loss. Even if the consumers in question still purchase the standardised product, the perceived quality of the product for them can be reduced. There is **4.40**

---

[21] If there is no attempt to limit arbitrage, it will be easier to argue that the impact of the arrangement is not significant. If the parties do attempt to limit arbitrage, that fact may itself indicate that there is a significant impact on market power—otherwise why would they bother?

no impact on price and only a limited impact on output (where the consumers in question do not purchase the product), but the for those who do purchase the product, a welfare loss occurs because they pay the same price for a product that, from their perspective, is of lesser quality.

4.41 Agreements between competitors can lead to a reduction in dynamic efficiency where they limit the opportunity or incentive for the parties to engage in innovative activities. Standardisation can have this effect, since the development of differentiated products is a principal motivating factor for innovation. Joint product development can have this effect, where independent development efforts could give rise to differentiated products in the future.

4.42 The difficulty with assessing restrictions that do not involve price or output is that assessment of the current market position is often not particularly useful in determining where there is likely to be competitive harm. Where the potential harm involves a reduction in choice, it is difficult to quantify the value put on choice by consumers.[22] The market shares of the firms involved in standardisation or joint innovation exercises is, at best an indirect indicator of the likely impact of the arrangements (to the extent that current market position provides an indicator of the ability of a party to develop or market an innovative product on their own). But there are many cases where relatively small players (or new entrants) have been able to build a strong market position on the basis of a genuine innovation. Rather than focusing on market shares, the appropriate focus in cases where there is an allegation of non-price effects will often be on barriers to entry or expansion by firms that do not participate in the arrangements.

(e) Competitor harm in horizontal agreements—joint purchaser power and foreclosure

4.43 The classic manifestation of collusion as described above involves joint effects that *increase* consumer prices. It is also possible, however, for competitors to engage in joint conduct that has the effect of *reducing* prices—the prices that they pay for their necessary inputs. This may occur where a group of rivals enter into an agreement for joint purchasing. This may also occur (more rarely) where rivals in a concentrated market 'collude' to reduce the price that they are willing to pay for an input. In this case, the possibility of collusion will depend on the same factors that allow successful collusion in a price-fixing case. In either case, however, the effect on consumer welfare is not clear because the joint conduct

[22] Unless a differentiated product either can command a substantial premium or leads to expansion in total sales, in which case there will normally be a provider if it is possible to produce the differentiated product on an efficient basis.

may not lead to higher consumer prices and may in fact lead to lower consumer prices, if the purchasers reduce their own output prices to reflect all or part of their saving.

Arrangements between competitors that lower input costs can have a negative effect on consumer welfare in two ways. First, they can harm rivals that do not benefit from the lower input prices, particularly where such competitors can only deal with suppliers that have higher costs and must therefore demand higher prices. Secondly, they can lead to a reduction in dynamic efficiency. This may occur where the effect of reducing input prices is either to drive suppliers out of the market that might otherwise generate better products or more efficient processes or to eliminate the margin needed to sustain investment in creating such improvements. These effects, which are sometimes treated together under the label 'foreclosure', are essentially the same as those which could be asserted by a single firm with market power in a vertical agreement, as described in more detail in the section that follows. **4.44**

(2) Vertical agreements and consumer harm

(a) Competitive harm in vertical agreements—general concepts

In the general discussion earlier in this chapter (para 4.24), we described a horizontal agreement as an agreement between competitors and a vertical agreement as an agreement between a firm and its suppliers or customers. These definitions are not mutually exclusive. It is possible for competitors to enter into a vertical agreement as long as one of them is vertically integrated—active in a market that either supplies inputs to the market where they compete or obtains inputs from the market where they compete. Where a vertically integrated firm enters into an agreement with a competitor for the supply or purchase of a product, that agreement can be described as vertical.[23] It makes sense to treat these agreements differently from other horizontal arrangements because: (i) the competitive harm may be less significant,[24] and (ii) the efficiency benefits of **4.45**

[23] This is reflected in the definition of vertical agreements at Article 2(1) of the Vertical Agreements Block Exemption which covers agreements 'entered into between one or more undertakings, each of which operates, for the purposes of the agreement, at a different level of the production or distribution chain, and relating to the conditions under which the parties may purchase, sell, or resell certain goods or services'. The definition is based on the function of the agreement itself, which depends in turn on the functional relationship of the parties in the transaction. Thus the agreement must perform the function of defining the conditions for purchase, sale, or resale of defined products, and the parties must be at a different level of the production or distribution chain in respect of those products.

[24] If, for example, a manufacturer of branded goods is vertically integrated into distribution, that manufacturer has no interest in colluding with other distributors to raise the price of the

supply agreements are different from those of other agreements between competitors. Nonetheless where firms *are* competitors, the competitive harms identified with horizontal agreements may be present. In considering the potential harm of vertical arrangements, it is therefore necessary to differentiate between the effects of vertical arrangements where the firms involved are competitors and other vertical arrangements. Where the firms involved are not competitors (or only compete in selling the products that are provided pursuant to the agreement), the agreement itself is unlikely to cause competitive harm. The focus of assessment in such a case is on any restrictive clauses in the agreement.

4.46 A second feature of vertical arrangements that is crucial for analysis of their effects derives from the fact that vertical agreements, by definition, involve firms at different levels in the production chain. This means that market power may occur at the supplier level, at the purchaser level or at either level or at both, and, as result, unilateral or coordinated effects (or a reduction in dynamic efficiency) could occur at either level. The facts necessary to support a theory of competitive harm will vary depending on where the competitive harm could emerge.

(b) Competitive harm in vertical agreements—theories of competitive harm

4.47 The Commission's Vertical Guidelines[25] identify four categories of economic harm that may result from vertical agreements:

> '(i) foreclosure of other suppliers by raising barriers to entry;
> (ii) reduction of interbrand competition between the companies operating on a market, including facilitation of collusion amongst suppliers or buyers, by collusion is meant both explicit collusion and tacit collusion (conscious-parallel behaviour);
> (iii) reduction in intra-brand competition between distributors of the same brand;
> (iv) the creation of obstacles to market integration, including, above all, limitations on the freedom of consumers to purchase goods or services in any Member State they may choose.'

4.48 This list gives rise to three observations:

1. It will be noted that the final category reflects the presumption in EU law against arrangements that lead to a partitioning of the market within the EU. In the same way that horizontal agreements that lead to market division are presumptively suspect, any agreement between non-competitors that

---

product. Any benefits that the vertically integrated firm would gain as a distributor would be outweighed by losses due to lower sales at the input level.

[25] At point 103.

restricts the ability of a party to produce or sell in part of the EU will be regarded as having some negative effects.

2. The Guidelines draw a distinction between 'interbrand' competition, and 'intrabrand' competition. An agreement affects intrabrand competition where it could lead to competitive harm in respect of the supply of the specific products provided by the seller pursuant to that agreement. Thus the agreement between a car manufacturer and one of its dealers may affect intrabrand competition if it 'reduces competition' in respect of that manufacturer's cars—the products supplied to the dealer pursuant to the agreement. Rivalry in a downstream market in respect of products supplied by a single firm is described as 'intrabrand' competition because most of the law and literature on vertical restraints involves the distribution and resale of branded consumer products.
3. The first three effects identified in the Guidelines: foreclosure; reduction of interbrand competition, and reduction of intrabrand competition, do not identify forms of competitive harm. Instead, they describe processes that lead to anticompetitive harm. In order to work out whether a 'restriction' in a vertical agreement in fact is likely to lead to competitive harm in the context of a specific case, it is necessary to understand how that process works. It is necessary to go behind the labels to identify the potential competitive harm of vertical restraints.

(c) Competitive harm in vertical agreements—foreclosure

'Foreclosure' describes the effect on competition that occurs where a firm engages in conduct that limits the access of its competitors either to inputs or to customers. A classic case of foreclosure would arise where a competitor purchased the only source of a scarce input. Because the rivals of the purchaser in the downstream market would be 'foreclosed' from access to this input, they could no longer compete effectively in the downstream market. The second classic case of foreclosure would occur where a competitor purchased all the potential retail outlets for its product. In this case, the competitors of the purchasing firm at the manufacturing level would be 'foreclosed' from access to necessary distribution channels and could no longer compete effectively. The first case involves unilateral effects in the downstream market, while the second case involves unilateral effects in the upstream market, but in either case, the acquiring company would gain market power by removing competitive constraints on its ability to raise prices. Both of these examples involve mergers and both involve total foreclosure of either a necessary input or a distribution channel. Article 81, however, is applicable to agreements that lead to foreclosure, where foreclosure is often less than total. It is therefore necessary to determine where a foreclosure effect can lead to an increase in market power and thus to competitive harm. **4.49**

4.50 The traditional approach to this question involved an assessment of the percentage of the market 'foreclosed'. If the 'foreclosees' (the firms subject to restrictions) taken together, constituted a substantial percentage of either the input market or the relevant distribution channels, and if the 'foreclosor' (the firm imposing the restriction) had market power, then the foreclosure effect would make it more difficult for rivals of the foreclosor to compete, strengthening the foreclosor's position. In effect, the foreclosure of inputs or outputs would act as a barrier to entry by new rivals and a barrier to expansion by existing rivals.

4.51 The introduction of economic analysis to antitrust policy led to a challenge to the traditional approach. Economists noted that customers or suppliers would only agree to deal exclusively with a putative foreclosor in return for some advantage. Since equally efficient rivals could match the advantages offered by the foreclosor, it was argued that foreclosure could only occur where rivals were less efficient. If a barrier to entry is defined as a cost that a new entrant would bear that was not borne by an incumbent firm, then a foreclosure effect could not be deemed a barrier to entry, since it would not affect an equally efficient entrant. As a result of this analysis, foreclosure effects were dismissed as significant competition policy concerns.

4.52 The foreclosure concept has been rehabilitated, however, under the label 'raising rivals' costs'. Some scholars noted that even efficient rivals could be foreclosed, if the foreclosor succeeded in obtaining preferential access to low cost inputs. By relegating rivals to the higher price inputs, the foreclosor could raise their costs and thus obtain a competitive advantage that could translate into increased market power. The same analysis could also be applied to downstream foreclosure, if exclusive access to distribution channels raised the costs to rivals of distributing their products the foreclosor could reduce their ability to impose a competitive constraint. The extra cost of distribution through the remaining channels would constitute a barrier to entry or expansion.

4.53 The 'raising rivals' costs' concept thus provides a theoretical basis for finding that certain vertical agreements can lead to competitive harm. It does, however, require some focus on the nature of the rivals' costs and whether they are in fact 'raised'. Where input costs are involved this question may be straightforward as it may be possible to identify the costs associated with alternative inputs. For distribution costs, in contrast, the level of foreclosure and the type of foreclosure that is likely to raise the costs of rivals is more complex. In some cases the inability of a rival or potential rival to achieve efficient economies of scale because of the number of outlets foreclosed may equate to an increase in costs. In other cases, the need to invest more heavily in new outlets may be a cost (where the foreclosor has foreclosed existing specialised outlets). In other cases, relegation to smaller volume outlets may lead to higher distribution costs. In all of these cases, the percentage of the market foreclosed is only the beginning of the enquiry. A reasonably high 'foreclosure percentage' will usually be a necessary prerequisite for a real foreclosure effect. Actual proof that foreclosure is likely requires some assessment of the actual market conditions, including the extent of any costs imposed on rivals.

The 'raising rivals' cost' theory of foreclosure is in essence a theory of unilateral effects.[26] It therefore also requires some finding of market power on the part of the foreclosor. Here again a reasonably high market share will normally be a prerequisite. Indeed, the market share threshold for successful foreclosure will probably be much higher than for other unilateral effects. (A supplier with 40% market share can only realistically tie up more than her inputs or distribution by paying suppliers or distributors to keep capacity idle. Unless there are really major cost differences, the uncommitted 60% would provide substantial opportunities for rivals). **4.54**

The 'raising rivals' costs' concept focuses on how a foreclosure effect could lead to a loss in static efficiency and thus to a reduction in consumer welfare. Some commentators remain concerned, however, at the possibility that foreclosure of less efficient competitors could lead to consumer harm either by reducing consumer choice or by reducing dynamic efficiency. (Dynamic efficiency would be reduced if smaller rivals or new entrants are important sources of innovative processes or products). As already noted in the context of horizontal agreements, the difficulty with these claims is that they are not readily subject to empirical proof. While it cannot be excluded that foreclosure of less efficient competitors could lead to economic harm, there should need to be some evidence that this would be likely in a particular case. Thus, where there is a claim that reduction in consumer choice is relevant, there should be some basis for concluding that a foreclosed provider does provide a differentiated product that meets a strong preference of a significant subset of consumers and that foreclosure will lead to elimination of that product from the market. If there is a claim that foreclosure of less efficient competitors will lead to a reduction in innovation, the history of innovation in similar industries and the nature and history of the foreclosed firms would be relevant. **4.55**

(d) Competitive harm in vertical agreements—interbrand competition

The possible impact of a vertical agreement on interbrand competition depends on whether the parties are competitors outside the scope of the vertical agreement itself. Where firms that are already competitors enter into a vertical supply **4.56**

[26] Arguably there could also be coordinated effects where foreclosure effects make higher cost competitors more willing to cooperate in collusive activities.

arrangement, obligations by the buyer to sell only the contract product or to purchase only from the seller can lead to unilateral effects in interbrand competition.[27] Even in the absence of explicit restrictions on the buyer or the seller, vertical agreements may lead to coordinated effects by facilitating collusion between the parties. This could occur if negotiation of the agreement and subsequent administration of the commercial relationship leads to regular contacts between the management of competing firms. It may also give the firms access to otherwise confidential information that makes collusion easier. (This effect may occur on the market where the cooperation occurs or in other markets.) Where a firm 'outsources' production of a key component to a rival, the common cost base for the two firms may also facilitate collusion.

4.57 The possibility of collusion is also the principal concern where the parties to a vertical agreement are not competitors in respect of other products. This can occur in a number of ways. An agreement between a manufacturer and a supplier fixing retail prices can facilitate collusion because it increases transparency regarding the consumer price. Where manufacturers work with exclusive distributors, the limitation on the number of distributors in the market may facilitate collusion at the distributor level. Where a dominant producer requires customers to purchase on an exclusive basis, this may facilitate collusion among the customers, since they will all share a common cost.[28] A vertical agreement between non-competitors can also lead to unilateral effects for the supplier by reducing the competitive constraint imposed by other brands on the beneficiary's pricing or production decisions. Thus an exclusive purchasing commitment by a distributor may increase market power for the supplier by reducing the ability of consumers to make a direct price comparison at outlets supplied by that distributor.

(e) Competitive harm in vertical agreements—intrabrand competition

4.58 Since intrabrand competition, by definition, involves competition in respect of the products supplied pursuant to a vertical agreement (and similar products sold by that supplier to other customers), the potential competitive harm of a

---

[27] These effects may be increased if the agreement is reciprocal—i.e. if firm A each agrees to purchase an input from firm B and firm B agrees to purchase a separate input from firm A.

[28] It is important to note that the contribution of a vertical agreement to collusive effects, regardless of whether that agreement is between competitors, is only part of a broader analysis. As in any case of potential collusion, this theory of restrictive effects requires assessment of whether the markets conditions lend themselves to coordinated effects, following the analysis laid out in the discussion of collusion in horizontal arrangements (see para 4.25). If the other conditions for finding likely collusive effects are not met, the agreements or restrictions cannot be deemed to lead to competitive harm.

vertical agreement involves competition in the downstream market.[29] The competitive harm may involve coordinated effects (the agreements may facilitate the ability of the customers to collude to raise the prices that they charge to their customers). The competitive harm may involve unilateral effects, because the restrictions in a vertical agreement reduce the competitive constraint that the purchaser would otherwise impose on the supplier's other customers. In either case, the potential competitive harm must be assessed in accordance with the general principles applicable to competitive harm of this nature.

Vertical agreements can facilitate intrabrand collusion on the same basis and subject to the same limitations that vertical agreements can facilitate collusion more generally. Vertical agreements may limit the number of competitors in the downstream market. Vertical agreements may lead to uniform conditions applicable to all intrabrand resellers, thus increasing transparency.[30] Where the vertical agreement provides for provision of confidential information to the seller and that information is shared with other resellers, that practice can also increase transparency. The presence of these features does not, however, mean that there is likely to be competitive harm in the form of coordinated effects, for two reasons. **4.59**

First, the relative significance of common inputs is crucial for effective collusion. This may be the case where the downstream competitors are reselling a finished product. (Thus for a car dealer the costs of purchasing cars from the manufacturer and of maintaining the necessary display and service facilities will represent a large proportion of total costs.) It will not always be the case where the customers are purchasing an input that is incorporated in another product. (Thus agreements for the sale of tyres to car manufacturers are unlikely to facilitate collusion in the market for new cars.)[31] Secondly and more importantly, collusion among resellers of the products of one seller is unlikely to create significant competitive harm where customers could react to a significant price increase by switching purchases to other 'brands'. In other words, a coordinated effect for the resellers of a 'brand' can only occur in circumstances where a *single seller* of that brand could benefit from unilateral effects.[32] **4.60**

[29] This contrasts with the effect of vertical agreements on interbrand competition or with foreclosure effects, discussed above (see para 4.49), where an agreement can lead to unilateral or coordinated effects that involve either the level of the supplier or the level of the purchaser.

[30] For example, the requirements in a car dealer agreement regarding maintenance of display or service facilities impose uniform costs on all dealers—each dealer is aware of the costs that other dealers must bear.

[31] The Commission acknowledges in the Vertical Guidelines that intrabrand effects are unlikely where the product is sold for incorporation in another product. In addition to the factors discussed in the text, all of the conditions for effective collusion must be present—in particular the ability to detect and punish deviation.

[32] If colluding intrabrand competitors could interact with other competitors to raise overall prices the vertical arrangements would contribute to interbrand collusion, already discussed above.

4.61 To assess whether competition from other brands would be sufficient to constrain the unilateral exercise of market power, the relative share of the branded product as part of the total market will be significant, as will, *inter alia*, the extent of barriers to entry, the ability of existing producers of rival products to expand production, and the degree of product differentiation. The extent to which resellers of other brands are subject to vertical restraints may also be relevant. If the rival 'brands' are properly part of the same market with the product sold in the vertical agreements, however, the share of the market subject to intrabrand collusion will need to be fairly high before intrabrand collusion becomes a concern.[33]

4.62 Vertical agreements can create unilateral effects where restraints accepted by one reseller reduces the competitive constraint that this reseller imposes on other resellers. This can occur where the seller imposes minimum prices on resellers. (Indeed the direct effect of this kind of restriction in eliminating price rivalry as a constraint creates an irrebuttable presumption of competitive harm.) Vertical agreements can also reduce competitive constraints by implementing market allocation between resellers of the same brand.[34] Market allocation will result, in particular, where vertical agreements limit the territory where resellers can sell or the customers to whom they can sell.[35] The extent that market allocation causes competitive harm in an individual case will depend, *inter alia* on the extent of the vertical constraints in question. Thus, the Commission has traditionally regarded a restriction on active sales outside an assigned territory as likely to lead to less significant competitive harm than a restriction on active and passive sales.[36] Indeed, a restriction on passive sales creates a presumption of competitive harm, reflecting the general concern in Community law, already noted above, with measures that impede integration.

4.63 Since market allocation is a unilateral effect, the effect of intraband market allocation depends on the market position of the beneficiary—if none of the beneficiaries possesses or will gain real market power as a result of the

---

[33] As we will see when we consider market definition, if a monopoly seller of the branded product would have sufficient market power to impose a significant price increase that may mean that the product is sufficiently differentiated to constitute a separate market.

[34] See the discussion of market allocation above (see para 4.31) for a description of the factors that determine the competitive harm resulting from market allocation. As that discussion indicates, the possibility of arbitrage is particularly important for assessing the unilateral effects of market allocation.

[35] The Vertical Guidelines refers in this context to creating a possibility of 'discrimination'. Where the unilateral effects of market allocation for one reseller differ from those for other resellers, price differences for customer groups or territories will result. The competitive harm remains the creation of unilateral effects; discrimination is only a description of the resulting effects.

[36] A restriction on passive sales would bar a response to unsolicited orders from outside the territory. The effect of a passive sales restriction is to limit the possibility of arbitrage.

agreements there will be no unilateral effect. In assessing unilateral market power by a reseller, factors specific to that reseller may also be relevant. Thus, e.g., if a reseller is the leading seller of electronic goods in a region, the strength of the reseller in the sector as a whole will carry over to sales of particular branded products, and vertical restrictions on other resellers could strengthen that position. Here again, constraints from other brands play a key role in limiting the ability of a firm to benefit from unilateral effects.[37] If the market as a whole is competitive, then unilateral intrabrand effects should not occur.

In assessing unilateral intrabrand effects it is important to isolate the real effects of the agreement. The Article 81 analysis requires that the position under the agreement be compared with the position that would apply if the restrictions were not in place. Where the restrictions limit the ability of resellers to sell to customers or territories within the EU, the assumption will be that, in the absence of these restrictions, the resellers would continue to resell the product. The result would be an increased constraint on unilateral exercise of market power by other resellers.[38] The position changes where there is no reseller currently under a restriction on sales into the EU, e.g. where a single reseller has been granted the whole EU as an exclusive territory.[39] The exclusion of *potential* intrabrand competitors does not necessarily signal an increase in unilateral market power, since the seller can achieve the same result by declining to supply competing resellers.[40] If the seller viewed the grant of an exclusive territory for the whole EU as profit-maximising, the reasonable assumption is that it will continue to supply that reseller alone as long as the reseller is effective in **4.64**

[37] It should be noted that the role of interbrand competition as a constraint on intrabrand competition is different here than in the consideration of coordinated effects. For coordinated effects it is necessary to consider constraints of interbrand competition on intrabrand sales as a whole. For unilateral effects it is necessary to consider the effects of interbrand restraints together with residual intrabrand constraints. Thus, the impact of a territorial restriction will be different if there are four intrabrand resellers selling in that territory, than if there is only one. Even limited interbrand constraints will further reduce the possibility of unilateral effects.

[38] The Commission has consistently rejected the argument that restrictions imposed on a party that was not a competitor in the resale market prior to the agreement should be excluded from Article 81 because the agreement enabled the competition to take place. '*Ex post*' restrictions (restrictions on competition made possible by the agreement containing the restrictions) have always been subject.

[39] A restriction on a customer outside the EU reselling into the EU may have restrictive effects if sales would otherwise be practical and the clause contributes to unilateral effects, there is, however, no presumption of such effects where a territorial restriction does not create separate territories within the EU.

[40] As we note later in this work (Chapter 10), even if the seller holds a dominant position, this refusal to supply would not be an abuse unless the product constitutes an essential facility. If the purchaser/reseller is dominant, on the other hand, an exclusive supply commitment imposed on the supplier may constitute an abuse.

promoting the goods at the retail level.[41] Under these circumstances, the restriction on sales by the seller may reduce the uncertainty for the reseller, but may not significantly increase the reseller's market power.

4.65 The same analysis applies where a vertically integrated manufacturer 'competes' against its own customers as a wholesaler or a retailer. An agreement by the manufacturer not to seek business from customers or in territories served by a customer will usually reflect the manufacturer's judgement that overall sales will be more profitable if the customer acts as an intermediary with respect to those customers.[42] In the absence of the agreement, it cannot be assumed that the manufacturer will engage in active sales efforts in respect of those customers, the main effect of agreement by the supplier not to compete with the customer will be to reduce uncertainty. A restriction on the customer, in contrast, could lead to unilateral effects by promoting the ability of the manufacturer to exploit a high price territory or customer group.

F. Assessing Economic Benefits under the New Paradigm

(1) General concepts

4.66 The new paradigm functions on the basis that restrictions that cause or could cause competitive harm in the form of higher prices or lower output can, at the same time, produce economic benefits in the form of improved efficiency. Where the benefits of improved efficiency exceed the cost in terms of economic harm the restrictions should not be prohibited by competition law, provided that the other requirements of Article 81(3) are satisfied. To determine whether a restriction satisfies this rule, it is necessary to identify and, to the extent possible, quantify the economic benefits claimed for an agreement. The Commission Guidelines therefore demand that parties asserting that an agreement qualifies under Article 81(3) should explain the nature of claimed efficiencies, show how these efficiencies are related to the agreements under review, and quantify the likelihood and magnitude of the supposed efficiencies. Where the restrictive effects of specific provisions are identified, it is necessary to show how those provisions contribute to the supposed efficiency gains.

4.67 It is evident that both static and dynamic efficiency gains may be relevant to

[41] If, on the other hand, the grant of exclusivity was the result of an exercise of market power by the reseller, so that the seller would otherwise view sale to multiple resellers as profit-maximising, the issue becomes one of foreclosure—not one of market allocation.

[42] In other words, where the vertically integrated manufacturer accepts a market allocation restriction with a customer, its economic interest as a seller in the overall market outweighs its desire to maximise profits in its distributor role.

assessment of a restrictive agreement or provision.[43] The process of assessing claims of improved efficiency differs significantly, however, depending on the nature of the efficiencies claimed. Where improvements in static efficiency (in terms of production efficiencies or short-term improvements of allocative efficiency) are in issue, it should be possible to identify with some precision the nature of the efficiency gain.[44] In many cases, it should also be possible to quantify the extent of the benefit, at least where the efficiency gain is identified with a reduction in costs.[45] Where gains in dynamic efficiency are claimed, in contrast, it will not usually be possible to quantify the benefit and it may be difficult to describe the nature of the benefit in detail. In this context, the proponents of an agreement must still identify the nature of the anticipated efficiencies and establish that the agreement is reasonably likely to achieve those efficiencies.[46]

Since Article 81 is based on a consumer welfare test, it is necessary to link the claimed efficiencies to consumer benefits—lower costs or increased product quality. The Article 81(3) Guidelines identify two types of 'efficiencies' which correspond to these benefits—cost efficiencies and qualitative efficiencies.[47] Cost efficiencies primarily involve factors that contribute to the productive efficiency of the firms involved in an agreement. As regards qualitative efficiencies, the Guidelines focus on dynamic effects, recognising that agreements can make technological advances available and speed the process at which technological advance proceeds.[48] The nature of cost efficiencies and qualitative efficiencies can be illustrated by reference to the examples of efficiencies set out in the Commission's Horizontal and Vertical Guidelines. **4.68**

43 See Horizontal Guidelines point 32.

44 See Article 81(3) Guidelines.

45 Where an efficiency gain is reflected in improvements in quality, it may be more difficult to quantify the benefits.

46 This leaves open an issue that is not addressed in the Commission Guidelines—what happens if the claimed efficiencies do not materialise? Under the old regime an agreement that had received an exemption under Article 81(3) was enforceable for the life of the exemption decision, unless the Commission took active steps to terminate that decision. It did not matter whether the agreement actually led to benefits in practice. Under the new paradigm, in contrast, Article 81 remains applicable through the life of the agreement; the assessment of restrictive effects and economic benefit must take place on the basis of the information available when the issue is raised.

47 Although the Guidelines refer to types of efficiencies, the categories actually describe two types of consumer benefit—benefits in cost and benefits in quality—that may be short-term or long-term.

48 The Commission also, however, identifies forms of 'qualitative efficiency' linked to productive efficiency—combining existing assets to make a better product or enhancing the ability of a specialist distributor to provide quality service by giving the distributor exclusive sales rights.

(2) Economic benefits of horizontal agreements

4.69 The Commission's Horizontal Guidelines identify efficiencies associated with common categories of horizontal cooperation as production, joint purchasing, commercialisation, standard-setting, and research and development agreements.

(a) Production agreements

4.70 The Guidelines classify joint production agreements,[49] and reciprocal specialisation agreements[50] together as 'production agreements'.[51] The economic benefits of a production agreement are economies of scale or better production technologies.[52] These benefits are treated as particularly likely where the cooperation will increase total output in the market.[53]

(b) Joint purchasing

4.71 Where rivals join together to purchase common inputs, they may achieve 'economies of scale in ordering or transportation'.[54]

(c) Commercialisation agreements

4.72 The Guidelines describe arrangements between competitors regarding 'selling, distribution, or promotion of their products' as 'commercialisation agreements'. Joint commercialisation can lead to both economies of scale and scope from the integration of the economic activities of the firms involved.[55]

---

[49] Joint production agreements include joint ventures for the production of a common product or input (e.g. an agreement among car manufacturers to produce batteries).

[50] Specialisation agreements involve reciprocal agreements between competitors by which each competitor specialises in a particular product or products and sells that product to its partner. (For example, two manufacturers of tableware could reach an agreement by which manufacturer A will make tea cups and manufacturer B will make saucers, so that each can sell cup and saucer sets. Each manufacturer can benefit by closing down the unnecessary manufacturing facility and increasing the usage of its remaining facility.)

[51] The Horizontal Guidelines also include subcontracting agreements in this category. Since the parties to a subcontracting arrangement are at different levels of the production chain for purposes of a subcontract agreement these arrangements are better analysed as a form of vertical agreement between competitors.

[52] Horizontal Guidelines point 102.

[53] Ibid. point 106, example 1.

[54] Ibid. point 132. Lower input prices due to purchaser power do not, however, qualify as an efficiency gain according to the Guidelines, even though they result in an increase in consumer welfare if the market where the purchasing partners operate is competitive.

[55] Ibid. at point 152. Thus, transporting the products of two or more firms in the same delivery van, and thereby allowing use of a larger vehicle to reduce the number of deliveries for each outlet, would be an efficiency gain from integration. A reduction in transport costs as a result of an allocation of customers would not, however, qualify as an efficiency gain according to the Guidelines, even if it was passed on to consumers.

(d) Standard-setting

The Guidelines define 'standardisation agreements' as arrangements that 'have as their primary purpose the definition of technical or quality requirements with which current or future products, production processes or methods may comply'.[56] The Guidelines suggest that such agreements can 'promote economic interpenetration in the common market or encourage the development of new markets and improved supply conditions'.[57] **4.73**

(e) Research and development

The Guidelines identify 'R&D agreements' to include 'outsourcing certain R&D activities', joint improvement of 'existing technologies', and 'cooperation concerning the research, development, and marketing of completely new products'. This includes joint production agreements and joint commercialisation agreements (as defined above, paras 4.70 and 4.72) 'provided that the gravity lies in R&D'.[58] The Guidelines suggest that 'most R&D agreements . . . bring about economic benefits by means of cost savings and cross fertilisation of ideas and experience, thus resulting in improved or new products and technologies being developed more rapidly than would otherwise be the case'.[59] The Guidelines recognise that these benefits may accrue even where the cooperation creates 'considerable market power', but suggests that in such circumstances the parties would need to show 'significant benefits of carrying out the R&D, a quicker launch of new products/technology or other efficiencies'.[60] **4.74**

Most of the potential economic benefits identified here concern productive efficiencies. The most often mentioned benefit is *economies of scale*—the ability to use the same assets to produce more products, leading to lower costs per unit, that are associated with production agreements, joint purchasing, or even joint research and development.[61] *Economies of scope*—use of the same asset to support activity in different fields—can occur as the result of production agreements, commercialisation agreements, or R&D agreements.[62] *Cost synergies*—where integrating the existing assets of the firms allows use of the most **4.75**

[56] Horizontal Guidelines point 159. [57] Ibid. point 169.

[58] Ibid. point 39. [59] Ibid. point 68.

[60] Ibid. point 69. The application of this test is addressed in more detail in Chapter 8 below.

[61] Presumably the use of a single research facility to accommodate more researchers or different research projects in the same area are examples of economies of scale in R&D.

[62] Use of the same facilities to support research in different fields would be an example of economies of scope.

efficient assets of either partner—can result from production agreements or commercialisation agreements.[63]

4.76 Secondly, some agreements (production agreements or commercialisation agreements) can also lead to *quality synergies*—where combining existing assets can make a better product.[64] The integration benefits of standardisation is a form of *direct allocative efficiency*—by reducing barriers between national markets the efficiency of the entire industry is enhanced. The principal benefits of standardisation or research and development, however, involve contributions to *dynamic efficiency*. The direct contribution of joint research and development to dynamic efficiency may be obvious where the combined 'ideas and experience' of the partners lead to new ideas—quality synergy.[65] The role of standardisation is particularly interesting in this regard, because it contributes to dynamic efficiency by resolving 'hold-up' problems that could delay or prevent the emergence of new products.

4.77 Neither the Horizontal Guidelines nor the more recent Article 81(3) Guidelines give much assistance in determining how these efficiencies should be proven. The Article 81(3) Guidelines make it clear that efficiencies must be 'quantified'. The Horizontal Guidelines underscore the need to distinguish the efficiency gains from cooperation (which may be reflected in lower costs and higher profitability) from gains attributable to an increase in market power. Thus, for example, where participants in a cooperative venture are able to reduce costs by reducing output, that cost reduction cannot be considered an efficiency gain. Economies of scale, economies of scope, and cost synergies are usually capable of identification and quantification. However, the ability of firms to produce solid quantification of even these benefits is limited. The most that any firm can do is project the likely scope of productive efficiencies[66] based on historical experience, the nature of the assets, and the market situation. Quality synergies are even more elusive when it comes to quantification (because it is difficult to measure the value put on quality by consumers) and dynamic efficiency benefits are, as already noted above, inherently speculative.

---

[63] Thus, e.g. if two firms entering into a joint distribution arrangement have strong distribution networks in different territories, combining their distribution in a single network will make distribution for both partners more efficient.

[64] Thus, e.g. a joint venture between two vehicle manufacturers that combined the technology of one manufacturer for off-road vehicles with the technology of a second manufacturer for luxury interior design could create a luxury off-road vehicle.

[65] This is a principal benefit associated by the Commission in connection with R&D cooperation. See Horizontal Guidelines point 68.

[66] Even where the benefits of an existing agreement are at issue, it will not be possible to measure the benefits with real precision, since the position that would apply today without the agreements cannot always be known with any certainty.

(3) Economic benefits of vertical restrictions

The Vertical Guidelines identify eight potential benefits of vertical restraints:[67] **4.78**

(a) Preventing 'free rider' problems.[68]
(b) Funding initial investments for new markets.[69]
(c) Avoiding the 'certification free rider issue'.[70]
(d) Avoiding 'hold-up' problems—to allow either the supplier or the customer to make 'relationship-specific investments.[71]
(e) Avoiding the specific hold-up problem for know-how—firms that possess substantial know-how may be reluctant to provide it to another firm without limits of the ability of the other firm to use the know-how or transfer it to third parties.[72]
(f) Achieving economies of scale in distribution.[73]
(g) Addressing capital market imperfections.[74]
(h) Achieving uniformity and quality standardisation.[75]

In addition, the Article 81(3) Guidelines identify both 'cost efficiencies' and 'efficiencies taking the form of new and improved products', which result from **4.79**

---

[67] Vertical Guidelines point 116.

[68] A distributor who faces direct competition from other resellers of the same product will be reluctant to invest in promotion of that product or in facilities to support that product if the rival resellers can benefit from that investment without making an equivalent investment themselves. Restrictions that limit the ability of rivals to take a 'free ride' on this investment (such as territorial or customer restrictions) may provide the beneficiary reseller with the incentive to make to appropriate investments.

[69] A distributor may be motivated to make the investment necessary to introduce a new product by a promise of protection from competition for a limited period.

[70] This arises where a specialist distributor is important for initial introduction of a new product, because the reputation of that specialist distributor will 'certify' the product as meeting standards of style and quality.

[71] Thus, if a customer needs to invest in specific equipment or facilities in order to market the supplier's product, the customer may require some level of exclusivity for the time required to depreciate its investment. Likewise, if the supplier needs to make specific investments in order to supply the customer, then an obligation to protect the supplier (e.g., exclusive purchase obligations) may be justified.

[72] The Vertical Guidelines suggest that where the know-how was not readily available, and was indispensable to operation of the agreement, exclusivity requirements 'normally fall outside Article 81(1)'.

[73] By limiting sales to a single distributor, that distributor will be able to cover its fixed costs associated with distribution of the product over a larger number of units—thus reducing per-unit costs.

[74] Where a customer needs funds for investment, the supplier may be able to provide finance on a more attractive basis than financial institutions because it has better information regarding the risks of the customer's business and the commercial opportunities open to the customer. Restrictions to protect the seller such as minimum quantities or exclusive purchasing may be justified while the loan remains outstanding.

[75] Franchise or selective distribution arrangements may include restrictions on purchasing or on the goods sold in outlets in order to maintain a uniform image that adds value to all participants.

vertical restrictions. The cost efficiencies include both synergies from integration of existing assets[76] and efficiency from better planning of production.[77] For new and improved products, the Article 81(3) Guidelines refer to licence agreements that allow 'more rapid dissemination of new technology in the Community and enable the licensee(s) to make available new products or to employ new production techniques.'[78]

4.80 Taking the benefits identified in the two Guidelines together, it will be noted that only three items—economies of scale in distribution, integration of assets, and better production planning—involve a direct productive efficiency. The other examples involve 'quality efficiency' (uniformity and quality standardisation); direct contributions to allocative efficiency (by correcting disincentives to investment); and contributions to dynamic efficiency (by either allowing new products to come on to the market or allowing them to come on to the market more quickly). Thus for each of these examples the elusive nature of quantification, already noted in the context of horizontal cooperation, is evident. Here again the challenge facing the firms involved in an agreement is to create a convincing theoretical basis for the assertion of significant economic benefits.

4.81 A particularly interesting aspect of the discussion of benefits in both the Article 81(3) Guidelines and the Vertical Guidelines involves the treatment of 'hold-up' problems. The Guidelines recognise that contractual restrictions play an important role in making suppliers more willing to make their products available and customers more willing to make the commitment of effort and resources needed to market a product sucessfully. In some cases the hold-up in the absence of restraints is absolute—the seller will stay out of the market or the customer will not market the products at all. In other cases, the hold-up is relative—in the absence of restrictions the product will come on the market at a later time. This distinction between absolute and relative hold-up issues is significant. 'Restrictions' necessary to address absolute hold-up problems may fall outside Article 81(1) altogether.[79] Restrictions that address relative hold-up problems, in contrast, are beneficial only to the extent that they shorten the period needed to bring the product to market. Long periods of territorial

---

[76] Point 57, '[I]f one undertaking has optimised one part of the value chain and another undertaking has optimised another part of the value chain, the combination of their operations may lead to lower costs.'

[77] Article 81(3) Guidelines point 60 (example of agreements allowing for 'just in time purchasing.').

[78] Ibid. point 62.

[79] In making this assessment it is still necessary to consider whether the restriction is indispensable, if a less restrictive alternative would still make entry possible. Then the more extreme provision would be restrictive. Thus if a distributor will make the necessary investments to introduce a new product line, based on the promise of three years' exclusivity, a five year agreement could be caught by Article 81.

exclusivity or exclusive purchase obligations will be difficult to justify for products where the hold-up appears relatively limited.

G. Economic Assessment under the New Paradigm—Conclusion

Central to the new paradigm as described in Chapter 3 is the proposition that it is possible to balance the competitive harm associated with an agreement or 'restriction' against the economic benefits that may be generated by that agreement or restriction. Yet the discussion above of competitive harm and economic benefits shows that it is rarely possible to quantify either in a statistically significant way. Even the most sophisticated economic analysis will not generate a figure in euros for the cost of an agreement, or for its benefits, let alone reliable amounts for both in relation to the same agreement. The emphasis of the Guidelines on quantification is thus misleading to the extent that it suggests that benefits can be precisely measured in some scientific way. **4.82**

This does not mean that the economics-based approach of the new paradigm cannot work. What it means is that the economic assessment must be based on consideration of theories of competitive harm and theories of economic benefit. In considering competitive harm, the role of the economist is to identify whether the restrictions at issue are likely in the context of the relevant markets to lead to competitive harm. This requires the economist to identify a robust theory of competitive harm based on accepted economic theory and explain how this theory applies in the context of the market.[80] While this may not allow precise quantification of the harm, it should usually be possible to categorise the harm on a sliding scale—minimal, appreciable, significant, very significant (sometimes the last two may be described as 'serious' or 'very serious'). As regards economic benefits, the role of the economist is not usually to deliver a precise estimate of the benefits of an agreement, but rather to verify both whether benefits are likely in the commercial context of the agreement and the significance of those benefits. Here again the economist must develop a coherent theory and ensure that this theory is consistent with the facts. The result of the analysis will again often be a judgement on a sliding scale of 'minimal' up to 'very significant'. **4.83**

---

[80] This assessment can never be completely accurate, since it will rarely be possible to obtain complete information regarding the market structure. More important, the competitive impact will depend on the reaction of customers and actual or potential competitors, which cannot be predicted with certainty. Moreover, factors outside the market (such as significant changes in input prices, legal changes, or changes in overall levels of economic activity) may alter the impact of conduct in the market. The goal of the Article 81 analysis cannot therefore be to *show* that conduct *will* result in competitive harm. The goal is rather to show that significant anticompetitive harm is *likely*.

4.84 The nature of the process outlined above for identifying both harms and benefits suggests Article 81 will become a balance of partly subjective conclusions. A 'significant' benefit would outweigh an 'appreciable harm', but a 'very significant' harm would outweigh even a 'significant' benefit, while an 'appreciable benefit' would be justified even it caused some 'minimal harm'. This is not necessarily a bad thing. Even though the final determination is inevitably subjective to some degree, the structured analysis of both harm and benefit and the need both to identify the appropriate theory and to show that the theory justifies conclusions regarding likely effects in the factual context provides the basis for competition authorities and courts to test the economic conclusions submitted to them. The experience of courts in the US applying a structured rule of reason suggests that significant benefits rarely co-exist with significant harms, so that in the vast majority of cases a balance of probable results does not create insuperable administrative burdens for courts or enforcement agencies.

4.85 This analysis is reflected in the block exemption regulations and the various Guidelines. The block exemption regulations typically use a market share screen to identify agreements where restrictions are not likely to lead to significant competitive harm. Restrictions that are likely to lead to significant competitive harm even at lower market share levels (such as price-fixing or market allocation) are excluded from the Regulation. Since the block exemptions cover agreements or restrictions that are likely to lead to some appreciable benefits, this structure ensures that the agreements subject to the block exemptions would almost always conform with Article 81, without need for individual assessment.

4.86 Outside the scope of the block exemptions, the various Guidelines all employ the rhetoric of significant benefits and/or serious restrictions in discussing the treatment of different agreements or restrictions under Article 81. The Guidelines use market shares as the principal indicator for determining where a competitive issue is likely outside the safe harbour of the block exemptions.[81] In providing 'top level' advice this is appropriate—since market share will often be the single most important factor in determining whether there is market power. It thus makes a certain amount of sense to use market shares as a filter for determining whether full economic assessment is warranted. Yet the discussion above shows that the probative importance of market share will depend on the nature of the market and on the nature of the theory of harm or benefit to be invoked. The Guidelines and the block exemptions together create an initial

[81] Market share serves both as an indicator of competitive harm (since high shares may indicate market power) and as an indicator of lack of benefit (where there is market power it cannot be assumed that restrictions in agreements are intended to increase efficiency).

filter for review that can 'clear' the great majority of agreements. The Guidelines also perform an extremely useful function in identifying the theories of competitive harm or economic benefit that may be applicable to different types of restriction. However, in difficult or complex cases, application of Article 81 will require assessment of the underlying economic theories as they apply to the markets in question.

5

IP LICENSING UNDER THE NEW PARADIGM

A. Introduction

This chapter explains how the new Article 81 paradigm applies to agreements that involve intellectual property. Following the structure described in Chapters 3 and 4, the discussion below outlines how each of the elements in the Article 81 analysis applies to intellectual property transfer. This outline will then be 'fleshed out' in the chapters that follow by reference to specific types of 'restriction' and in the context of different commercial relationships. The step-by-step analysis that is outlined in this chapter is not, however, necessary for assessment of most intellectual property related issues that emerge in day-to-day practice. In **5.01** the great majority of cases a detailed formal assessment will not be necessary, because a preliminary assessment shows that risks under Article 81 are limited. This chapter therefore concludes by describing the structure of 'self-assessment' under the new paradigm, explaining how practitioners can engage in an appropriate preliminary assessment and discussing what that assessment can achieve.

B. Applying the Paradigm to IP Licensing—Step-by-Step

5.02 The final section in Chapter 3 describes application of the new paradigm as a 10 step process. In this section we address each step in that analysis in the context of intellectual property licensing.

(1) Step one—is there an 'agreement' or 'concerted practice' between 'undertakings'?

5.03 Most cases involving intellectual property involve a contract for transfer of rights in intellectual property—a licence or an assignment. Even where the licensor is a private individual, that individual will still usually be considered to constitute an 'undertaking' for purposes of Article 81, because he or she will be operating in a commercial capacity.[1] There may, however, be situations where intellectual property raises competition law issues in the absence of a formal agreement. There could, for example, be an informal agreement between competing firms not to challenge their respective intellectual property rights. Furthermore, parties to agreements that are limited in scope may informally agree to further commitments that raise issues under Article 81. Thus participants to a standardisation exercise could agree informally not to produce products that are incompatible with the standard, even though they do not include such a commitment in their formal agreements.

5.04 There is a significant body of law under Article 81 that deals with the requirements for proving an agreement or concerted practice in the absence of a formal agreement. A detailed explanation of these rules is outside the scope of this work.[2] In principle, however, the rules for determining the existence of an agreement or concerted practice are no different for arrangements involving intellectual property than for any other commercial transactions.

---

[1] *Reuter/BASF* [1976] OJ L 254/46.

[2] See e.g. Faull, J and Nikpay, A, *The EC Law of Competition*, 2nd ed, 2006, Oxford University Press.

(2) Step two—is the agreement 'restrictive' for purposes of Article 81(1)?

(a) The relationship between 'restrictions' and 'restrictive effects'

The next stage in the Article 81 analysis is to determine whether the agreement itself, or any provisions in the agreement, constitute a 'restriction of competition' for purposes of Article 81(1). As explained in Chapter 3, this analysis remains grounded in the approach to defining restrictive agreements under the old antitrust regime. Under the new paradigm, however, this assessment is preliminary to the central analysis of economic effects. This means, that it is unimportant if an agreement is not a 'restriction' and if no clause in the agreement is a 'restriction', then the agreement falls outside Article 81, without the need to consider economic effects. **5.05**

(b) When does an agreement restrict competition?

The key issue regarding an agreement restricting competition is whether the parties to the agreement are actual or potential competitors.[3] Parties to an agreement that involves intellectual property may be actual or potential competitors at two levels: the level where the intellectual property is used (e.g., in manufacturing products using licensed technology) or at the level where the technology is marketed. Where the parties compete at the level of marketing technology, the Commission considers that an agreement between them to license technology to each other could constitute a restriction on competition, even if it does not contain any further restrictions on their conduct in respect of the technology.[4] This would be the case if the result of the licence could be to remove the incentive for one party to continue developing its own technology. Where the parties compete only at the second level, in contrast, the position appears to be that the agreement as such does not constitute a restriction of competition. Article 81 is only applicable where the agreement contains *provisions* that constitute a restriction on competition. **5.06**

(c) Restrictive provisions—application to 'ex post' restrictions

Some commentators have suggested that this analysis is too interventionist. According to this argument, where two firms that are not competitors enter into **5.07**

[3] Quote from definition of actual or potential competitors in Horizontal Guidelines. The Commission defined a 'potential competitor' in the horizontal guidelines, as a firm that 'would be likely to undertake the necessary additional investments . . . [to] enter the relevant market in response to a small and permanent increase in relative prices'. Horizontal Guidelines, n. 10 at para 9.

[4] The Commission refers in this context to competition between 'technology poles' Horizontal Guidelines point 81. This concept, which is borrowed from the US Guidelines, represents an application of the distinction between interbrand and intrabrand competition familiar from cases involving distribution. Within one technology pole, agreements between IP owners can only restrict intrabrand competition—these become significant only if the number of other technology poles is limited.

an agreement, restrictions on the parties in that agreement cannot 'restrict competition' because 'competition' has only been made possible as a result of the contract. They argued that competition law should only be concerned with restrictions on '*ex ante*' competition (restrictions on competition that pre-existed the agreement) and not where an agreement contains restrictions on '*ex post*' competition (restrictions on competition that exists as a result of the agreement).[5] This argument has been asserted with particular vigour in respect of restrictions in intellectual property licensing.[6]

5.08 From a legal perspective, the *ex ante–ex post* argument is of limited significance, since the application of Article 81 to *ex post* restrictions is clearly established. Indeed, limiting Article 81 to *ex ante* restrictions would severely compromise the Commission's policy in respect of market integration, since many territorial restrictions would fall into the *ex post* category. Moreover, a systematic limitation of Article 81 to *ex ante* agreements would appear questionable from a broader policy perspective, particularly in the context of technology licensing. The suggestion that 'competition' in respect of a technology or products using that technology 'would not take place' in the absence of an agreement seems too deterministic to form the basis of policy. While it may be true in the short term, in the longer term the determination of a licensor to exploit technology alone may well lead to more rapid development of alternative technologies or of alternatives to products that use the technology. Particularly where a licence involves a product with long-term protection like industrial know-how or copyright-protected software, the likelihood that in the longer term rival products will emerge that could exert a competitive constraint on products using the technology in question must be high. Under these circumstances, the 'restrictive' nature of provisions that maintain territorial or customer restrictions between competitors 20 or 30 years after transfer of the original technology cannot be dismissed, even if the firms involved were not competitors at the time that the technology was originally transferred.[7]

[5] See e.g. Korah, V, *An Introductory Guide to EC Competition Law and Practice*, 2004, Hart Publishing, Oxford, at sections 2.4.1 and 9.5.1.

[6] See e.g. Charles River Associates, *The European Commission's Draft Technology Transfer Block Exemption Regulation and Guidelines: A Significant Departure from Accepted Competition Policy Principles*, 2004, www.crai.co.uk/pubs/pub_3563.pdf.

[7] An example of possible restrictive effects resulting from licence agreements where the parties were not competitors prior to the agreement may be *US* v *Pilkington Plc* 1994–2 Trade Cases (CCH) at para 70,842 (Ariz 1994). That case involved hybrid patent/know-how licences for Pilkington's float glass technology. According to the US DoJ, these licences included 'stringent territorial, use, and sub-licensing restrictions' that continued in force as long as any of the licensed know-how remained confidential. (See Speech by Asst Attorney General Anne Bingaman, *Competition and Innovation: Bedrock of the American Economy*, 19 September 1996 (on DoJ website). DoJ argued that these restrictions prevented US-based competitors from expanding into foreign markets, particulary China. The case was settled by consent decree.

(d) Restrictive provisions—restricting competitive autonomy

A different analysis applies to assessment of a contractual *provision* (as opposed to assessment of the agreement that includes that restriction). Whether an individual provision constitutes a 'restriction' depends upon whether it restricts the autonomy of a party in respect of a significant competitive parameter. In other words, the provision must either explicitly limit competitive conduct, or must have the direct effect of limiting competitive conduct. A provision that limits the customers to whom a licensee can sell, the territory or locations where the licensee can sell, the price at which the licensee can sell, the quantity that the licensee can sell, the trademark that the licensee can use, the ability of the licensee to use other technology, or the ability of the licensee to sell competing products are all restrictions on the licensee's competitive autonomy. Likewise a provision that limits the licensor's ability to license others constitutes (according to the Commission) a limitation on the autonomy of the licensor and a restriction on the ability of third party potential licensees to compete with the licensee. **5.09**

In considering what constitutes a significant competitive parameter, special considerations arise in the licensing context. In particular, efforts to limit or avoid the application of intellectual property rights may be regarded as a legitimate form of competitive behaviour, since they reduce costs (by eliminating the need to pay royalties) and avoid the possibility that competitive behaviour could be blocked altogether. Accordingly, an agreement not to challenge the validity of intellectual property rights could be classified as a 'restriction' since it limits this competitive parameter. **5.10**

(e) Indirect restrictions

Some provisions may become restrictive because of their specific commercial context. This applies in particular to provisions regarding payment of royalties. For example, it is normally not a restriction of competition to require a licensee to pay a royalty for each unit sold by the licensee of the product that would or could incorporated the licensed technology. Where, however, the level of the royalty payment and nature of the products mean that the provision effectively prevents the licensee from using a rival technology, the provision would be regarded as commercially equivalent to an agreement to use only the licensed technology. This would render the royalty provision a 'restriction' for purposes of Article 81. **5.11**

(f) Restrictive provisions—the 'limited licence' exception

The existence of a direct or indirect 'restriction' under Article 81 must be distinguished from the possibility of a limited licence. To the extent that intellectual property rights are governed by national law it is possible to grant a **5.12** licence valid in one EU Member State while withholding rights covering the same technology in other Member States. If conduct by the licensee in an EU territory not covered by the licence would infringe the licensor's intellectual property rights in that territory, enforcement by the licensor of its rights in the second territory is not a restriction of competition. The limitation on the licensee's conduct results from the nature of the intellectual property right itself and not from the provisions of the licence agreement.[8]

(g) Restrictive provisions—limits accepted by licensor as part of essence of right

5.13 A similar argument applies to restrictions on licensing by the licensor. There is an argument that where intellectual property rights are created by law, the ability to assign or license the right is part of the essence of the right. On this basis, an 'open exclusive licence' that gives the licensee exclusive rights in a territory without the promise of contractual restrictions on sales into the territory by other licensees may not constitute a restrictive provision.[9] An agreement by a licensor regarding the extent to which the licensor will grant licences to third parties can, however, constitute a restriction for purposes of Article 81. Thus an agreement with a licensee not to grant a licence in that territory to a third party or an agreement with licensors of similar technology to license solely on a national basis may constitute 'restrictions' for this purpose.

(3) Step three—is a restrictive provision 'ancillary' to an otherwise non-restrictive agreement?

(a) General

5.14 Many common 'restrictions' in licensing agreements are arguably ancillary to a broader relationship and therefore not in themselves restrictive of competition for purposes of Article 81. In some cases the ancillarity principle applies to the licence agreement itself, where the licence is part of a broader cooperation

[8] The same analysis would apply if the licensor had separate patents covering different fields of use of the same technology—a licence for one field of use may not confer a licence for the second field of use.

[9] The Commission decision in *Maize Seed* [1978] OJ L 286/23 found that restriction on marketing by the licensor in an exclusive territory allocated to a licensee was caught by Article 81(1) along with a restriction on the granting of licences to third parties. In the proceedings before the Court on the application for annulment of the decision, the applicants argued that the restriction on marketing by the licensor was not restrictive, since the ability to transfer the licensor's own rights in a territory formed part of the specific subject matter of the right in question Case 258/78 *Nungessor* v *Commission* [1982] ECR 2015 at 2037). This submission was supported by the UK and German governments. The Court did not, however, reach this issue, holding that in the circumstances of that case an 'open exclusive licence' (which did not purport to limit the activities of licensees or customers in other territories) would be be subject to Article 81.

between the parties. Some 'restrictions' in licence agreements are justifiable on the basis that they are objectively necessary to protect the legitimate interests of the licensor. Thus a requirement that the licensee use the licensor's trademark or a requirement that the licensee meet specified certain quality standards would protect the licensor's interest in maintaining the reputation of the technology.[10]

(b) Hold-up problems in licensing

In the intellectual property context, many 'restrictions' may be regarded as ancillary because they are necessary to resolve 'hold-up' problems, on the side either of the licensor or of the licensee.[11] Where contractual protection for either the licensor or licensee is necessary to ensure either that the technology transfer takes place or that the technology transferred is actually used, then these restrictions should fall outside Article 81. This standard can apply to hold-up on the side of either the licensor or the licensee. As regards the licensor, one example of a restriction needed to avoid hold-up problems is a limitation on disclosure or use of secret know-how.[12] Another example may be the right of a licensor to a licence under improvements developed by the licensor, which ensures that the licensor can continue to compete in markets using the technology that it has developed. As regards the licensee, a requirement that the licensor give the licensee a free licence on any improvements to the technology protects the licensee from exploitation by the licensor once it has become locked into the licensor's technology. **5.15**

The most difficult hold-up issues typically arise where protection of the licensee is arguably necessary to induce the licensee to introduce or use the technology. Thus where the technology is being introduced on a market for the first time, some contractual protection for the licensee may be objectively necessary to compensate the licensee for its investment in introducing the technology and for the risk that the technology may be unsuccessful in the market. Similarly, where the market for a technology is limited, the contractual protection needed to ensure that exploitation takes place may exceed those objectively justified where the potential market is broader.[13] The TT Guidelines make it clear that **5.16**

---

[10] This type of ancillary restriction is similar to the restrictions discussed in the Vertical Guidelines designed to ensure maintenance of uniform appearance and product quality in a franchising or selective distribution context. See Chapter 4 above at para 4.78.

[11] Hold-up problems can also occur in joint product or cross-licensing contexts. As the Horizontal Guidelines recognise, standardisation may be necessary in order to make it possible for products based on new technology to reach the market. Where further contractual restrictions are necessary for this to occur, these restrictions could also be ancillary under this analysis.

[12] As noted in Chapter 4, the Vertical Guidelines indicate that a provision of this nature is not restrictive because it is objectively necessary to avoid hold-up problems.

[13] For example, in the case of 'orphan drugs'—pharmaceutical products that treat a condition affecting a limited number of patients, both the licensor and the licensee may have requirements that are not applicable for licensing products with a broader application.

the legitimate interests of the licensor and licensee depend on an objective consideration of the technology and products involved.[14] The fact that either party would not have agreed to the licence without these terms is not decisive.[15]

(c) The ancillarity of a restriction depends on the commercial context

5.17 The ancillarity principle may also apply because of the broader relationship between the parties. Thus where technology is licensed on a sub-licence basis to allow the licensee to produce products on behalf of the licensor, the sub-licence relationship may justify restrictions on the licensee that could be regarded as restrictive in other commercial contexts. In cases where technology is licensed in the context of joint products or joint sales, even direct agreement on pricing, customers, and territories may be necessary for effective implementation of the broader arrangements. This analysis is particularly important when we come to consider pooling arrangements. Since it is necessary that a patent pool set the price for a licence to third parties in order to function, agreement by the pool members on that price level is not in itself restrictive of competition.

(d) Restrictions must be necessary

5.18 The ancillarity principle only applies to restrictions that are *necessary* for the implementation of the agreement. Where a provision with less restrictive effects would be sufficient to deal with a perceived hold-up problem, the broader restriction will not be ancillary. Thus a prohibition on both active and passive sales will not be justified if a passive sales restriction would be sufficient to resolve the hold-up issue. This rule can also apply to the duration of a restrictive provision. Particularly where territorial or customer restrictions are concerned, a restriction that is objectively justified to allow introduction of a new technology may no longer be justified after the technology is established or the licensee has recovered the costs of the technology's introduction.

(e) Ancillary restrictions may be relevant in assessment of agreement as a whole

5.19 Classifying a provision as ancillary does not mean that the possible competitive harm associated with that provision is irrelevant. The impact of the provision must be assessed in the context of considering the anticompetitive effects of the agreement as a whole. Thus where a joint venture involves licensing of technology to the joint venture company by the joint venture partners, discussions

[14] This is a specific application of the difference between absolute hold-up problems and relative hold-up problems discussed in Chapter 4 at para 4.81. As noted in that discussion, the assessment must involve both whether there is an absolute hold-up and whether the restriction at issue is necessary to resolve that hold-up.

[15] TT Guidelines para 12(b).

of pricing between the partners would not be in themselves restrictive. If these discussions had spillover effects on other markets, those effects would need to be considered in evaluating the application of Article 81 (or merger rules) to the joint venture. Likewise, in a licence agreement between two competitors, a territorial or customer restriction may well be necessary in order to resolve a hold-up problem, the impact of that restriction on the market where the parties compete will nonetheless be part of the broader analysis of the competitive impact of the agreement as a whole.

(4) Step four—does the agreement or restriction have the object *of restricting competition?*

Certain provisions are regarded as having the 'object' of restricting competition. Where a provision falls in this category, the analysis immediately goes to economic benefits because, in the absence of significant benefits, the provision will infringe Article 81. Under the old regime, this assessment was easy because certain types of restrictions were regarded as restrictive by their nature. In particular, agreements that determined the price at which a party could sell, how much it could sell, the customers to whom it could sell, and the territory where it could sell were regarded as restrictive by object because they limited the essential aspects of competitive rivalry. Under the new paradigm, in contrast, these provisions are those that, from experience, are so likely to lead to competitive harm that detailed economic analysis is not required. This leads to a significant limitation on the scope of the 'restrictive by object' category in the context of intellectual property licensing. Territorial and customer restrictions that might formerly have been regarded as having the object of restricting competition are now evaluated solely on the basis of their economic effect, as long as they do not involve agreements between parties that were not actual or potential competitors prior to the licence in the markets where the technology is used.[16] **5.20**

The Commission has taken a step towards resolving uncertainty regarding the extent of restrictions by object in the context of intellectual property licensing by incorporating a list of 'hardcore' restraints in the TTBER.[17] This list starts with the classic restrictions by object: price, quantity, territory and customers. The list then, however, presents a series of exceptions to these rules that differ depending on: (i) whether the parties to the agreement were actual or potential competitors at the time that they entered into the licence agreement; and (ii) **5.21**

[16] TTBER Article 4(3).

[17] TTBER Article 4. In the Guidelines para 75 the concept of 'hardcore restraints' is described as 'based on the nature of the restriction and experience showing that such restrictions are almost always anticompetitive'—the same language used in the Article 81(3) Guidelines to define restrictions by object. The TT Guidelines make a cross-reference at para 75 to the cases cited in support of para 14, which provides an abstract definition of the restriction by object concept.

whether restrictions in an agreement between competitors are reciprocal or non-reciprocal. The clear implication is that provisions that fall within the Article 4 exceptions are not 'hardcore' and therefore should not be regarded as restrictive by object as that concept is understood under the new paradigm. Although these restrictions may still be subject to Article 81, application of Article 81 will require an affirmative showing of competitive harm.

5.22 Classification of a provision in a licensing agreement as restrictive by object does not mean that the restriction is not compatible with Article 81. Application of Article 81 still requires a balance of competitive harm and economic benefit under Article 81(3). It means, however, that it will be necessary to determine whether the restriction contributes to a significant economic benefit that could not be achieved by some means that is not restrictive by object. In practice, it will be rare that a quantifiable economic benefit can be identified that requires imposition of one of the hard core restrictions listed in Article 4 of the TTBER.[18]

(5) Step five—does the agreement or provision have the effect *of restricting competition?*

(a) General

5.23 Competition law treats the technology covered by intellectual property as an input that a licensee can acquire for use in making its own products for sale in a downstream market. The starting position is that this input is treated like any other input (raw materials, electricity, manufactured components) that the licensee requires to conduct its downstream business.

5.24 By treating technology as an input, the analysis outlined in Chapter 4 becomes directly applicable. The market for providing technology is an upstream input market (*technology market*) and the market where the licensee sells products using the technology (*goods market*) is downstream from that market. Where a firm holds technology of its own, so that it does not require a licence, that firm is vertically integrated. A licence for use of technology is thus an agreement between two firms at different levels of the production chain for purposes of the agreement. The theories of competitive harm applicable to vertical agreements are therefore also applicable to licence arrangements.

[18] This is reflected in the comment in the TT Guidelines para 75 that 'the Commission considers that in the context of individual assessment hardcore restrictions of competition will only in exceptional circumstances fulfil the four requirements of Article 81(3)'. As already noted above (para 5.14), however, the position is different where the ancillarity standard is applicable.

(b) Competitive harm in the Commission IP Guidelines

The relationship between the general approach to competitive harm outlined in Chapter 4 and the application of Article 81 to intellectual property licensing is reflected in the Commission IP Guidelines. The Guidelines suggest that restrictions in licensing agreements can lead to competitive harm in the following ways:[19] **5.25**

(i) reduction of intertechnology competition between the companies operating on a technology market or on a market for products incorporating the technologies in question, including facilitation of collusion, both explicit and tacit;
(ii) foreclosure of competitors by raising their costs, restricting their access to essential inputs or otherwise raising barriers to entry; and
(iii) reduction of intratechnology competition between undertakings that produce products on the basis of the same technology.

These categories of competitive harm follow closely the categories identified in the Vertical Guidelines.[20] The major difference (aside from the failure to give separate mention to the concern for market integration), is that the twin concepts of 'interbrand' and 'intrabrand' competition in the Vertical Guidelines have been replaced by 'intertechnology' and 'intratechnology' competition. The clear intention of the Commission is to apply the general concepts applicable to competitive harm identified in the previous guidelines to the specific context of intellectual property. Licence arrangements are treated as a form of vertical agreement, while technology pools and related R&D or standardisation efforts are treated as forms of horizontal cooperation. **5.26**

The Guidelines provide, however, only limited assistance in determining whether a 'restriction' could contribute to competitive harm. The purpose of this section is to identify the factors that may be relevant to finding competitive harm in the context of licensing arrangements by reference to the theories of competitive harm identified in Chapter 4. The implications of these factors are then discussed in more detail in subsequent chapters that consider the application of Article 81 to specific types of individual restraints. **5.27**

(c) Facilitating collusion as a theory of competitive harm

In the discussion below, reference is made in a number of contexts to the possibility that licence arrangements could 'facilitate collusion'. As already noted in Chapter 4 above, it is well established that agreements that facilitate **5.28**

[19] IP Guidelines point 141.
[20] Compare Chapter 4 above at para 4.47.

collusion may be restrictive where coordinated effects are more likely.[21] It is important to recognise, however, that for Article 81 to be applicable to an agreement there must be a direct causal link between the allegedly restrictive agreement or clause and a significant anticompetitive harm. The fact that a licence agreement facilitates collusion may not be sufficient to invoke Article 81 in respect of that agreement where the licence is only one among many factors that cumulatively lead to coordinated effects.[22]

C. 'Reduction of Intertechnology Competition'

5.29 A licence agreement is most likely to have an effect on competition involving rival technologies where the licensee either currently uses a rival technology or itself controls a rival technology. In either case, the licence will tend to make use of the rival technology less likely (an effect that may be increased by restrictive provisions in the licence agreement). Where the licensee is currently a competitor (using a different technology) of the licensor in the downstream market, the potential for competitive harm is particularly significant because a licence agreement could directly affect the constraints that the licensor and licensee exert on each other in that market. In light of the way that the prior competitive relationship between the parties affects the analysis of competitive effects, the discussion below treats separately:

1. effects where parties were previously competitors in the technology market;
2. effects where parties were previously competitors in the downstream market; and
3. effects where parties were not previously competitors in either market.

(a) Parties previously competitors in technology market[23]

(i) General

5.30 Where the parties were previously competitors in the technology market, agreement by one competitor to use the rival technology may create unilateral effects in favour of the licensor in the technology market, by withdrawing

[21] This applies in particular to agreements for exchange of confidential information.

[22] Where, however, a wider agreement or concerted practice is proven, alterations in licensing arrangements may be an appropriate subject for remedies even if those licence agreements were not themselves subject to Article 81.

[23] The discussion in the text involves cases where the parties have *competing* technologies. The issues that emerge where parties that license each other have *complementary* technologies do not involve intertechnology competition. The difference between competing and complementary technologies is addressed in Chapter 6.

the competitive constraint of the licensee's technology.[24] The extent of these unilateral effects may depend on a number of factors. These could include:

- the extent to which the licensee's technology was the closest substitute for the licensed technology;
- the scope for new licensing;
- the extent to which alternative viable technologies exist; and
- whether the technology market can support multiple technologies.

The assessment of the position in the technology market is complicated because, in contrast to the position in markets for other inputs, market share is an unreliable guide to market power where technology is concerned. In contrast with the position for manufactured products or physical services, the provision of technology is not limited by capacity, access to resources, or established trading patterns, all of which are reflected by market shares. Where a rival technology is a full substitute, the ability of that technology to restrain the exercise of market power by an incumbent is not limited by a small current market share, since a rival licensor can license as many licensees as may apply.[25] The assessment of market power in technology markets will often therefore focus on whether rival technologies are in fact full substitutes, which may depend on issues such as technical functionality, interoperability, and reputation among downstream users. (These factors are addressed in greater detail in Chapter 6.) **5.31**

(ii) Parties previously competitors in technology market—scope for new licences

The scope for new licensing is a factor that is specific to intellectual property licensing. Licences of technology are usually made on a long-term basis with limited scope for the licensor to terminate the agreement or obtain an increase in royalty rates. Thus for the committed licensees, the removal of constraints on the licensor due to the removal of a technology from the market does not lead to unilateral effects, since the licensor cannot raise prices or reduce output. **5.32**

(iii) Parties previously competitors in technology market—tipping markets

Where a substantial number of potential licensees remain uncommitted, the extent of unilateral effects may depend on the degree to which rival technologies remain available. Where rival technologies are available, the TTBER creates a safe harbour where there are three other technology 'poles'. Even where only one 'pole' is present, there may be no real unilateral effects if the remaining **5.33**

[24] This effect may inevitably follow from the decision of the licensee to accept a licence. It may also follow or be reinforced by contractual commitments to use the licensed technology or forgo the rival technology.

[25] Limits on the supply of other inputs needed to exploit a rival technology may, however, constitute a barrier to widespread exploitation. As discussed below in Chapter 6, this affects the extent to which the technology is a full substitute.

technology is a viable alternative. Since technology licensing is not subject to production or capacity restraints, there is nothing to prevent a licensor offering a genuine alternative technology from serving all remaining potential licensees, if the first licensor attempts to implement a significant increase in royalty rates.

(iv) Parties previously competitors in technology market—scope for multiple technologies

5.34 The assessment of market power in technology markets can also be complicated by the fact that technology markets can often be subject to 'tipping'.[26] After an initial phase of development, interoperability concerns and other network effects mean that one technology will become a *de facto* standard. At that point rival technologies cease to provide a competitive constraint. Where a rival 'accepts the inevitable' and takes a licence from the successful licensor, there is no unilateral effect and thus no real competitive harm.

(v) Parties previously competitors in technology market—coordinated effects

5.35 The withdrawal of one technology due to licensing between competitors could also lead to coordinated effects between the remaining licensors in the technology market. For example, if there were three technology poles prior to the agreement, the scope for coordination between the remaining licensors could become greater. The conditions for effective collusion will, however, be difficult to achieve in most technology markets, even if there are only two licensors.[27]

(vi) Parties previously competitors in technology market—reduction in dynamic efficiency

5.36 There is also a possibility that removing a licensor from a technology market could have an effect on future development of technology. This could occur if the licensee's technology might otherwise be the basis for development of new technologies, or where the licensee itself might otherwise develop new technologies. As noted in Chapter 4 this theory is inherently speculative. A theory of competitive harm of this nature should require solid evidence that new technology based on the excluded technology was likely and that the licensee was a particularly important source of innovation.

(vii) Parties previously competitors in technology market—quantifying likely effects

5.37 In light of the limiting features outlined above, cases where a licence agreement leads to genuine competitive harm in a technology market are likely to be rare.

[26] See e.g. *Commission* v *Microsoft* (COMP/C-3/37.7920) 24 March 2004.

[27] See the discussion of the conditions for effective collusion in Chapter 4 above at paras 4.27ff. The non-price terms in a licence may limit transparency. The commitment by licensees (so that competition to reach a licence agreement becomes competition for that licensee's future business) means that each potential licensor has scope and motivation to 'cheat'.

However, even where licensors gain market power by licensing potential competing licensees, the consumer welfare loss associated with that competitive harm may be limited. Since licence costs will be a small share of total costs for many licensees, even a substantial increase in licence costs may not lead to increases in prices at the consumer level. Thus, where there are genuine efficiency gains associated with a licence agreement, it cannot be assumed that unilateral or coordinated effects in the technology market outweigh those benefits, even if the conditions for such effects are present.

(b) Parties previously competitors in the goods market

(i) General

In principle, the possibilities for competitive harm where licensor and licensee are competitors in the downstream product market are the same as those that apply in any situation where a competitor supplies an important input to a rival. The licence arrangement itself may lead to coordinated effects by facilitating collusion among downstream rivals. Furthermore, restrictions in the licence agreement can also aid in facilitating collusion and can also create unilateral effects benefiting the licensor. Where rivals in a downstream market license each other on a reciprocal basis, the possibility for collusion may increase, and restrictions in the agreements may lead to unilateral effects benefiting either or both parties. **5.38**

The theoretical possibility of competitive harm in licensing between competitors does not mean, however, that licensing between competitors is necessarily harmful or even likely to be harmful. The nature of the technology, the market position of the licensor and licensee (or licensees), and the nature of any contractual constraints require careful examination before the existence or extent of competitive harm can be assessed. **5.39**

(ii) Parties previously competitors in the goods market—coordinated effects

Where a competitor licenses most or all companies in the market, any restrictions on price or quantity may create a direct consumer welfare effect.[28] Where the parties to a licence agreement (together with other rivals that are licensed by the same licensor) account for a substantial proportion of the goods market there is a possibility that those agreements could facilitate collusion in the market as whole. For licensing among competitors to facilitate collusion, however, the conditions in the goods market must meet the other requirements for **5.40**

[28] Because of the potentially serious nature of this effect, such commitments will be treated as restrictions by object. Where there is an argument that these restrictions are nonetheless necessary to achieve a consumer benefit, the extent of the competitive harm must be assessed by relation to the position of the firms subject to the restrictions in relation to the market as a whole, following the approach to intertechnology effects discussed below.

successful collusion.[29] The most likely contribution of licence agreements among competitors to successful collusion is probably a reduction in product differentiation—increasing transparency. Although other vertical arrangements among competitors may contribute to collusion by creating common cost structures, licensing arrangements by themselves may play only a limited role in this regard, since the cost of licensing is usually a relatively small part of the costs of inputs for the downstream product. Restrictions in licence arrangements may also contribute to collusion by increasing transparency.[30]

(iii) Parties previously competitors in the goods market—unilateral effects

5.41 The unilateral effects of licensing arrangements like those in other vertical arrangements between competitors typically result from restrictions in the agreements.[31] Thus territorial restrictions, customer restrictions, quantity restrictions, or other contractual restrictions may create unilateral effects benefiting the licensor or other licensees (or the licensee itself where the licensor accepts a restriction). These arrangements have the effect of market allocation These effects can be serious. Indeed, many of the classic 'cartel' cases involved market allocation by a central licensor who was also a downstream competitor in some geographic markets.[32] The actual effects in a given case, require an assessment of:

- the market position of each firm that benefits from the restriction; and
- the effect of the restrictions in question in reducing competitive constraints on that firm.[33]

(c) Licensor not competitor in downstream market[34]

5.42 Where the licensor was not previously a competitor in the downstream market, the scope for competitive harm on *intertechnology* competition is limited. There is a theoretical risk that restrictions in licence arrangements by different licensors could facilitate collusion on the downstream market, by limiting the number of players or increasing transparency. (Thus, where there are two or three

---

[29] See Chapter 4 at 4.27–4.30.

[30] For example, quality control restrictions could lean to greater homogeneity in downstream products. It is also theoretically possible that where competing licensors grant competing licensees exclusivity in respect to customers and territories the result may be to facilitate collusion by limiting the total number of competitors on the downstream market.

[31] See discussion in Chapter 4 at 4.18.

[32] See e.g. *United States* v *National Lead* 332 US 287 (1997); *United States* v *United States Gypsum* 333 US 364 (1998).

[33] See discussion in Chapter 4 at para 4.65. Thus the market share of the beneficiary firm, the extent of barriers to entry, the nature of the restriction (e.g. passive sales v active sales) and the scope for arbitrage may be relevant to assessing the extent of any unilateral effects.

[34] The case of the new entrant in the downstream market where the licensor is already a competitor is covered in the context of intratechnology competition 5.48–5.50, as is the case where the licensee was previously active in the downstream market using another licensor's technology.

licensors, each of whom gives exclusive licences for a territory, their parallel licensing policies would limit the number of players to three. Likewise where each licensor charges similar royalty rates, those rates (if known) would constitute a common cost-base.) The limitations on collusion effects in markets where the parties were previously competitors are, however, equally applicable here. Royalty payments are still unlikely to constitute a sufficient proportion of costs to make collusion more likely. The big difference is that where the licensor is not vertically integrated, the licensor's interest in facilitating a downstream cartel that includes rival technologies is limited. (Because licence revenues usually vary with output, but not with the downstream price, the licensor will prefer lower prices and higher output.)

D. Foreclosure Effects

(a) General

It will be recalled that 'foreclosure' describes the effect on competition that occurs where a firm engages in conduct that limits the access of its competitors either to inputs or to customers.[35] Foreclosure is a form of unilateral effects. As with other forms of unilateral effects it requires an assessment of the market position of the beneficiary (the 'foreclosor') and the impact on that position of reducing competition from the competitor subject to foreclosure (the 'foreclosee'). Foreclosure effects can occur on input markets (where access to customers is foreclosed) or on downstream markets (where access to inputs are foreclosed). In the context of technology markets, this means that foreclosure can, in principle, occur in either a technology market or in a downstream goods market. **5.43**

(b) Foreclosure effects on technology markets—the Technology Transfer Guidelines

The Technology Transfer Guidelines give the following description of foreclosure on technology markets: **5.44**

> 'Restrictions on the ability of licensees to use other technologies could limit the ability of rival technologies to establish themselves in the market. If the excluded technology presents technical advantages over the licensed technology for some consumers, the exclusion would create a direct consumer harm by reducing product quality without a corresponding reduction in price. If a rival technology, once established, would reduce the ability of the licensor or its licensees to demand monopoly rents, the exclusion of that technology could give the licensor and its licensees the ability to increase or maintain high prices.'[36]

[35] See Chapter 4 above at para 4.49.

[36] It will be noted that both of these effects are unilateral—they increase or maintain the market power of each licensee.

Thus, in theory a licensor can 'foreclose' other licensors by entering into agreements with licensees providing for exclusive use by that licensee of the licensor's technology. If enough potential licensees are committed to the licensor on this basis, providers of competing technologies may be forced to exit the market and providers of new technologies may not be able to market those technologies.

5.45 The conditions for finding this effect in practice may, however, be difficult to satisfy. First, it will be necessary to show that the licensor has real market power. Yet as explained above, it will be difficult to show real market power if true alternative technologies exist—current share of the licensing market cannot be a reliable guide. Secondly, it should be necessary to identify the competing technologies that are or will be foreclosed—since Article 81 under the new paradigm requires a restrictive effect by comparison to a realistic counterfactual. It will be noted that this second requirement may make the first difficult to satisfy, since the existence of a real competitor may negate the existence of significant market power. Third, it will be necessary to show how the restrictions in question operate to impede development or use of better or less expensive technology.[37] This may also be difficult because, in contrast with the position in a traditional distribution context, economies of scale in distribution do not apply for licensing.[38] There may, however, be minimum revenue streams that are necessary to support an effective programme of R&D.[39]

5.46 In light of these limitations, foreclosure effects arising from intellectual property licensing should be relatively rare. The real foreclosure problems that do emerge will often concern the development of a rival technology that was not a realistic alternative when the agreement was signed, but which becomes an alternative during the life of the agreement. Where this occurs the licence condition is preserving or protecting the market power that the licensor possessed at the time of the original agreement, meeting the first requirement identified above, while the existence of a viable alternative meets the second

[37] In this context it will be necessary to distinguish the barrier to entry imposed by the contractual restrictions from the switching costs inherent in the choice by the licensee of the licensed technology.

[38] The suggestion in the IP Guidelines that depriving the licensee of better technology or lower royalty rates represents an inherent welfare loss may be simplistic. Presumably a licence agreement that did not require use of the licensed technology would have either entailed higher royalty rates or other commitments that would have been accepted by the licensee (since the value of the technology to the licensee does not change). To determine the consumer welfare cost of the provision, the subsequent lower price would need to be offset against these initial benefits.

[39] To be effective in depriving the licensor of the revenue stream needed to fund product development, foreclosure would need to occur across all the product and geographic markets where the technology is used. Thus where technology is provided on a global basis, but downstream markets are local, total foreclosure in one geographic market could not have a foreclosure effect where sales in other markets are sufficient to fund research and development.

requirement. If the owner of a rival technology faces significantly higher costs or is truly excluded by the incumbent firm's licensing practices, then a case for foreclosure may be present.[40] Where profitable entry by a rival licensor is possible, however, competition in the downstream market should ensure that the benefits of the better technology and lower prices are available to consumers. Courts and competition authorities should be alert to ensure that an Article 81 foreclosure effect cannot be invoked too readily by licensees of the incumbent in order to escape the consequences of poor decisions made in agreeing to the original licence terms.

(c) Foreclosure effects on downstream markets

A licensee may obtain an advantage over rivals in a downstream market by securing an exclusive license for vital technology. In principle, this kind of foreclosure needs to be assessed on the same basis as any other claim involving foreclosure of a necessary input.[41] The key issues will be: (i) the market power of the licensee, and (ii) the 'essentiality' of the input. Where the licensee has significant market power and the technology is an essential input, the possibility of an increase in the market power of the licensee is real, for the reasons discussed in Chapter 4. The difference between this case and a foreclosure case in a non-technology environment, however, is that market share in the upstream market does not provide useful insight into the likelihood of foreclosure. The central question is whether the licensed technology constitutes an essential input. **5.47**

E. Intratechnology Competition

The analysis of intratechnology competition closely follows the analysis of intrabrand competition in Chapter 4. Like effects on intrabrand competition, it involves possible restrictive effects on the downstream market—the goods market. Like effects on intrabrand competition, effects on intratechnology competition can be manifested in either coordinated effects or unilateral effects. Like effects on intrabrand competition, effects on intratechnology competition are only likely where competition from outside (i.e. from competitors using rival technologies) do not impose a sufficient competitive constraint on the intratechnology rivals. As in the case of intrabrand competition, the focus of **5.48**

[40] The IP Guidelines suggest that bundling of rights could create a foreclosure issue on this basis. It is difficult, however, to see where the Article 81 'restriction' lies in an agreement to license a bundle of rights between firms that are not competitors. The competitive harm (if any) derives from the refusal to license parts of the package separately. The appropriate context for considering 'bundling' allegations is therefore Article 82. See Chapter 10 below at paras 10.31–10.32.

[41] See Chapter 4 above at para 4.54.

assessment is on the restriction of current licensees, not on the prospect that additional firms might obtain licences in the absence of a restriction. As in the case of intrabrand competition, the presumption that territorial restrictions that partition the EU lead to economic harm means that territorial restrictions can be presumed to create some economic harm where a licensee is allowed to sell in the EU but is restricted to certain territories within the EU.

5.49 Intratechnology effects differ from intrabrand effects to the extent that coordinated effects in the form of collusion between licensees are unlikely to be the result of technology arrangements. To the extent that licence arrangements facilitate collusion by reducing the number of competitors, the resulting collusive effects are likely to manifest themselves in collusion involving all competitors on the market. Insofar as uniform licensing terms facilitate collusion, this is unlikely to be an issue unless royalties are a substantial proportion of cost for the licensees.

5.50 As in the case of intrabrand effects, concern in the intratechnology context focuses on unilateral effects. A full assessment of those effects in a specific case therefore requires consideration of:

(i) the position in the downstream market of the firm that benefits from the restrictions (which may be the licensor or another licensee; and
(ii) the extent to which the firm or firms subject to the restrictions would impose a competitive constraint on the benefiting firm if the restrictive provision were removed, but contractual relations were otherwise unchanged.

Thus, with the exception of territorial restrictions within the EU (which are presumed to lead to competitive harm), unilateral effects in intrabrand competition only become an issue where the beneficiary firm has a reasonable level of market power and other existing licensees would be the principal source of competitive constraint in the absence of those restrictions. Where competitors using rival technology can constrain the exercise of market power by the beneficiary firm, restrictions on other licensees of the same technology are unlikely to be important.

(1) Step six—does the agreement generate economic benefits?

5.51 At a general level, the economic benefits of a licence agreement are clear. As the various guidelines recognise, technology licences foster dynamic efficiency by allowing new products, techniques or ideas to reach consumers more rapidly. Furthermore, since the firms that develop new technology are not necessarily the firms best placed to develop and market products using that technology, a licence agreement facilitates the integration of complementary assets. In addition, the combination of technology already available to the licensee with technology provided by the licensor may generate real synergies. Finally, even where the licensee does not wish to use the licensed technology, licences may also play a key role by allowing firms to pursue their own development strategies without the risk of suits for infringement.

The more difficult problem is quantifying the economic benefits for a specific agreement. As discussed above in Chapter 4, paras 4.68 and 4.77, there is an inherent difficulty in quantifying benefits associated with increases in dynamic efficiency because of the need to speculate about future benefits. This does not mean that it is impossible to quantify benefits. It does mean that the conclusion relating to benefits cannot necessarily be expressed as a monetary value. The result of the economic benefit analysis is more likely to be descriptive—on a range from 'minimal' through 'significant' to 'very high'.[42] **5.52**

In assessing the significance of benefits in a particular case, the starting point must be the nature of the technology itself and any advantages that this technology has over other technologies that are currently available or likely to become available.[43] An incremental improvement in existing processes may be a 'limited benefit', while a new process that could generate substantial savings in other inputs might be a 'significant benefit'. The second step in assessing the benefit is to consider the scope for benefits. A new medication for a disease that is suffered by 4,000 patients worldwide is less important (in a purely economic sense) then a medication with similar efficacy for a disease suffered by 400 million patients. The third step in assessing the benefits may be to consider the position of the licensor and licensee.[44] If the licensor has the resources and facilities to exploit the technology fully on its own, the benefits may be limited. Likewise the benefits of the agreement may be limited where the licensee already has access to technology that is comparable to the licensed technology. Where the primary benefit of the licence is creation of space for innovation, the extent to which the licensed technology was objectively likely to constrain the licensee's design decisions would be relevant in this context. **5.53**

---

[42] This does not exclude the possibility of more exact quantification in some cases. It would be unrealistic, however, to require such quantification as a prerequisite for application of Article 81(3).

[43] There is an apparent paradox here. As noted earlier in this chapter, the nature of market power and therefore the potential for competitive harm in technology markets is related to the degree to which rival technologies are full substitutes. Yet the benefit of technology licensing depends largely on the unique features of the technology.

[44] The position of the parties is important because Article 81(3) applies to *agreements*. The value or importance of the technology that is licensed is not the issue, but rather the extent to which the agreement between the parties will pass on the benefits of that technology.

(2) Step seven—do the benefits 'outweigh' the restrictive effects?

5.54 The concluding section of Chapter 4 explained that the 'economic analysis' under the new paradigm is not a completely scientific exercise. It is really an analytical framework for assessment of benefits and harms. The benefit of the new paradigm over the old Article 81 regime is that the analytical framework allows appropriate consideration to be given to the relevant economic considerations at each stage in the analysis. In the end, the assessment of competitive benefits and harms will have a subjective element. The subjective nature of the assessment is troubling, however, when it comes to balancing competitive harm and economic benefit. Where both sides of the balance include a subjective element, the ultimate conclusion from the balancing exercise cannot be particularly reliable. From a policy perspective, it might be appropriate to resolve the balance in favour of validity of the agreement where the balance is genuinely uncertain.[45]

5.55 The problem of balancing does not pose a problem in most cases involving intellectual property licensing, however. In almost any case where the licensor and licensee were not competitors before the agreement and that does not involve a restriction by object, the result of the balance will be the same; the benefits of the agreement will normally outweigh the restrictive effects. This follows because of the apparent paradox noted earlier—the greater the market power (and thus the greater the competitive harm) the greater the benefits. For example, in a case where competitive harm is based on foreclosure (which will usually be due to exclusivity favouring the licensee), the degree of foreclosure is based on the essential nature of the input. If the exclusive nature of the agreement is 'necessary' to overcome a hold-up problem (under the test applied in the following section), the significant benefit of the new technology will kept from consumers. For the same reason where benefits are limited, competitive harm will also be limited. The real issue in cases involving agreements between firms that were not previously competitors is not balancing, but indispensability.

5.56 Agreements between parties that were competitors prior to the agreement raise more complex issues that will be explored in more detail in Chapter 8. For present purposes, however, three points are of interest. First, in some cases the balance will be clear because the restrictive impact will outweigh the economic

[45] The IP Guidelines are silent on how this issue should be resolved. Since Article 2 of Regulation 1/2003 puts the burden of showing that Article 81(3) is satisfied on the proponents of the agreement, the burden may be on the proponents to show that benefits are greater, since the balancing requirement emerges under Article 81(3). The US Guidelines, in contrast, resolve a 'tie' in favour of validity. Antitrust Guidelines for Collaboration among Competitors, 2000, section 3.36 ('the Agencies consider whether cognizable efficiencies likely would be sufficient to offset the potential of the agreement to harm consumers . . .').

benefit. This will usually be a case where the parties have significant market power and the licensee has only limited need for the licensed technology (or for space to innovate). Secondly, in other cases the balance will be clear because the economic benefit outweighs the restrictive effect. This will usually be a case where the parties have limited market power. Thirdly, the balance may change depending on the extent of restrictive conditions in the agreement. (This is a separate issue from the issue of whether the restrictions are necessary to achieve those benefits.)

(3) Step eight—are restrictive provisions 'indispensable' to obtain the benefits?

The most important competition law issue raised by most intellectual property licences is whether any provision or provisions that restrict competition are indispensable to generate the benefits of the agreement in which they are contained. The principal justification for many restrictive provisions is to deal with hold-up issues. As explained in Chapter 4, para 4.81, some hold-up problems create an absolute block on effective dissemination of the technology. As noted earlier in this chapter, the restrictions necessary to resolve these problems are treated as 'ancillary' and should not be considered individually under Article 81. Other hold-up issues, however, are relative. They do not prevent use of the technology, but rather lead to a slower take-up or less extensive use of the technology. As with absolute hold-up problems, these issues can affect the licensor or the licensee. **5.57**

As regards hold-up problems for the licensee, these primarily concern the investments necessary to use a new technology or to promote the purchase of products using the new technology. The licensee may be unwilling to make these efforts without some protection from competition by rival users of the technology in respect of either the customers that they serve or the territories where they sell. It will usually not be difficult to justify some restrictions of this nature as indispensable, particularly where the agreement is made between firms that are not competitors in respect of the products produced using the technology. The indispensability of the restriction, however, is linked to the nature of the hold-up problem. A hold-up problem related to the speed at which technology is introduced will not necessarily justify protection for an unlimited period. Thus, particularly where the intellectual property rights protecting the technology are, as a practical matter, not limited in time (as in the case of know-how or copyright) restrictions that were justified at the point of the original licence may no longer meet the indispensability requirement at a later date. **5.58**

As regards hold-up problems for the licensor, these could include the need to protect the reputation of the technology (which might justify provisions on **5.59** quality control or use of the technology), the need to preserve the licensor's own ability to compete in a technology market (which might justify license-back provisions), or the need to protect the licensor in a goods market where it currently competes (e.g. where the licensor would not license the technology to a competitor but is willing to license for a non-competing use or territory). In many cases there will be a trade-off between the royalty fee that a licensor would demand for a non-restricted licence and the lower fee it would accept where the licence is restricted. Where the licensee intends to use the technology for only a limited purpose, the licensor may be willing to grant more favourable terms, either because the licensee's revenue stream from that activity will be more limited than that for other uses, or because the licensee will not be competing with the licensor in respect of that activity. In these circumstances, the benefits of the agreement are increased as a result of the restriction, since the licensee receives the benefits that it wants at a lower price than it would otherwise pay. (Considerations of this nature might justify *inter alia* territorial restrictions, customer restrictions, or restrictions on the number of products produced using the technology.)

5.60 In agreements between non-competitors, hold-up problems are usually genuine and justify restrictions, at least for an initial period. This follows because otherwise the licensor and licensee would not have an economic incentive to agree to these restrictions. In agreements between competitors, however, hold-up problems may be exaggerated to justify a market-sharing arrangement. (An agreement by an existing competitor to stay out of a territory or to refrain from selling to a specified consumer group under the cover of an IP licence could be a cover for a genuine cartel.) This problem will be particularly acute where competitors enter into reciprocal restrictive commitments in licensing each other. This explains the less favourable treatment accorded reciprocal licensing in the TT Guidelines and the TTBER.

5.61 A central component of the indispensability test is the requirement that a less restrictive alternative not generate the same benefits. In the licensing context this means that the necessity of absolute territorial protect or protection in respect of specific customers will require particular scrutiny, since the same hold-up problem might be addressed with a more limited restriction on active sales or limitation to a specified field of use.

(4) Step nine—do consumers share in the benefits?

5.62 Because of the role of technology transfer in fostering dynamic efficiency (and thus leading to better products, processes, etc. over time), an agreement that passes Steps One to Eight will almost inevitably benefit consumers.

(5) Step ten—does the agreement 'eliminate competition'?

It is important to recall that the final requirement of Article 81(3) refers to elimination of competition as *a result of the agreement*. In this sense an agreement between firms that were not competitors previous to the agreement is unlikely to eliminate competition, since an agreement cannot both create and eliminate competition at the same time. Where an agreement involves existing competitors in either the technology or goods market, there is a possibility that an agreement *could* eliminate existing effective competition either in the market where the parties are competitors or in a downstream market. The Commission has cited this risk as a reason for requiring that firms in a patent pool be free to license independently, and also as a reason for requiring that a patent pool license third parties on 'fair, reasonable, and non-discriminatory terms'. **5.63**

F. The Block Exemption, the IP Guidelines and the New Paradigm

The Commission's application of the new paradigm to technology licensing is embodied in the Technology Transfer Block Exemption Regulation and the Guidelines on Technology Licensing. The TTBER and the Guidelines reflect the model that the Commission has applied in other sectors (and discussed in Chapter 2)—a block exemption to provide a safe harbour for agreements that are clearly lawful under Article 81 and guidelines on application of Article 81 outside the scope of the block exemption. Both the TTBER and the Guidelines. reflect the analytical approach to technology licensing issues set out in the first section of this chapter. **5.64**

The TTBER creates its safe harbour by defining a class of agreements 'which can be assumed with sufficient certainty to satisfy the conditions of Article 81(3)'.[46] This class is defined on the basis of four principles. The first principle is that technology licensing usually leads to economic benefits.[47] The second principle is that where licensing occurs between firms that were not previously competitors, substantial competitive harm is highly unlikely if the market share in the affected markets does not exceed 30%.[48] The third principle is that where firms **5.65**

[46] TTBER Recital 9.

[47] Ibid. Recital 5: 'Such agreements will usually improve economic efficiency and be pro-competitive as they can reduce duplication of research and development, strengthen the incentive for the initial research and development, spur incremental innovation, facilitate diffusion, and generate product market competition.'

[48] The market share thresholds apply to both the technology market and the downstream product market or markets. Since the vast majority of technology licences involve technologies

were already competitors in the relevant markets, competitive harm is highly unlikely if the previous market share did not exceed 20%. The fourth principle is that certain contractual provisions are by their nature 'severely anti-competitive' and thus could cause competitive harm even where market shares are low.[49]

5.66 The first principle reflects Step Six in the analytical model laid out earlier in this chapter. Since benefits can be assumed, Article 81(3) is potentially applicable. The second principle reflects the general conclusions of an analysis of competitive harm under Step Five—for most licence restrictions potential effects depend on market power and market power is unlikely where market shares are low. The Step Five assessment also supports the conclusion that agreements between competitors in a downstream product are more likely to cause competitive harm, which is reflected is in the third principle underlying the TTBER. The fourth principle reflects the strand of analysis that starts with Step Four—certain agreements or provisions may have the 'object' of restricting competition and are likely to lead to competitive harm.[50] In effect, the TTBER applies the balancing test prescribed in Step Seven. If restrictive effects are unlikely (principle two or three) and benefits are 'usual' (principle one) then Article 81(3) is normally satisfied.[51] Where 'severely anticompetitive' restrictions are involved (principle four), it cannot be presumed that the benefits of licensing agreements below the market share thresholds outweigh their competitive harm. Thus the TTBER is designed to ensure that an exemption will be available only in conditions where a more complete assessment would lead to the conclusion that Article 81 is inapplicable.

5.67 The extent of the safe harbour created by the TTBER is extremely limited. Since the TTBER cannot apply where the share of the product markets using the licensed technology exceeds 30%, many legitimate pro-competitive licence arrangements fall outside its scope. The recitals to the TTBER itself clearly indicate that there 'can be no presumption' that agreements above the thresholds are either 'restrictive' under Article 81(1) or cannot satisfy the requirements of Article 81(3).[52] Thus the TTBER is peripheral to application of Article 81 in cases where there are real policy issues.

5.68 The Guidelines are therefore the real key for understanding the new paradigm as it applies to IP licensing. If the TTBER is a 'safe harbour', then the Guidelines can be viewed as a form of 'sailing instructions'. They provide the information

where market shares as computed under the regulation could exceed that threshold, the TTBER **is almost never directly applicable in practice.**

49 TTBER Recitals 10 and 11.

50 This principle also reflects the Step Eight analysis of 'indispensablilty'. See TTBER Recital 13.

51 Ibid. Recitals 10 and 11.

52 Ibid. Recital 12.

needed for safe navigation, including some indication of good anchorages in most weather, as well as navigational hazards (*per se* restraints). In general, the Guidelines are an impressive and coherent body of work. They reflect the overall approach of the new paradigm, making consumer welfare the guiding principle for competition policy. The Guidelines set out both the general approach to applying Article 81 in the intellectual property context, and an assessment of the application of Article 81 to a range of individual provisions.

The Guidelines suffer, however, from two inherent limitations. First, the discussion in the Guidelines of licensing outside the scope of the TTBER is relatively short (19 pages in the Official Journal). Secondly, the Commission is constrained by its limited experience in this area. Because of the prescriptive approach of the old regime, the Commission has had very few cases in which it has needed to consider the benefits and harms of intellectual property licensing in a practical context. The Guidelines are therefore understandably cautious in reaching firm conclusions on what should be allowed, where such guidance could limit the ability of the Commission to reach the appropriate result if an actual case comes before it. This means that the Guidelines do not always provide the kind of practical advice that practitioners require in assessing individual agreements. Many remain in a grey area—outside the TTBER safe harbour—not restrictive by object, where the Guidelines indicate that the extent of market power may determine the application of Article 81. **5.69**

G. Applying the New Paradigm in Practice—The Process of Self-assessment

The reliance of the new paradigm on economic theory and empirical evidence creates a real problem for practitioners who are called upon to give practical advice. The safe harbour of the TTBER is of little practical use. The Guidelines may provide a model for competition authorities to follow in investigating suspected infringements, and may provide benchmarks for courts in Member States to apply where cases raise competition law issues, but they do not necessarily make it easy for companies or their advisors to assess whether their agreements are in compliance with the law. Where there is no pending court case or official investigation, firms do not always have the ability to obtain access to the information required to perform the kind of analysis described in Part A of this chapter. Yet the new paradigm relies heavily on 'self-assessment' in place of the notification of agreements for exemption that featured in the old regime. **5.70**

One of the goals of this book is to assist firms in conducting this self-assessment. The key point is that a full economic analysis is seldom necessary in the absence of threatened or pending litigation. As long as advisors understand the analytical **5.71** structure applicable to a full assessment, it is possible to 'cheat'—to reach a reasonably safe result without the full assessment.

(a) The goals of self-assessment

5.72 In an ideal world, firms would not enter into agreements that are invalid or contain provisions that are invalid. A self-assessment exercise may support a legal opinion that an agreement or restrictive provision within an agreement are definitely lawful. In some cases, however, it may not be possible to reach a definitive conclusion on the legality of the arrangements, whether due to the limited information available to the parties or because the cost of a full enquiry cannot be justified in light of the economic value of the transaction. When the legal status of a commitment is not clear, the objective of the exercise becomes risk assessment. This starts with assessing the risk that the agreement or provisions infringe Article 81, but may also involve assisting the firms involved in determining how the risk of unenforceability, of eventual enforcement activity, or of private lawsuits could affect the commercial viability of a proposed commitment or of the arrangements as a whole. Finally, an important purpose of the self-assessment exercise is to create a compliance record. Even where the exercise does not lead to a definitive conclusion that an agreement is lawful, the fact that firms have assessed the legal status of the arrangements and concluded that they have reasonable arguments that the arrangements are lawful should help protect them from serious penalties if the agreements are eventually challenged and ruled unlawful.

(b) The structure of self-assessment

5.73 Self-assessment as described below is a three stage process. The first stage involves a preliminary consideration of the proposed arrangements on the basis of a working model of competitive conditions. The second stage requires review of any issues remaining after the preliminary assessment on the basis of a detailed economic analysis of restrictive provisions and their likely effects on the market. The third stage involves a risk assessment of remaining unresolved issues.

(i) Stage one—preliminary consideration of potential restrictions

5.74 The first stage of self-assessment can be further divided into three steps:

- developing a working model of the markets where the agreement could have restrictive effects to determine whether there are any likely market power issues;
- reviewing the agreements to determine which provisions could be regarded as restrictive; and
- classifying each potentially restrictive provision as either clearly prohibited, clearly allowed, or potentially problematic.

Develop a 'working model' for identifying market power The big shift between the old regime and the current self-assessment system is that it is now necessary for advisors actually to understand something about the market in which their clients operate. This does not mean that it is necessary to bring in economists to perform detailed econometric assessments. It means that the advisor needs to understand the nature of the products and technology involved in a transaction, who the customers for the products ultimately are, and how the firms involved in a transaction see the competitive landscape in terms of the position of actual or potential rivals. It means that, at a minimum, the advisor must understand the competitive relationship between the parties to the transaction—are they actual or potential competitors or not? **5.75**

With a basic understanding of market conditions, it is usually possible to create a working model of the position of the parties, to determine whether there are likely to be market power issues. Sometimes this means identifying a worst case scenario (where narrow markets lead to findings of market power) and a most likely scenario. It also may require assessing, with the aid of the client, how likely it is that the worst case scenario will be applicable. In the great majority of cases, this working model, with a rough idea of probable outcomes, will provide the basis for actually assessing the proposed arrangements. The discussion of market definition in Chapter 6 is meant to assist in this process, while also addressing the theoretical issues relevant to a more detailed assessment. **5.76**

Identify potentially restrictive provisions The next step is to review the agreements to determine whether there are potentially restrictive provisions. This analysis follows Step Two in the step-by-step approach outlined earlier in this chapter. The discussion in Chapters 7 and 8 of specific restrictions provides a useful checklist, but ultimately a provision is potentially restrictive where it limits either a licensor or a licensee in what is sold, how much is sold, to whom it is sold, or where it is sold. Provisions that describe the technology licensed or that establish the method of payment are usually not in themselves restrictive. Many licence agreements will not include any restrictive provisions. It is not unusual for an agreement hundreds of pages long to include restrictive language covering less than a page. **5.77**

Identify 'hard core' restrictions Where an agreement includes restrictions that fall within the list of hard core restrictions in the TTBER, there is always a serious risk that Article 81 will be infringed. If these provisions cannot be excluded from the agreement, a full assessment will be required. (See above, para 5.74.) **5.78**

Classifying each remaining restrictive provision Each remaining 'restrictive' provision needs to be assessed in the context of the working model of market conditions (including assessment of whether the parties are actual or potential **5.79**

competitors (which itself depends on the market assessment). The discussion of individual restrictions in Chapters 7 to 9 is intended to facilitate this process by indicating how the analysis of each restriction varies depending on the market position of one or the other party and on the competitive relationship between the parties. In most cases that do not involve agreements with competitors, this will conclude the process, because the identified restrictions do not raise serious issues. In other cases, the analysis will reveal restrictions that could not satisfy Article 81 in any defensible market analysis, which also leads to a clear result for the self-assessment exercise.

(ii) Stage two—detailed economic analysis

5.80 Relatively few agreements in practice involve provisions the substantive status of which remains in doubt after the preliminary review. If such provisions are identified, it will often be sensible to determine how important these clauses are to the overall transaction. Where the clauses are not really important or where alternative contractual provisions could meet the needs of the parties with less legal risk, the cost-effective option may be to change the agreement rather than pursuing a more detailed legal assessment.

5.80A Where a more detailed assessment can be justified, that assessment will require careful consideration of both the relevant market and the likely impact of the restrictive provisions on that market. The detailed discussion of market definition principles in Chapter 6 is intended to assist in the market definition assessment. Likewise, the description of the analytical framework for assessing restrictions under Article 81 in Part A of this chapter cross ref and the application of that framework to specific restrictions common in licensing arrangements in Chapters 7 and 8 are intended to facilitate the substantive assessment.

5.81 In some cases it may be useful to engage economists to assist in this assessment. In other cases, however, it may be sufficient for legal advisors to develop a better understanding of the markets with the help of personnel from the firms involved. In many such cases, investigation of the competitive context in which the agreement will be implemented will show that competition policy concerns are not warranted. It will not, however, always be possible to reach a definitive conclusion based on available evidence. It is particularly important that external advisors manage the expectations of their clients so that they do not expect a definitive legal opinion in an area where some uncertainty is often inevitable.

(iii) Step three—risk assessment

5.82 Where provisions in an agreement involving intellectual property do raise serious competition law issues, the purely legal assessment is only part of the equation. The real challenge for legal advisors is to put their legal analysis in a context that allows the firms involved to manage the legal risks. Risk assessment involves in part the uncertainties in the competition law analysis: How likely is a narrow market definition? How likely is it that a particular clause will be deemed to constitute a lawful field of use restriction or a (more questionable) customer restraint? Risk assessment also involves understanding the enforcement priorities of competition authorities—are they likely to take an aggressive view of how Article 81 applies to a particular provision? In addition, risk assessment involves the relationship of the parties and of interested third parties: Is the clause likely to become an issue? Who will complain? Would it matter to the client whether a particular clause were eventually ruled unenforceable? Finally, risk assessment involves evaluating the relationship of the clauses at issue to the whole transaction under review—if one or more clauses should eventually be struck down, would that make the whole arrangement between the parties untenable?

(iv) Step four—keep a record of the assessment

Good lawyers will, of course, normally record their legal advice to clients in writing and keep a copy of that advice on file. Licence agreements may be in effect for many years, however, and it is surprisingly easy for advice provided at the time the agreements were signed to disappear. Even where the original advice can be found, it may be difficult to understand the conclusions when the drafters of the advice are no longer around to explain them. In order to obtain the full benefits of the self-assessment process, it is particularly important that any review of provisions raising significant substantive issues be drafted in a way that can be understood by future advisors unfamiliar with the background of the transaction and that such advice be preserved along with the agreements themselves. **5.83**

6

MARKET DEFINITION

A. Introduction

The focus of the EU's modernised enforcement regime on the identification and control of market power makes careful definition of the relevant markets the essential first step in a competition law assessment. Where intellectual property rights are involved, this analysis often appears difficult due to the need to consider markets at different levels of the production chain and the technical complexity of both the products involved and the underlying technology. This need not be so. Market definition in technology-driven markets nearly always involves the application or adaptation of well-established principles of market definition in competition law. Pragmatic application of these principles can resolve most market definition issues. In particular it is vital to keep in mind that market definition is not an end in itself—its purpose is to allow identification of market power. **6.01**

In the discussion below we consider: **6.02**

(i) general principles applicable to product market definition in EU competition law;

(ii) the special issues raised by product market definition in technology-driven markets;
(iii) geographic market definition on technology-driven markets; and
(iv) the analytical steps that should be followed in identifying and defining the relevant markets.

We conclude with a review of the criteria used for assessing the existence and extent of market power in these markets.

B. General Principles for Defining Markets in EU Law

6.03 The first step in the transformation of EU competition rules into a system based on economic principles was the Commission Notice on Definition of the Relevant Market for purposes of Community Competition Law [1997] OJ C 372. Drawing on the Commission's experience in defining markets in merger cases and relying heavily on the analytical framework developed by the US authorities in their merger guidelines, the Market Definition Notice imposed an economics-based structure on a body of law that had previously lacked doctrinal rigour. This does not necessarily mean that the cases from before 1997 are irrelevant or that the Notice itself is inconsistent with prior cases. The Community courts have always recognised that economic analysis was relevant to market definition, but the structure of appellate review in competition cases did not facilitate imposition of an analytical structure on the Commission.[1] More important, while the modern approach makes explicit reference to economic concepts, the overall purpose of the exercise remains unchanged—to provide the basis for assessing 'economic power'. The discussion below relies heavily on the approach set out in the Market Definition Notice, making reference where appropriate to the case law.

(1) The purpose of market definition

6.04 In defining markets for applying competition rules, it is important to remember *why* it is necessary to define markets at all. Even in the days before economic concepts attained their current pre-eminence in applying Community competition law, the Court of Justice warned that mechanical application of market definition 'rules' was not appropriate; the application of these rules must be

[1] In most cases the courts have endorsed the Commission's own approach to defining markets, particularly where appellant parties have sought to overturn a market definition on 'technical' grounds (see e.g. Case 322/81 *Michelin* v *Commission* [1983] ECR 3461 paras 23–45). In cases where the Courts have overturned the Commission (e.g. Case 6/72 *Europemballage and Continental Can* v *Commission* [1973] ECR 215) the direction has generally been in favour of more economic analysis rather than less.

guided by awareness of their purpose.[2] For market definition, the clear purpose is to allow an assessment of market power. The rules that have evolved for defining markets are designed to assist in this assessment, and new issues can only be resolved on the basis of how market power can best be assessed.

Where a market is properly defined, the dominant firm should be able to achieve monopoly power if all of its rivals (and thus all sources of constraint) are removed. This leads to a fundamental feature of markets for competition law purposes—a 'market' is something that could be monopolised. If a firm could eliminate all rivals in a 'market' and still could not raise prices significantly above the competitive level—the 'market' is not a true market for competition law purposes.[3] A 'product market' for competition law purposes thus comprises a product or products of the firm whose conduct or agreements are under review, plus the products produced by firms that exert a constraint on the ability of the first firm to raise prices for its products significantly above a competitive level. In the remainder of this section, we consider the rules that have developed for defining markets and the increasingly sophisticated economic tools available to assist in this process. **6.05**

(2) The analytical framework for market definition

The starting point for market definition is always to identify the product or products produced by each firm under consideration. Then for each product it is necessary to apply a three step analysis: **6.06**

(a) *Demand substitutability*—What products currently available constitute functional substitutes for the product in question?
(b) *Supply substitutability*—Could rival producers quickly switch production of other products to making products that are substitutes for the products in question?
(c) *Geographical scope*—What constitutes the territory where competitive conditions are the same or similar for the firm or firms under investigation and their principal rivals?

---

[2] See *Michelin*, n 1 above. As a general matter, interpretation of the EC Treaty (including the competition rules) is meant to follow a 'teleological' approach; the goals of the Treaty must inform the meaning of each provision. This general approach is also applicable to interpretative questions arising within the Treaty framework.

[3] There are two circumstances where a monopolist in a properly defined market may lack market power. The first is where there are many partial substitutes for a product, none of which by itself could shift sufficient demand to make a price rise unprofitable. The second is where barriers to entry are very low, so a large range of firms could enter the market in response to a price increase. In the second case, the market is said to be 'contestable' and the firm lacks real market power. In each of these cases the real problem is that the market definition methodology has failed to capture all of the competitive constraints that might properly be contained in the 'market'.

(a) Demand substitutability

(i) The general concept

6.07 The starting point for defining a product market is 'demand side' substitutability. This means that two products belong to the same market if the potential purchasers for those products view them as in competition with each other. Customers will normally regard products as in competition if they are functionally interchangeable; the function of product A can be performed by product B.[4]

6.08 Markets rarely if ever involve competition between perfect substitutes.[5] The interesting issues in product market definition tend to arise where products are *differentiated*—perform the same basic function but with differences in features or quality—or where products are *partial substitutes*—where a product is a substitute for some customers or for some functions. In these cases the issue becomes whether the products are sufficiently substitutable to provide a significant constraint on the ability of the firm under consideration to exercise market power.

(ii) Assessing demand substitutability—different customer needs

6.09 The essence of demand substitutability is an assessment of customer needs. It is common for a product to be used for different purposes or for a similar purpose by different groups of customers. The general rule applicable in these circumstances is to assess demand substitutability separately for each group of customers. This tests whether the products that are substitutable for those customers' needs and are available to that group of customers could form a different product market. This reflects the possibility that a firm may have market power in respect of a specific group of customers, who lack alternatives to the firm's products, even though the firm does not enjoy market power on a more general basis.

6.10 This rule only applies, however, where the firm is capable of *discrimination* between customers. If the firm cannot maintain higher prices in respect of its supposedly captive customers, then its ability to exercise market power is no different in respect of those customers. This will typically be the case where *arbitrage* is possible, i.e. where products sold at a lower price to other customers could make their way to the 'captive' customers. Where discrimination is not possible, then interchangeability needs to be considered for all customers for the product, even though they themselves are in different markets.

[4] By focusing on the *function* of a product, the interchangeability test expands the range of products in a market to include products that differ in form from the products sold by the firm whose agreements or practices are under investigation.

[5] In economic theory, the prices of two products that are perfect substitutes will increase or decline together, because any increase in the price of one product will lead to a shift in demand for the other product.

(iii) Proving demand substitutability—the pragmatic approach

Determining whether two products are 'substitutes' may appear straightforward enough, at least where consumer products are concerned. As consumers, we think we know what products are substitutes, at least from our personal perspective. Practical experience can indeed lead to useful insight into product markets, particularly where reflected in standard industry classifications or categories used by third party collectors of data. Survey information can also provide insight into whether customers regard products as substitutes. Particularly in merger cases, competition authorities may send questionnaires to customers seeking their views as to whether products are functionally interchangeable. **6.11**

In the great majority of cases, these sources of information, along with a firm's own subjective assessment of what products form part of its 'market', provide a sufficient basis for performing a 'self-assessment' exercise.[6] The usefulness of this pragmatic approach to market definition is limited, however, because it can often understate the range of substitutes that are relevant for assessing market power. From an economic perspective, two products may be substitutes even if most of the customers for the first product would not regard the second product as an option. The key question is whether the prospect of a switch to the second product is sufficient to constrain an increase in the price of the first product. To determine whether markets are properly defined from an economic perspective, it is sometimes necessary to have recourse to economic tools. **6.12**

(iv) Proving demand substitutablity—'cross-elasticity of demand'

The central economic concept for defining markets is 'cross-elasticity of demand'. This concept describes the relationship between the price of one product and sales of a second product. If an increase in the price of the first product leads to a significant increase in sales of a second product, i.e. higher in percentage terms than the increase in price, then there is high cross-elasticity of demand. This will usually mean that the second product is in the same market as the first product. The major difficulty with applying this concept in practice inevitably involves proof; it is not usually commercially practical to simply raise prices and see what happens. From a legal perspective it is also necessary to consider how much cross-elasticity is enough to establish that products are in the same market. Finally, there is also an issue that cannot be readily resolved where the firm under investigation already has market power. If prices are already uncompetitively high, then cross-elasticity may exist with products that have not constrained the ability of the firm to raise price.[7] **6.13**

[6] It is usually possible to make a 'worst case' scenario based on the narrowest realistic market. If there is no market power on the basis of that 'market' a more detailed assessment is superfluous.

[7] This is the so called the 'cellophane' fallacy.

(v) Proving cross-elasticity of demand—economic tools

6.14 The economic tools available to assess cross-elasticity of demand include price correlation analysis and shock analysis. Price correlation analysis involves studying the movement of the prices of two products over time. Where prices tend to move together, this may be indicative of the fact that the products are close substitutes. It may also, however, simply indicate that many of the input costs for producing the products are the same. An important feature of economic analysis of price movements is identifying and testing for alternative explanations for price correlation other than cross-elasticity of demand. Because of the difficulties inherent in price correlation analysis, a shock analysis can provide more useful results when the necessary data is available. Shock analysis involves a situation where external factors have caused an increase in the price of a product or a decrease in its availability. Studying the actual shifts in demand following a sudden price or availability change can provide a powerful indication of what products are in fact economic substitutes.

(vi) Proving cross-elasticity of demand—the SNIP test

6.15 The Market Definition Notice uses the 'SNIP test' as its central analytical tool for assessing whether products are in the same market. This 'test', which originated in the US 1992 Merger Guidelines, requires that a competition authority consider what would happen if a monopolist in respect of the first product imposes a Small but Non-insignificant Increase in Price (hence 'SNIP'), which is defined as an increase of 5%, while the price of the second product remains constant. If sufficient customers would switch to the second product to defeat this price increase, the second product forms part of the same market. The analysis is then repeated until the point is reached where no additional product provides a sufficient constraint to prevent a putative monopolist from imposing a SNIP.

6.16 Although this analysis is almost invariably referred to as the 'SNIP test', it is not really a 'test' in the sense of a legal rule or an economic theory that can be applied to a set of facts. Although stated as a rule, it involves three conditions that are almost invariably hypothetical and therefore not subject to empirical evaluation:

- the monopoly in respect of the first product;
- the 5% price increase; and
- the effect on sales in the second product.

The SNIP test is thus really a framework for analysing anecdotal information about products and the data from economic analysis of demand. The important insight that this framework provides is that cross-elasticity must be related to market power—the ability to raise price.

The SNIP test creates this link by making the inclusion of the second product in the market contingent on its ability to defeat a price increase. In doing so, the SNIP analysis also deals with the question of how much cross-elasticity is necessary for two products to share a market. While economists can grade the level of cross-elasticity between two products, the SNIP approach tells us that the absolute amount of cross-elasticity of demand is not important, but rather the impact of cross-elasticity on the ability of the producer to raise prices. Furthermore, the 5% benchmark for a SNIP provides guidance on what constitutes a competitively significant effect. This effect can occur even when cross-elasticity is relatively low. In appropriate conditions, switching to a different product may defeat a SNIP even if less than 5% of customers were to switch. Thus the SNIP test focuses appraisal in difficult cases on the market perceptions of marginal customers.[8] **6.17**

(b) Supply-side substitutability

Once current substitute products and producers have been identified, it is necessary to consider whether possible market power on the part of the firms under investigation could be constrained by the reaction of firms that are not currently producing substitute products. If a firm could switch existing production to producing substitute products quickly and at relatively low cost, then that production capability exercises a potential constraint on the ability of the firm under investigation to raise prices above a competitive level.[9] The Commission Technology Transfer Guidelines illustrate this principle by reference to paper **6.18**

---

[8] An issue that complicates the assessment of demand substitutability is the so-called 'cellophane fallacy'. If a firm really does have current market power, it will increase its price to the point where further price increases are unprofitable. This may be because its customers are willing to do without the product rather than pay more, but it may be because more expensive substitutes become viable once the monopolist's price reaches a certain level. Thus either a pragmatic approach to demand substitutability or an economics-based SNIP approach may lead to a 'market' that does not provide a proper basis for assessing market power, since it includes 'substitutes' that do not constrain the creation or exercise of market power.

In the context of merger analysis, where the economics-based approach to market definition originated, this phenomenon does not really matter. In a merger case the real concern is whether the merger or acquisition will lead to an *increase* in market power or in collusive effects compared to the current position. Assessing interchangeability based on current commercial conditions provides a reasonable basis for assessing whether the transaction will lead to such effects. Likewise in cases involving conduct, to the extent that there is concern that agreements or abusive conduct will lead to a reduction in existing competition, then current conditions should provide the appropriate benchmark. If the concern is historical—that past practices have led to an increase in market power—then the analysis is the same but the interchangeability analysis needs to be conducted in light of the commercial position at the time that the alleged conduct occurred.

[9] This rule was established in EU law by the Court judgment in *Continental Can* v *Commission*, n 1 above. In that case the Court annulled a Commission decision that found a distinct market for metal cans used for selling fish products, without considering whether glass bottle producers could adapt their products to serve the same purpose. It is interesting to note that in US law supply-side substitutability is theoretically at least not part of market definition, but is taken into account in assessing whether market power actually exists in a market defined on the basis of demand substitutability.

manufacturers that make different grades of paper, which are not fully substitutable for each other, but who can easily switch production in accordance with shifts in demand or pricing patterns. Under these circumstances, a manufacturer of a particular type of paper is constrained in its pricing by the possibility that rivals could shift production. Thus defining a market for specific types of paper and its direct substitutes is not appropriate—there will tend to be a broader market for 'paper'.[10]

6.19 Supply-side substitutability is different from *potential competition*. Any entry by a new firm in response to an increase in price will restrain the ability of a hypothetical monopolist to raise prices. Where entry will only occur following a time delay, the entrant's production is not considered part of the market. Likewise where entry requires significant additional investment or significant adjustment of existing assets, the potential entrant's production is not considered part of the market. The apparent reason for this distinction is that the likelihood of entry following a time delay or of entry requiring a decision to reconfigure assets is more speculative than the likelihood of entry that can occur quickly without major costs.[11] Because potential entry is speculative, including potential entrants in the market may make the market larger than appropriate, overstating the competitive constraint that controls a firm's conduct. This does not mean, however, that potential entry is ignored in the analysis of market power. The identification of potential entrants and the assessment of potential barriers to entry by potential entrants are important parts of the substantive market power assessment discussed at the end of this chapter.

(c) Defining geographic markets

(i) General considerations

6.20 The classic test for geographic market definition in EU competition law was set out by the Court in *United Brands*:

> The conditions for the application of Article [82] to an undertaking in a dominant position presuppose the clear delimitation of the substantial part of the common market in which it may be able to engage in abuses which hinder effective competition and this is an area where the objective conditions of competition applying to the product in question must be the same for all traders.[11A]

---

[10] The Court reached a similar result in *Michelin I*, see n1 above, where the Commission defined the relevant product market as including all tyres for heavy goods vehicles. The Court rejected the criticism that specific sizes of tyres were not substitutable for each other because of the ease of substituting production of one tyre for another. The Court found that it was not necessary to include car tyres in the market because supply substitutability from car tyres to larger tyres was not as simple.

[11] These costs may include *opportunity costs*—the loss of the benefits that may be obtained by using productive assets for their previous purpose.

[11A] *United Brands* v *Commission* [1978] ECR 207.

This approach to geographic markets incorporates several important elements. First, it requires consideration of whether conditions are *sufficiently* homogeneous to allow assessment of market power. Thus geographic market definition, like product market definition, is not an end in itself, but rather a tool for addressing the real issue—whether the firm or firms under investigation have significant market power. Secondly, there is no requirement that competitive conditions be the same across a geographic market. 'Sufficient' homogeneity may allow significant differences in price, regulatory constraints, or consumer preferences within the same geographic market. Thirdly, the reference to conditions being the same for all competitors indicates the real focus of the exercise—determining whether a firm or firms that may possess market power have competitive advantages within a definable geographic area that they do not possess to the same extent in other areas. Where the same competitors face the same constraints, the geographic market may be EU-wide or even worldwide, despite differences in customers, suppliers, and regulation. **6.21**

This approach is in accord with the modern emphasis on economic assessment. In the same way that a product market must be 'worth monopolising', a geographic market must be an area where a sole supplier could have significant market power, even if conditions outside that area remain competitive. Where this condition does not apply, market shares and other structural indicators cannot provide reliable insight to the presence of market power. Thus to establish a separate geographic market, it is necessary to show *both* that different conditions apply in the relevant geographic area *and* that those differences give a significant competitive advantage to one or more firms. **6.22**

(ii) Regulatory barriers

Legal constraints that apply in one country or region can make the affected area a separate geographic market.[12] These constraints may take the form of special or exclusive rights that limit the number of competitors (e.g. licence requirements for telecoms operators) or legal requirements applicable to sales in that country that do not apply elsewhere (e.g. requirements regarding safety or environmental features of a product). Tariffs constitute a classic form of regulatory barrier, since they impose a cost on certain competitors that does not apply to domestic producers. In the absence of a direct limitation on the number of competitors, however, different regulatory conditions only create a barrier to entry where they provide a significant advantage to one or more firms. Where the costs of complying with the regulatory regime are comparable for all **6.23**

[12] Thus in *United Brands* (see para 6.22 above) the existence of separate national law regimes governing banana imports in the UK, France, and Italy justified exclusion of those countries from an EC-wide banana market.

competitors, the existence of that regime does not make the affected region or country a separate geographic market.[13]

(iii) Transport costs

6.24 Where transport costs constitute an important proportion of the costs for supplying a product, firms may enjoy a competitive advantage in the area close to their production or shipping facilities.[14] As in the case of regulatory barriers, however, transport costs are only relevant where they give a firm or firms a competitive advantage—higher costs incurred in supplying a region do not create a separate market where competitors bear comparable costs.[15] Even where transport costs mean that each producer can only operate on a local or regional basis, firms may not have real market power in their local 'market' if there is a significant overlap between their sales area and adjacent sales areas. Where overlaps exist, a reduction in price by sellers in one local market can have a 'ripple' effect from one adjacent market to the next, leading to a price reduction in another region where the firms initiating the price reduction are clearly not active. Under these circumstances, even very high transport costs do not justify narrow geographic markets.[16]

(iv) Other barriers—reputation, supplier or customer relationships

6.25 Transport costs and regulatory restrictions constitute objective external factors that can make competitive conditions different on a local, national, or regional basis. Sometimes, however, the nature of commercial relations in a region may itself have evolved in a way that gives an advantage to long-established firms in that market. Since consumer preferences are shaped by experience, an incumbent supplier with a national or regional identity may enjoy a significant 'reputational' advantage in a country or region that may be transferable in part to new products and services. An incumbent supplier may also have the advantage of established distribution channels and favoured relationships with suppliers in a country or region. Where these supply and distribution networks

---

[13] Thus in Case T-83/91 *Tetra Pak* v *Commission* [1994] ECR II–755 ('*Tetra Pak II*') at para 97, where the Commission defined a western Europe market for liquid packaging machines, the CFI rejected the argument that tariffs then applicable to importing machines into Spain and Portugal invalidated the market, observing that these requirements affected all producers equally since there were no domestic manufacturers in those countries.

[14] In Cases 40–48, 50, 54–56, 111, 113, 114/73 *Suiker Unie* v *Commission* [1975] ECR 1663 (the '*Sugar Cases*'), the cost of shipping sugar was regarded as sufficient to create regional markets. Transport costs have been particularly important in cases where sea crossings were involved.

[15] In *United Brands* (para 6.20 above), all banana producers faced the same transport costs in shipping bananas from the main entry ports. See also Case T-65/96 *Kish Glass* v *Commission* [2000] ECR II–1885 (Ireland not a separate geographic market for flat glass because there was no domestic producer).

[16] See Market Definition Notice (para 6.03 above) at para 57. This phenomenon is described as a 'chain of substitution'.

cannot readily be duplicated on a short- or medium-term basis (or where incumbent firms have significant reputational advantages), the incumbent firm or firms may enjoy an advantage over rivals that justifies treatment of the relevant country or region as a separate market.[17]

(v) Methodology—relevance of price differences—price correlation—SNIP test

Differences in prices between countries or regions do not necessarily indicate that the countries or regions in question are different geographic markets. Such differences may reflect differences in costs or differences in consumer perception that affect all competitors equally. Persistent regional price differentiation does, however, suggest the possibility of narrower geographic markets. Price correlation can provide a useful test for assessing this possibility. Where price relationships are constant over time (so that price increases or decreases together across the various territories), separate markets are less likely.[18] The Market Definition Notice suggest that a SNIP test can also be useful for assessing geographic market issues—where a small but significant local increase in price would attract additional sales from outside the territory, that territory should not constitute a separate geographic market.[19] **6.26**

(vi) Geographic market definition in practice—national markets for distribution—broader markets upstream

The Commission has often defined markets in competition cases on a national or regional basis. Most of these cases, however, have involved markets for final sale to end-users, particularly of consumer products or markets for sale of complete products to distributors. In these cases, barriers based on reputation or distribution/supplier relationships have often been central to the assessment. For intermediate products, in contrast, including most inputs for industrial production, markets have been defined on a European or even global basis, in the absence of regulatory barriers or significant transport costs.[20] **6.27**

---

[17] This will often be the case where the local advantages are the legacy of previous State action—e.g. where the incumbent firm is a former State monopoly.

[18] This assessment may also depend on the reason for the price decrease or increase. Changes in the price of a raw material input may have a uniform effect across different markets.

[19] The comments on the SNIP test in respect of product market definition (paras 6.15–6.17) are equally relevant here. The possible relevance of a SNIP test does, however, underscore that the purpose of geographic market definition is to identify areas where a firm might have market power due to local characteristics that do not apply elsewhere.

[20] These would include barriers based on infrastructure—electricity markets continue to be national.

(3) Special issues

(a) Defining markets 'from the bottom up'

6.28 A practical consideration in defining markets derives from the fact that market definition typically proceeds from the 'bottom up'—starting at the consumer level and working up. This results from the fact that the central question in market definition is demand substitution. To assess demand substitution, it is often necessary to know something about the customers. This is not a major issue where the product sold at the downstream level is a processed version of the input product. (A banana remains a banana, even if it is ripened at the distribution level.) For other customers of intermediate products, the ability to substitute different inputs will often depend on their purpose in purchasing those inputs. The market or markets where they are active as sellers will therefore define the inputs that are potential substitutes for them as purchasers. Furthermore, the possibility that different uses will lead to different levels of substitutability makes consideration of the downstream market the logical first step in assessing whether different customer groups required identification of different markets.

6.29 The need to define markets from the bottom up also applies in the context of defining geographic markets. Where the purchasers of an input operate in a market that is EU-wide, the market for the input will often also be EU-wide, even if transport costs or regulatory restrictions would otherwise lead to narrower markets at the upstream level. This will be the case where producers in the downstream market could source the input elsewhere (either by shipping the input themselves or by shifting production to the lower cost territory). If a sufficient number of customers could shift the source for the input and continue to supply the downstream market, then the upstream market will at least correspond to the size of the downstream market.[21]

(b) Derivative markets and switching costs

6.30 A recurring feature in competition cases is the identification of product markets linked to the products of a particular manufacturer or group of manufacturers. These cases often involve spare parts for a manufacturer's products. Even where the market for the products themselves is competitive, it is sometimes argued that each manufacturer may have market power in respect of the spare parts for

[21] Following the SNIP analysis (see para 6.15 above) the amount of production that needs to shift may actually be quite small. This phenomenon is most applicable where the input in question is primarily of use for a specific downstream purpose. Inputs that are used for a wide variety of purposes, where the ability of providers to discriminate is limited (e.g. electricity or natural gas) will not be as susceptible to changes in demand from a specific customer group and may therefore constitute narrower geographic markets than the downstream markets where their customers operate.

its products. This has led to identification of separate product markets for each manufacturer's parts. Since product market definition is based on supply-side substitutability this makes some sense, since for most purchasers of spare parts, the parts for one product are not substitutes for parts for another product. You cannot repair a Volvo with parts from Mercedes. To the extent that the purpose of market definition is to identify market power, however, spare parts markets are more problematic. It is often pointed out that if the market for the main product is competitive, a manufacturer could not exercise market power over parts. Any attempt to raise the price of parts above a competitive level would lead to a loss of sales for the main product. (This assumes that customers for the main product have accurate information about servicing costs.) The current position in EU law is that the existence of 'markets' for parts does not depend on the competitiveness of the main market. If the main market is competitive, however, then there may not be market power in the derivative market and neither Article 81 nor Article 82 may be applicable, regardless of the market shares or barriers to entry.[22]

Similar issues arise where products are developed to be compatible with other products. Thus, for example, the Commission's case against IBM in the early 1980s was based on the argument that IBM was dominant in the market for peripheral devices that could operate with IBM mainframe computers.[22A] Here again the logic of a system of market definition based primarily on demand substitutability favours the definition of a narrow market. Once customers have opted for a 'system', products that are not compatible with that system will no longer be viewed as substitutes, because to use rival products the customers will have to bear the cost of switching systems. The definition of separate markets for products compatible with individual manufacturers is, however, subject to limitations. If the compatible products are also sold to new users then it may be important to determine if the seller can maintain higher prices for the 'captive' users—if not, the market may need to include the competitive options for new customers. (Whether a user is 'captive' depends on switching costs, as described below.) Supply substitutability may also be an important factor where compatibility requires relatively minor changes in rival products. Finally, as in the case of parts markets, if the market for the main product is competitive and if the costs for compatible products are transparent, then there may be separate markets for compatibles but there will be no market power in those markets. **6.31**

An important tool in assessment of such cases is the concept of 'switching costs'. **6.32**
These are the costs that a customer would incur if it wished to switch to a

[22] *Pelikan Kyocera*, Twenty-fifth Report on Competition Policy, 1995, point 87.
[22A] *IBM*, Fourteenth Report on Competition Policy, 1984, points 94–95.

different product. The existence of switching costs inhibits the willingness of firms to switch suppliers in response to small changes in price. The higher the switching costs, the less likely the switch, and the less likely it becomes that demand by those customers could inhibit the exercise of market power. These costs may involve investment in different equipment—switching from oil to natural gas for heating may require investment in new boilers. These costs may involve changes in the way that the purchaser relates to its own customers—where a restaurant changes its beer supplier it may need to reprint its menus. These costs may involve the loss of economies of scale in distribution where the product currently purchased is obtained in connection with other products.

(c) **One-way substitutability**

6.33 Where competing firms sell differentiated products, the substitutability of the products sold by the firms may not be the same for each firm's customers. Thus customers who would usually purchase a standard family car might consider purchasing a more luxurious model if the price differential was small, but the customers for the luxury model would be less likely to step down to the family car. In some cases these asymmetrical demand constraints must be taken into account in assessing the likelihood of unilateral effects within a broader market. But in many cases, the difference in substitutability between the two customer groups will be reflected in the way that we define the markets. This will be particularly relevant in industries where differences in substitutability are the result of technical change. Thus the Commission used a market for containerised maritime shipping, excluding bulk shipping since shippers that use containers are unlikely to switch back to bulk shipping.[23] The Commission has also identified a separate market for internet connections (compared to dial-up) and has also suggested that ADSL could be separate from ISDN.[24]

6.34 In cases of one-way substitutability the nature of the market depends on the nature of the possible competitive harm. A merger between the last two providers of bulk maritime shipping may not raise significant competition issues if the merged firm will be constrained by the ability of customers to switch to containerised services. Where the harm involves bulk shipping the 'market' is all maritime shipping. The same applies to a merger between a bulk shipping supplier and a supplier of containerised services. Even though there is *also* a separate market for containerised services, the competitive harm (if any) can only manifest itself on the wider maritime services market. It is thus possible

[23] Case T-212/98 *Atlantic Container Lines* v *Commission* 10 October 2004, not yet officially reported.

[24] '*Deutsche Telekom AG*' [2003] OJ L 263/9. See also '*Wanadoo Interactive*', Commission Decision of 16 December 2003, not yet officially reported ('narrowband' not substitutable for broadband).

for the same product to be sold to the same customers at the same time in different 'markets' for competition law purposes. A competition law 'market' is an analytical tool for assessing market power. It should not be confused with 'markets' in a commercial sense.

(d) Potential markets

The distinction between commercial markets and markets for competition law purposes means that the competition law markets may exist even where the products involved are not currently sold at all, or are only sold as part of a larger package. The European Court has ruled that the competition rules can apply as long as there is a *potential market*.[25] A potential market may be said to exist where a product that fills a separable function can be identified and where customers would pay to purchase that product. If a shipping company owns port facilities, there will be a separate market for providing port facilities, even if the shipping company is not currently supplying services to third parties.[26] As a practical matter, the possibility of 'potential markets' can lead to confusion for commercial clients, since it can mean that the competition law assessment diverges from the way that a firm perceives the 'market' in which it operates. **6.35**

(e) Identifying potential relevant markets

A competition law assessment is not necessarily limited to a single 'relevant' market. In theory, a full assessment requires the identification of each market where the transaction or conduct under review could have anticompetitive effects. Furthermore once the 'potential affected markets' are identified, it is necessary to consider whether a party to a transaction (or a party engaged in questionable conduct) has market power in a market *related* to any of these potentially affected markets. This can mean that a merger case involving firms with extensive product lines can require consideration of hundreds of markets. **6.36**

As a practical matter, however, the number of possible markets that require consideration is more limited. In the usual Article 81 case, the starting point is the agreement or practice under review. The products or services that will be purchased, sold, or developed pursuant to that agreement are clearly products where the parties could have market power and where the arrangements could strengthen that market power. Accordingly, product markets need to be considered for each product that the agreement covers. Where either party purchases that product (either for resale or for incorporation in its own products), the downstream market or markets where the party is active may also be a **6.37**

[25] Case C-418/01 *IMS Health* v *Commission*, judgment of ECJ, 29 April 2004, not yet officially reported, at para 44.

[26] See e.g. *Sea Containers/SeaLink* [1994] OJ L 15/8.

relevant market. Likewise, where either party provides an input that is used for making the product, the upstream market for that input is potentially a relevant product market. It is often useful to get a rough idea of the position of the parties in each of these markets as a basis for making a preliminary assessment. A detailed assessment will only be needed, however, where the preliminary assessment identifies significant competition law issues.

C. General Principles of Market Definition—Conclusion

6.38 In applying these rules, it is important to recall two points. The first is that market definition in competition law is a means to an end—identifying market power. Mechanical application of legal precedents without regard for the purposes of the exercise will lead to false results. The second point, which is related to the first, is that there is often no 'correct' answer as to the relevant market in a competition case. Even where a regulatory authority can use its investigative powers to obtain information from all the relevant sources, the information available will often be capable of varying interpretation. At the same time, some answers will clearly be 'more correct' than others. The goal of a regulatory authority must be to make the best assessment of the market possible on the basis of available facts.

6.39 For firms involved in a self-assessment exercise, the challenge is even greater, since they must not just evaluate the hard information in their possession, but must also make their 'best guess' as the information available to their suppliers, customers, and rivals (actual or potential). The goal of the market definition exercise in self-assessment is to work out what the definition of the market is likely to be, if a full investigation were to take place. Even a detailed analysis may not provide a complete answer to this question. It can however, narrow the range of possible outcomes, and thus provide the basis for an informed assessment of the competition law risk associated with an agreement or transaction.

D. Defining Markets and IP Rights

(1) Product markets, technology markets, innovation markets

6.40 The Technology Transfer Guidelines and the Horizontal Restrictions Guidelines employ a tripartite division between 'product markets', 'technology markets', and 'innovation markets'.[27] As regards technology and product markets, this

[27] This division appears to be inspired by the US Guidelines on IP Licensing. It will be noted, however, that the US approach is slightly different because it does not focus on the question of whether the IP right itself is a 'product'.

division does no more than reflect the difference, described in the section above, between the market for the sale of technology by licence and markets for sale of products that are either protected by intellectual property rights or are produced using licensed technology. Following the general approach applicable to any agreement to supply a product or service, assessment of a technology licence requires at least a preliminary assessment of both the input market (the technology market) and the market where the input will be used or resold (the product market). The Guidelines essentially use these terms as shorthand to refer to the input market and the downstream market in the context of technology development, exploitation, and licensing.

The 'innovation market' raises different issues. This concept emerged as a tool for analysing merger cases involving technology-based firms, particularly pharmaceutical companies. To assess a merger between pharmaceutical companies, it is not sufficient to consider the products that they currently have on the market or the technology that can be used to produce those products. It is equally important to assess products that are 'in the pipeline'—in development but not yet ready for production—that could either replace or supplement existing products or will provide the basis for new products that are not yet in existence. If the parties to a merger are both developing products that could compete in some years' time, the relationship between these products is clearly relevant to the competitive assessment of the merger. **6.41**

It will be evident that an 'innovation market' is not a market in a competition policy sense, since even a putative monopolist does not have market power in an innovation market.[28] The innovation market is instead a device for focusing attention on the competitive relevance of ongoing development efforts. It is particularly useful in considering issues in the pharmaceutical sector, where the requirements for regulatory approval stages in the development process makes it relatively easy to identify products in development to treat the same condition. In addition to the significance of innovation markets for merger analysis, assessment of pipeline products may be important for assessing the impact of horizontal collaboration, including research and development. Depending on their level of development, the existence of pipeline products may be relevant to assessment of market power in technology or product markets. Identifying these effects is an important part of the competitive assessment, but it is not really part of the market definition process. **6.42**

The following sections address specific issues that apply to definition of product markets and technology markets. In keeping with the principle that markets **6.43**

[28] It is, of course, possible to license or assign technology that is still in development. To assess such a licence under Article 81, however, it is necessary to treat the licence as affecting a technology market and a downstream product market. If the licensee is not active in the technology market, market power in that market will normally not be an issue in the assessment.

should be assessed 'from the bottom up', the discussion begins with product market issues.

(2) Defining downstream product markets where IP rights are involved

(a) General

6.44 In principle, defining downstream markets where intellectual property rights are involved is no different from defining product markets generally. Indeed the ubiquity of technology and branding means that few product markets are unaffected by intellectual property rights in the broadest sense. There are nonetheless some aspects of market definition for downstream markets of particular significance for assessment of intellectual property related issues.

(b) Does technology define the product?

6.45 A preliminary question that arises in many intellectual property related cases concerns the competitive relationship between products incorporating different technologies. In the *IBM* investigation in the early 1980s, the Commission focused on 'IBM-compatible' peripheral products. In the recent *Microsoft* decision, the Commission defined markets for personal computer operating systems (not technology-specific) and for 'Microsoft-compatible' server products (technology-specific).[28A] The difference between markets where products using different technologies compete and narrower markets where competition is primarily between products based on or compatible with a specific technology generally depends on whether the customers for the product in question are committed to the product because of the 'switching costs' that they would face if they wished to use a rival product. Where the customers for a product are 'captive' (unwilling to change to products based on other technologies because of switching costs) then that product may constitute a separate product market.

6.46 The most common situation in which markets are defined on the basis of a separate technology involves products that are designed to be compatible with a product that is sold separately like a video recorder. Thus different formats for recording video programmes could constitute different product markets since they each require investment in different equipment. Likewise computer programs that work on one PC-operating system (e.g. Windows) may not be in the same market with programes that work on another operating system (e.g. Macintosh). Markets fitting this pattern are sometimes described as 'derivative markets' similar to the markets for spare parts or consumables discussed earlier in this chapter.

[28A] *Microsoft/W2000*, Commission Decision of 18 March 2004, not yet officially reported.

Markets may also be defined on the basis of a specific technology where they are constructed on the basis of a standard design to allow interoperability with a common infrastructure. (This may be a standard agreed among industry participants or a *de facto* standard.) Thus mobile phones may be in different markets if they apply incompatible standards. Even where there is no common infrastructure, the investments made by customers to adopt their internal operations to a technology may create switching costs. Thus customers may design their own internal data systems to work with particular types of computer software, or vehicle manufacturers may design cars to use a particular type of anti-pollution device. **6.47**

In all of these cases, the questions are: (i) whether the non-captive group of customers (customers purchasing a system for the first time or customers that need to replace an existing system) are sufficiently important to limit the ability of sellers to sustain a rise in prices;[29] and (ii) if so, whether the sellers can discriminate in price between captive and non-captive customers. If the answer to the first question is negative, then the products compatible with the technology constitute a separate product market. If the answer to the first question is positive and the answer to the second question is negative, then the captive customers would constitute a separate market. **6.48**

(c) Product groups

The *Microsoft* case[28B] also illustrates another characteristic of product markets in technology cases—the definition of a product market based on a group of products. In *Microsoft*, the Commission defined a market for Windows-compatible server products that included products serving different functions such as file management or e-mail distribution. From the perspective of demand substitutability these products do not belong in the same market since they serve different needs—they do not constitute direct substitutes for each other. The common link between these products, however, is their ability to work with the common operating system. At the same time, the biggest barrier to entering the market for producing one of these products is getting the product to work with the common operating system. This in turn means that a firm that imposed excessive prices for a Windows-compatible server product could attract competition from a producer of different Windows-compatible server products, but not from firms outside the market. **6.49**

[28B] Ibid.

[29] Since the sellers need to sustain a price increase, they also need to consider how existing customers could react to an increase in prices today when they renew or expand their own needs in the future.

(d) Breakthrough products

6.50 Where products are technology-intensive, technological change plays an important role in determining how markets are defined. Innovation can allow a new product to perform a function significantly better than products already in the market. Thus, for example, a new medication with fewer side effects than existing drugs, or a medication that requires a less intrusive method of administration (e.g. oral instead of injection) would be a significant advance over an existing product. If the innovation is sufficiently important, consumers may not view the existing products as full substitutes for the new product. Where a technological 'breakthrough' of this nature occurs, the seller of the new product may be able to demand a premium price. If the existing products do not constrain the pricing of the new product, then the product is sometimes described as a 'breakthrough product' and the market for that product can be defined separately.

6.51 The 'breakthrough product concept' is, in part, an application of the more general concept of one-way substitutability. As explained earlier in this chapter, the 'one-way substitutability' concept is based on the premise that market power in respect of products performing similar functions may be assessed differently depending on the firm under consideration. Where the market power of a firm selling a breakthrough product is at issue, the producers of old technology products may be left out of consideration, since they do not impose a competitive constraint. Thus, in assessing market power for suppliers of 'broadband' internet access, existing suppliers of 'narrowband' dial-up services would not be viewed as a constraint—broadband would be a separate market. Whether this is the case is an empirical question that depends on whether existing broadband users would switch back in response to a price increase.

6.52 Where technological change leads to one-way substitutability, there remains a possibility that an incumbent in the broader market could retain some market power. Although the possibility of switching to the new product may constrain that power, that constraint may not be complete, depending on the nature of the technological advance and the extent of the price premium demanded for the breakthrough product. An acquisition of rights to the product by the incumbent in the larger market could therefore give rise to some concerns. The concerns are minimal, however, where the nature of the technological change in question dictates a switch over time of all users to the new technology as it becomes accessible and affordable. In these circumstances the economic harm associated with any acquisition of rights is minimal. Whether a product is a breakthrough product in this sense requires an assessment of the nature of the technology itself and a projection of future consumer demand.

(3) *Defining technology markets*

(a) Intellectual property rights are not 'products'

There is a clear distinction in EU competition law between the product protected by an intellectual property right (which is clearly a product) and the intellectual property right itself. The ECJ established this point in the 1968 *Serena* case.[30] This distinction is justified on the basis that allowing an intellectual property right to define the product market could lead to definition of markets where there was no real market power. Thus in *Serena* the court was concerned that defining the market for a pharmaceutical product on the basis of the extent of a specific patent would mean that the possible competitive constraint from competing medicines would not be taken into account. The distinction between the product that is protected by intellectual property and the intellectual property itself recurs in a range of competition law contexts. Thus in *Magill*[30A] a market was defined for television listing information, rather than for the copyright licence (which would give the party obtaining that information the right to use it). Similarly in the recent *Microsoft* decision,[30B] the Commission has defined broad markets for server software and identified an abuse based on refusing to supply 'interoperability information' for a whole range of server products, without reference to whether various aspects of this information might be covered by different copyrights and/or patents. **6.53**

---

[30] Case 40/70 *Serena* v *Eda* [1970] ECR 69 at 83:

'It should first be observed that the proprietor of a trade-mark does not enjoy a "dominant position" within the meaning of Article 86 merely because he is in a position to prevent third parties from putting into circulation, on the territory of a Member State, products bearing the same trade-mark. Since the article requires that the position in question should extend to at least a "substantial part" of the Common Market, it is also necessary that the proprietor of should have the power to impede the maintenance of effective competition over a considerable part of the relevant market, having regard in particular to the existence and position of any producers or distributors who may be marketing similar goods or goods that may be substituted for them'.

See also the Opinion of Advocate General Roemer in *Parke Davis* v *Centrafarm* (Case 24/67) [1968] ECR 55 at 78:

'[M]ost of the parties have rightly emphasized that this [existence of a dominant position] is an economic concept and that consequently it was not enough to prove the existence of a legal monopoly (like that enjoyed by the holder of a patent) but that it was necessary to determine the position that the holder of the patent occupies on the market, whether the product which he manufactures is of its nature subject to competition, or at least does not meet substantial competition, and whether therefore the person enjoying the monopoly is in a position to fix prices and terms.'

[30A] Case T-69/89 *RTE* v *Commission* [1991] ECR II–485; Cases C-241 and 242/91 [1995] ECR I–743.

[30B] See n 28A above.

6.54 The distinction between the intellectual property and the product protected by the intellectual property works deals with a case like *Serena* where the competition law issue involves possible market power in the protected product. This analysis can be difficult, however, to reconcile with the commercial reality that intellectual property rights are bought and sold all the time. If an intellectual property right is not a product, what is being bought and sold in a licence or assignment agreement? For many products, it can be argued that the licence is the means by which the underlying product is marketed. Thus broadcast media may function on the basis of copyright licences, but the product remains the actual content (e.g. the film, music, or sporting event). But in some cases intellectual property rights *are* bought and sold separately from the products that they protect. Indeed in some cases, the party selling the intellectual property may not have the intention or even the ability to manufacture the product or provide the service that will be sold by the licensee. In other words, there are 'markets' for intellectual property rights in the commercial sense. Furthermore, holders of intellectual property rights may have market power from an economic perspective; the price for an intellectual property right may vary depending on the alternatives available to the licensee.

6.55 EU law deals with this apparent dilemma by treating the '*technology*' covered by intellectual property rights as the product. Thus while a patent is not a 'product', the technology described in the patent *may* be a product. Similarly, 'know-how' may be 'technology' and thus a product that can be sold through a licence, even though it may not be protected by intellectual property in a conventional sense. The licence of an intellectual property right thus becomes the transfer of a technology product, and the licence agreement establishes the conditions for that transfer of technology. The intellectual property right does not, however, define the parameters of the 'market' where this transfer takes place. The relevant market will depend on the existence of substitutes for that technology and on the nature of customer demand.

(b) Blocking, complementary, and substitutable IP rights

6.56 The analysis of product markets for technology can be obscured by the separate considerations applicable to the infringement of intellectual property rights. The relationship between intellectual property rights is often described in the rhetoric of rivalry—do the rights compete with other or block each other?—or the rhetoric of efficiency—do the rights 'complement' each other? Where there are two valid patents, and each patentee cannot practise the innovation covered by his patent without infringing the second patent, then the patents are in a 'blocking' position. Where exploitation of an innovation would infringe two or more patents, those patents are complementary to each other. Patents may also be 'complements' if exploitation of one would lead to a better product than would be produced using only one patent.[31] Where a potential patentee can choose between different patents, those patents are described as 'substitutes' (although they may be complementary to other patents). These relationships are important for assessment of competitive harm and economic benefits. Assessing whether intellectual property rights are in a blocking, competing, or complementary relationship is, however, primarily a question of the extent (and validity) of the various patent claims. It does not necessarily involve the considerations of substitutability or potential market power that drive the definition of markets for competition law purposes.[32]

In some cases, the considerations applicable to definition of technology markets and those applicable to the relationship between intellectual property rights are the same. If a potential licensee requires a licence under only one patent to produce a product and the licensee can choose between a licence under patent A or a licence under patent B, then patents A and B are competing patents and patents A and B also define a relevant technology market. Where, however, the potential licensee requires a licence under a number of patents in addition to A or B, then the A/B 'technology' is not a separate product market. For a potential licensee the technology package is the 'product'. If the potential licensee cannot acquire the other necessary intellectual property rights, then the A/B rights have no independent value. The ability of patentees offering components of the package (such as A or B) to exercise market power depends on the availability of alternative technologies. Where a patent is an essential element of a technology, the owner of that patent will have market power, but the market in which that power is exercised is a market for the technology (and any competing technologies) not the 'market' for the bit of technology covered by the essential patent.[33] **6.57**

(c) Defining a 'technology'

If a 'technology' is a 'product' and intellectual property rights themselves do not define that product, the key question becomes how should the technology product be defined. Put another way, what do we mean when we refer to '*the* **6.58**

[31] When used in this sense a 'complementary patent' is contrasted with an 'essential patent' (necessary to exploit a technology) or a substitute patent.

[32] This terminology has evolved in the case law and literature under US antitrust law dealing with patents and antitrust. The traditional US analysis has viewed the IP rights themselves as products and assumed competitive relationships on the basis set out in the text. The influence of the US IP Licensing Guidelines and US practice on the evolving EU policy has led to adoption of the US rhetoric, sometimes with the simple substitution of 'technology' for 'intellectual property'. This approach has led to some analytical confusion, particularly in respect of cross-licences and pooling as discussed below in Chapter 9, paras 9.33–9.35.

[33] In economic terms there is a single demand curve for the technology package. Where there are multiple patentees holding essential patents, double marginalisation may force the price up the demand curve to a sub-optimal result from the perspective of allocative efficiency, but there is not a separate demand curve for the different essential patents (if they truly are essential).

technology? The Horizontal Guidelines state that 'Technology Markets consist of the intellectual property that is licensed and its close substitutes, i.e. other technologies which customers could use as a substitute.'[34] This definition is not helpful because it directly identifies the 'technology' with the intellectual property. This is inconsistent with the approach used in conduct cases such as *Magill*[34A] and *Microsoft*.[34B] More important, it is inconsistent with the goal of market definition—to identify circumstances where market power can be exercised.

6.59 The better approach would define the 'technology' on a functional basis from the perspective of potential licensees. For some licensees, the full bundle of technology required to produce a product might constitute 'the technology'. For other licensees, a product may require different 'technologies'. (For example, a digital camera may require different technologies for image capture, storage, transmission, etc.) In the software context, a separate 'application' would usually constitute a 'technology' on this basis. In the standard-setting context, the full bundle of rights needed to implement the standard might constitute a 'technology' in many cases. (A standard may also define multiple applications, which are not all required by all potential licensees.) This approach can generate its own difficulties in borderline cases. It does, however, provide a better basis for assessing competitive relationships and potential market power than an analysis based on the intellectual property rights held or licensed in a particular case.[35]

(d) Competing technologies—technology 'poles'

6.60 The Commission defines 'R&D poles' in the Horizontal Guidelines as 'R&D efforts directed towards a certain new product or technology.' An 'R&D pole' in this context could be described as the precursor in an innovation market of a technology in a technology market. As such, an R&D pole is not a product, just as an innovation market is not a market. It is, however, a useful construct for understanding the significance of different structures of cooperation in R&D. Consideration of R&D poles also help in assessing the extent to which technology 'in the pipeline' is likely to constrain suppliers or users of existing technology.

[34] Horizontal Guidelines para 47.

[35] The difficulty here arises in part because IP rights are from a legal perspective the right of the holder to block the potential licensee from doing something that they want to do. The right of the patentee to block the potential licensor is not defined by relation to the licensee's objective, but rather by reference to the licensor's innovation. Consider the position if an applicant to a club with 100 members requires the signature of the chairman, co-chairman, and 49 other members to join and each signatory expects a payment for signing the application. From the potential member's perspective, the price that matters is the cost of membership in total. The fact that the chairman and co-chairman are essential and can demand a greater payment than individual members or that a group of 30 members might provide a package with each of their votes for a set price does not affect the interest of the applicant in the package as a whole.

[34A] See n 30A above. [34B] See n 28A above.

Technology poles can also be useful in describing the relationship between technologies, families of technologies, and products using different technologies. 'Intertechnology competition', which the TT Guidelines set up as the equivalent of 'interbrand competition', is really a description of competition between products associated with different technology poles, both present and future. 'Intratechnology competition', in contrast, is competition within a technology pole. The technology pole concept does not, however, serve to define product markets. In some cases, the relevant market may involve competition between competing technologies and in others there may be a relevant market restricted to a specific technology. This will depend on the extent to which switching costs for dependent user groups create separate derivative technology markets. **6.61**

(e) **Dependent user groups, switching costs, and derivative markets**

The issues raised by switching costs and captive customers in technology markets mirror those encountered in technology-intensive product markets. This is not surprising, since from the customer's perspective, the ability to switch to a different technology is subject to the same constraints as switching to components based on a different technology. Furthermore, where potential licensees are themselves selling in markets that are properly defined on the basis of different technologies, this fact may itself make the licensee a captive customer for improvements in that technology. **6.62**

As in product market cases, the questions relevant to defining narrower technology markets include: (i) whether the non-captive group of customers (customers licensing a technology for the first time or licensees that need to update existing systems) is sufficiently important to limit the ability of the licensor to sustain a rise in prices; and (ii) if so, whether the licensor can discriminate in price between these captive licensees and new licensees. The answers to these questions will depend in part on the switching costs of existing licensees. They will also, however, depend on the ability of the licensee's customers to switch to products using a new technology. Thus the nature of the technology market depends on the nature of the downstream product market. **6.63**

(4) Geographic markets in IP cases

Since intellectual property rights are granted under national law, it is easy to fall into the trap of treating the national territory where a specific intellectual property right is valid as a relevant geographic market either for technology licensing or for the products incorporating the technology. Applying the basic principles for geographic market definition identified earlier in this chapter, however, it is evident that geographic markets for technology will seldom be limited to national territories. The national character of intellectual property **6.64** rights does not change the characteristics of the markets for products that incorporate the technology protected by those rights. Only in the case where intellectual property rights issued in different jurisdictions conflict should the existence of intellectual property rights affect the geographic scope of the downstream product market.

(a) Geographic scope of technology markets

6.65 Technology markets will usually be EU-wide or even global in scope. This follows because the usual factors that limit geographic markets for physical products or services do not limit intellectual property licensing. Transport costs are, of course, irrelevant to sale of rights in intangible property. Although the need to apply for patent protection might be regarded as a regulatory barrier, this barrier applies to all the actual or potential competitors in a technology market. At least within the EU it should not be the case that a local firm has an advantage in applying for and obtaining patent protection that would give it a significant advantage over other firms developing and licensing technology. From the demand side, where the downstream product market extends beyond national boundaries the ability of potential licensees to obtain their technology elsewhere and supply the broader market limits the ability of a licensor to exert market power in respect of a single territory.

(b) Geographic scope of downstream product markets

6.66 As regards the downstream product market, the general approach to defining geographic markets remains valid where those products are technology-based. The one circumstance in which intellectual property rights may affect geographic markets is where firms established in different territories have conflicting rights. This occurs most often where firms have conflicting trademarks.[36] Patents or similar rights can, however, also lead to separate geographic markets where they allow the incumbent in one territory to block entry by a firm active in selling the same products in a different territory.[37]

E. Assessing Market Power in Technology Cases

6.67 Earlier chapters have discussed the concept of market power in general in competition law. This chapter has outlined how markets are defined in cases involving technology in order to provide the necessary basis for assessing the impact of

---

[36] See e.g. Case 119/75 *Terrapin* v *Terranova* [1976] ECR 1039.

[37] This implies of course that the two firms are the only seller of the relevant product.

agreements between firms on the exercise of market power. This final section considers how market power can be assessed in the context of those markets.

It is important to bear in mind that for competition law purposes, market power and possible competitive harm are not limited to firms that hold a dominant position for purposes of Article 82. The primary function of Article 81 is to limit the use of agreements that lead to unilateral or coordinated effects at levels of market power below dominance. Thus practice under Article 82 or old merger cases applying a dominance standard do not necessarily provide useful guidance on how to assess market power in other cases. It is also important to recognise that where products are at all differentiated, the existence of some market power is commonplace, as a matter of practice low levels of market power will be ignored. **6.68**

(1) Market power in downstream product market

The traditional competition law approach to determining whether a firm enjoys market power requires assessment of structural factors in the context of the defined market. This is reflected in the IP Guidelines[38] that list the following factors relevant to market power in technology-intensive product markets: **6.69**

- *The market position of the parties.* A low combined market share (20%) is unlikely to raise concerns—a relatively high combined market share (40%) is a strong indicator of market power.[39]
- *The market position of competitors.* A lot of small competitors may make the cumulative market share more significant as an indicator of market power, stronger competitors may make it weaker.[40]
- *The market position of buyers of the licensed products.* If buyers have market power, then market shares of sellers are discounted as indicator of market power.[41]
- *Entry barriers.* The Guidelines suggest that entry barriers are 'low' where 'effective entry, preventing or eroding the exercise of market power is likely to occur within one to two years.' The Guidelines cite economies of scale or scope, government regulations, state aid, import tariffs, intellectual property rights, ownership of scarce resources, first mover benefits, brand loyalty, and existing restrictive agreements as examples of entry barriers.[42]
- *Maturity of the market.* Where a market is evolving quickly, structural factors may be a less strong indicator of market power, particularly where demand is expanding.[43]

---

[38] IP Guidelines point 132. [39] Ibid. point 134–35.
[40] Ibid. point 136. [41] Ibid. point 137.
[42] Ibid. point 138. [43] Ibid. point 139.

- *Other factors.* (These include cumulative effects of other agreements, the duration of such agreements, the regulatory environment, or a history of collusive conduct.)[44]

6.70 The Guidelines observe that the importance of these factors 'may vary from case to case and depends on all other factors,' concluding that it 'is therefore not possible to provide firm rules on the importance of the individual factors'. This conclusion is certainly correct in terms of assessing the position where a given factor is *present.* In assessing technology markets, however, the key point will often concern the features that are *absent.* Thus while a high market share may not always show market power, a low market share (less than 15%) is very unlikely to be associated with market power. Likewise while high entry barriers are often associated with market power, a market with low barriers to entry and identifiable willing entrants is unlikely to give rise to significant market power. Furthermore, while a mature market is often associated with market power, a rapidly expanding market (like many technology-based product markets) seldom gives rise to competitive harm unless there is a very small number of competitors and high barriers to entry. Finally, where the conduct of firms in the market is currently competitive in terms of prices and product quality there should not be a concern about present market power (although that position could change as a result of agreements by firms in the market).

6.71 It will be noted that the Guidelines confirm that intellectual property rights themselves can constitute a barrier to entry in a product market. Whether intellectual property rights actually *do* constitute a barrier to entry in a given case clearly depends in part on the importance of the technology covered by those rights. Equally important, however, is the licensing policy of the firm that controls those rights. Where licences are available on reasonable terms or where the terms available to new entrants are no worse than those provided to incumbent firms, the existence of intellectual property rights may not be a barrier to entry at all.

6.72 There are some aspects of market power assessment that are important for technology-based products, but are not discussed in the Guidelines. The first involves competition from partial substitutes. Although the goal of market definition is to include in the market all products that exert a significant competitive constraint, there will often be products that are not sufficient in themselves to constrain market power—not enough customers would shift to prevent a price increase. Sometimes, however, the cumulative impact of a large number of partial substitutes may operate to limit market power. This can be a particularly important consideration in technology cases where product differentiation is common.

[44] IP Guidelines point 140.

A second factor relevant to market power assessment is the relationship between competition in primary and derivative markets. As already noted, the Commission has acknowledged that where the primary market for a product is competitive, a firm will not have market power in derivative markets (including markets for spare parts or consumables) as long as the costs in the derivative market are transparent. A 100% share of the derivative market will not create an inference of market power under those circumstances. **6.73**

A third factor involves the possible existence of 'bidding markets'. Where products are primarily sold in response to requests for competitive tender (whether formal or informal), the market share and other structural indicators of market power are not relevant. As long as there is a credible alternative bidder (and in the absence of collusion), there is unlikely to be real market power. **6.74**

Finally, where intermediate products are involved, competitive conditions in downstream markets can impose real constraints on exercise of market power. Even where customers may be 'captive' or dependent on a supplier, so that the supplier's product can be regarded as a distinct market, the ability of the supplier to raise prices will be constrained if the result would be higher prices for those customers and a consequent loss of sales in the downstream market. From the supplier's perspective the loss in quantity of sales could offset any benefit from higher unit prices. Thus, even though switching costs and technological dependence may lead to definition of separate markets based on different technologies, effective intertechnology competition in downstream markets may limit market power and deal with any concerns regarding competitive harm. **6.75**

(2) Market power in technology markets

The key point about assessing market power in technology markets has already been noted in Chapter 5—market share is not a reliable indicator of market power for licensors or potential licensors. This results from the fact that market share is indicative of nothing in itself, but rather serves as an indicator for barriers to entry or expansion, particularly those involving capacity and distribution constraints. For technology transfer capacity and distribution restraints are rarely an issue—one licensor can supply the demands for a whole industry. Where multiple technologies exist, the Guidelines suggest that the existence of four or more means that there is no market power on the technology market. In practice, however, even if there is only one viable alternative technology, it will be difficult to maintain market power since technology markets are seldom suited for effective collusion (new licences are relatively rare and existing users do not often switch). Thus the key question in assessing technology markets is whether there are viable alternative technologies. **6.76**

6.77 This analysis can change, however, depending on features specific to the technology market in question. One such feature involves the extent to which licensees for existing technologies may be able to switch at lower expense to follow-on technologies. In these circumstances the installed base of existing users give their supplier the ability to charge higher royalties. A particularly important feature for assessment of market power can, in technology markets, be the existence of network effects. Where the value of a technology increases because of the extent to which it is used by others, the supplier of the leading technology can soon obtain a decisive advantage. At this point the market is said to 'tip' in favour of the leading firm, which will obtain control over the market regardless of the technical advantages offered or the price charged for competing technologies.

Part III

APPLYING THE METHODOLOGY TO AGREEMENTS UNDER THE NEW RULES

7

'VERTICAL' AGREEMENTS FOR TECHNOLOGY TRANSFER BETWEEN 'NON-COMPETITORS'

A. Introduction

In the preceding chapters we have shown that the potential competitive harm of an agreement depends primarily on the competitive relationship between the parties and the nature of the agreement. In making this assessment, analysis under the new paradigm classifies agreements in two ways. First, there is a distinction between agreements between 'competitors' and other agreements. Secondly, there is a distinction between 'vertical' agreements and other agreements. The first distinction makes sense because, as shown in Chapters 4 and 5, the potential for competitive harm in any agreement clearly changes where the parties are competitors or potential competitors. The second distinction can, however, be confusing since agreements between competitors are often referred to as 'horizontal' agreements—suggesting that 'vertical' agreements should be understood as covering agreements between firms that are not competitors. As explained in Chapters 4 and 5, however, the 'vertical' nature of an agreement does not depend on the competitive relationship of the parties, but rather on the nature of their relationship for purposes of the agreement. Where one party is **7.01** supplying an input to another party for the second party to use in a downstream market, that agreement is a 'vertical' agreement for purposes of the Article 81 analysis. On this basis, any licence for transfer of technology qualifies as a 'vertical' agreement for purposes of Article 81.

7.02 Since any *bona fide* licence of intellectual property constitutes a vertical agreement—it follows that some licence agreements will constitute vertical agreements between competitors. In light of the general considerations set out in Chapters 4 and 5, it is not surprising that licence arrangements between competitors can raise serious concerns from a competition policy perspective. 'Restrictions' in licence arrangements that would not be problematic in most licence arrangements may raise concerns in arrangements between competitors. The special issues that are raised by such arrangements are discussed in detail in Chapter 8. Competitors may also collaborate on issues that involve licensing going beyond the simple transfer of rights to use existing technologies. In particular, collaboration in research and development, collaboration in standard setting, and the creation of patent pools will often involve agreement on licensing to third parties, including possible joint licensing. The special issues raised by these forms of collaboration are discussed in Chapter 9.

7.03 The purpose of this chapter is to address the application of Article 81 to licensing arrangements in the cases where the parties are not 'competitors'. These agreements can be regarded as 'simple' or 'pure' vertical agreements, since they do not raise the complex concerns found where existing competitors collaborate. The treatment of simple vertical licences provides a 'baseline' for the discussion in Chapters 8 and 9. While licences between non-competitors do not raise as many issues as the agreements discussed in the subsequent chapters, where such issues do exist, thy can arise in the context of any licence arrangement, including licensing between competitors or licensing in the context of a broader collaboration. Thus the discussion of specific restrictions in this chapter should be read as applicable to any licence arrangement, unless the special concerns raised in Chapters 8 and 9 warrant different treatment.

B. When is a Licence Agreement Deemed to be between Non-competitors?

7.04 As explained in Chapter 3, whether an agreement is between 'competitors' depends on whether the parties would have been actual or potential competitors *in the absence of the agreement.* If without the agreement, the parties would not have been competitors, they will be deemed to be non-competitors for purposes of Article 81.[1] Thus even though as a result of the agreement the parties may be competitors and the agreement clearly affects the way in which they compete against each other, the agreement nonetheless falls to be considered under the more relaxed standards applicable to simple vertical arrangements. This may appear anomalous for licence agreements where the need to assess application of Article 81 can arise many years after the event. It makes sense, however, for two reasons. First, the policy considerations outlined in the previous chapters dictate a different approach to agreements between parties that were not competitors at the point where the agreement was entered into.[2] Secondly, as the discussion in this chapter will show, characterising an agreement as one between non-competitors does not mean that the agreement is immune from scrutiny under Article 81. Where restrictive provisions in an agreement lead to competitive harm and the agreement does not generate significant benefits outweighing that competitive harm, the agreement can be caught by Article 81, even if the parties were not competitors when they entered into the agreement.[3]

Assessing the status of parties as 'competitors' at the time the agreement is entered into is particularly important for application of the TT Guidelines safe harbour. Where the agreement is deemed to be between non-competitors, the general 'exemption' under the TT Guidelines applies as long as the market shares of the parties do not exceed 30%. Furthermore, the list of 'hard core' restrictions under the TT Guidelines is more limited where the agreement is between non-competitors. Fixing the status of an agreement under the TT Guidelines in relation to the position of the parties at the time it was made provides some additional legal certainty, particularly where there are restrictions on the ability of the licensee to exploit its own technology or conduct separate research and development.[4] Even if changed circumstances subsequently allowed a licensor to enter the market, the TT Guidelines market share threshold would mean that Article 81 would still be applicable in any case where significant competitive harm could arise. **7.05**

Assessment of the competitive status of the parties to a licence agreement must be performed in the context of each of the markets that could be affected by **7.06**

[1] TT Guidelines Article 1(j)(ii).

[2] Where a licensee was not an actual or potential competitor prior to the agreement, both the assessment of competitive harm and the assessment of economic benefit under Article 81 are simply different, and remain different as long as the agreement remains in effect. Likewise, where a licensor that was previously not active in the downstream market subsequently enters the market (possibly as a result of subsequent agreements or acquisitions), the potential competitive harm is no different from that which could occur where an incumbent producer in the downstream licenses a new rival.

[3] The application of the TT Guidelines.

[4] These obligations are blacklisted for agreements between competitors under TTBER Article 4(1)(d).

the agreement. This requires definition of those relevant markets on the basis outlined in Chapter 6. Within these markets, the assessment of whether the parties to a licence agreement are actual or potential competitors follows the general approach applicable to any agreement under Article 81. The Relevant Market Guidelines refer to the familiar SNIP test, stating that a licensor and licensee will be viewed as *potential competitors* on the relevant product market and geographic market(s) if, in the absence of the agreement and without infringing the intellectual property rights of the other party, it is likely that they would have undertaken the necessary additional investment to enter the relevant market in response to a small but permanent increase in product prices within a short period such as a year or two.[5]

7.07 Four points should be noted regarding this assessment. First, the identification of separate markets for the licensing of technology and for the sale of products that use the technology is essential to this process. If, prior to the agreement, the parties were competitors in the technology market, in the downstream product market or in both markets, the agreement must be assessed under the more stringent rules detailed in Chapter 8. Secondly, for application of the safe harbour of the TT Guidelines, status as a competitor in the *technology market* depends only on whether the parties were *actual* competitors prior to the agreement.[6] This deviates from the general rule that the parties will be competitors for the purpose of this analysis if they were either actual or potential competitors prior to the agreement. Thirdly, in technology markets the parties will not be considered to be competitors where a party controls blocking intellectual property rights—where, in the absence of the agreement licensing those rights, the other party would not be able to exploit its own technology.[7] Fourthly, the parties will not be considered competitors in the downstream product market where the licensed technology involves a 'breakthrough' product, even though they may be competitors in respect of existing products that the new product will replace.[8]

[5] See Relevant Market Guidelines. The application of the SNIP test is discussed in more detail in Chapter 6. In individual cases longer periods can be taken into account. The period of time needed for undertakings already on the market to adjust their capacities can be used as a yardstick to determine this period. The parties are for instance likely to be considered potential competitors on the product market where the licensee produces on the basis of its own technology in one geographic market and starts producing in another geographic market on the basis of a licensed competing technology. In such circumstances, it is likely that the licensee would have been able to enter the second geographic market on the basis of its own technology, unless such entry is precluded by objective factors, including the existence of blocking patents.

[6] This provision provides a slight improvement in the legal certainty afforded by the TT Guidelines, since it is easier to determine whether a technology controlled by a party is an actual competitor for rival technology then to assess potential competition by technologies not yet on the market. Outside the safe harbour of the TT Guidelines, potential competition can be taken into account to some extent in technology markets. TT Guidelines para 66.

[7] For further discussion of this issue see Chapter 8.

[8] See generally Chapter 6.

(1) *Applying the new paradigm to agreements between non-competitors*

The general application of the new paradigm to licensing agreements is discussed on Chapter 5. It may, however, be worth recalling here some of the key points that relate to licensing between non-competitors. **7.08**

(a) Article 81 does not apply to the agreement itself

Where an agreement is entered into between firms that are not actual or potential competitors in a relevant market, that agreement does not in itself 'restrict competition' for purposes of Article 81. Article 81 can apply to restrictive provisions within the agreement. Accordingly, the discussion below focuses on specific provisions commonly found in licence agreements. **7.09**

(b) 'Ancillarity' can take a provision outside Article 81

Where 'restrictions' are necessary to avoid a hold-up problem, those restrictions may fall outside Article 81 without any need to assess whether they have the object or effect of restricting competition in the actual market context.[9] This concept is particularly important in the licensing context because many non-territorial obligations between licensors and licensees are so essential to purposes of a licensing agreement that unless included, there would be no point in transferring the technology and the know-how in the first place. Some examples would include confidentiality obligations, obligations on licensees not to sub-license, obligations not to use the licensed technology after the expiry of the agreement, provided that the licensed technology remains valid and in force, obligations to assist the licensor in enforcing the licensed intellectual property rights, and obligations to pay minimum royalties or to produce a minimum quantity of products incorporating the licensed technology.[10] **7.10**

(c) The cases where a provision will be deemed to have the 'object' of restricting competition are limited

The clauses deemed to be restrictive by object under the new paradigm will generally include those listed as 'hard core' restrictions in Article 4 of the **7.11**

[9] Where an agreement is between competitors, in contrast, the existence of such restrictions may still be relevant to the overall competitive assessment of the agreement.

[10] A similar analysis can be performed for contractual restraints such as field of use restrictions. What is important to note here is that the new methodology recognises that the IP owner can subdivide its powers of exploitation by contractual restriction and not be caught by Article 81(1) under the analysis of restriction on competition rather than the scope of the patent or limited licence doctrines which stem from IP law. Consequently, in the case of agreements between non-competitors, the new methodology gives a wide scope for contractual territorial restrictions between licensor and licensee not to be caught by Article 81(1) in the first place.

TT Guidelines. For agreements between non-competitors, the effect of the list in Article 4(2) of the TT Guidelines is to limit restrictions by object to agreements that fix the resale price of products produced using the licensed technology and certain restrictions on passive sales by the licensee.[11] For other restrictions in agreements between non-competitors, Article 81 is only formally applicable if the provision has the effect of restricting competition in its commercial context.

(d) Competitive harm in agreements between non-competitors will usually involve either effects on 'intrabrand' competition or 'foreclosure'

7.12 Although there may be circumstances in which a licensing agreement between non-competitors may affect 'intertechnology competition', these effects can usually be discounted in a self-assessment exercise.[12] The principal concerns that arise in respect of agreements between non-competitors involve (i) effects on 'intratechnology competition' and (ii) foreclosure effects in either the technology market or the downstream product market.

(i) Effects on intratechnology competition

7.13 As explained previously, effects on intratechnology competition are effects in the downstream product market.[13] Such effects are unlikely to lead to competitive harm unless there is insufficient competition from competitors using rival technologies to constrain the licensor and/or the licensee in that downstream market. Furthermore, even in the absence of interbrand constraints, there will

---

[11] The concept of a hard core restriction under Article 4 has changed from that of the blacklisted provisions in the 1996 TT Guidelines. Formerly the test whether the provision was restrictive in nature was elaborately defined and this resulted in seven blacklisted restrictions. Now the test has been modified so that it also asks whether the restriction is so likely to lead to anticompetitive harm that detailed economic analysis of effects is required. Guidelines to Application of Article 81(3). Cf. TT Guidelines para 75: 'based on the nature of the restriction and experience showing that such restrictions are almost always anti-competitive'. Based on this test, other provisions not specifically mentioned in Article 4 of the TT Guidelines could still be regarded as restrictive by object. For example, a provision requiring a licensor or licensee that is subject to a restriction on active or passive sales to impose the same restrictions on its own customers would be restrictive by object, on the same basis that such restrictions are classified in the application of Article 81 to distribution arrangements.

[12] The TT Guidelines indicate that such effects could occur where there is a network of agreements by competing licensors that lead to market rigidity and therefore to collusion in downstream markets. As explained in Chapter 5, these effects, while theoretically possible, require the presence of additional conditions that will rarely occur together. Even where this effect does exist, however, it would not necessarily mean that the licence agreements themselves are restrictive. The presence of the licence agreements may instead be relevant to determining whether there is a broader concerted practice present in the downstream market.

[13] Chapter 5.

only be significant intrabrand effects where the beneficiary of the restrictions enjoys some market power.[14]

(ii) Foreclosure effects

7.14 Foreclosure effects can occur in either the downstream market or the technology market. In the downstream market, they are only likely where the licensee or licensees do not otherwise face significant competitive constraints. In the technology market, foreclosure effects will only be likely if: (i) the licensor has significant market power; and (ii) there are viable alternative technologies that cannot effectively enter the market due to restrictive provisions in the licence (either alone or in combination with similar agreements with other licensees).

(iii) Excluding agreements that are unlikely to cause competitive harm from the Article 81 analysis—role of the TT Guidelines

7.15 Since these two types of competitive harm are only likely where there is significant market power in either the technology market or the downstream market, it makes sense to exclude from extended analysis agreements where the parties lack such market power. This function is performed by the 30% market share threshold of the TT Guidelines. Although the TT Guidelines purports to apply Article 81(3), its principal role is to 'exempt' provisions to which Article 81 probably did not apply in the first place.

(e) Most restrictive provisions in licences between non-competitors generate consumer benefits—the real issue is often 'indispensability'

7.16 Most 'restrictions' in licence agreements between non-competitors outside the hard core category can at least arguably be characterised as contributing to dynamic efficiency either by reducing hold-up problems or by addressing free rider issues. In some cases, however, the market power of the party benefiting from the restrictions means that the benefits of the restriction are outweighed

[14] Under EC law, since the *Consten/Grundig* case [1966] ECR 299, there have long been specific competition concerns with restrictions on intratechnology competition such as provisions placing obligations on licensees not to sell directly into the territories of other licensees. From the Commission's point of view such obligations are regarded as restricting the potential competition that could have existed between the licensees in different territories in their absence and therefore restrict competition for the purposes of Article 81(1). One universal example of restrictions on intratechnology competition would be a price restraint placed on all licensees by a licensor. See TT Guidelines Article 4(2). A second example of such a restriction might be an output restriction. A third example consists of restrictions on passive sales by licensees into the territories of other licensees of the same technology. Insofar as a licensing agreement between non-competitors contains such a restriction on intratechnology competition, it will be necessary to resort to the analysis under Article 81(3) to decide whether it is exemptible under Article 81 as a whole.

by the competitive harm. The requirement of 'indispensability' may, however, be more significant. A restriction on all sales into a territory might, for example, be defended as useful in overcoming hold-up problems. It would nonetheless be necessary to show why a restriction on active sales alone would not be sufficient to address those issues. This is the basis for the general rule disfavouring restrictions on all sales, but allowing restrictions on passive sales.

(2) Analysis of specific restraints

7.17 Since a licence agreement between non-competitors does not in itself restrict competition for purposes of Article 81, the focus of the Article 81 analysis is on the individual restrictions contained in the agreement. The discussion below focuses on the most common restrictions in licence agreements, which are discussed in the TT Guidelines and which have featured in previous block exemption regulations. We start by analysing restrictions on territories, customers, and output, which in other circumstances could be deemed restrictions by object because of their inherent impact on competition. The competitive harm resulting from each of these types of clauses is clear, although significant effects will be unlikely in most agreements between non-competitors. We then consider the broad range of restrictions that have traditionally been regarded as raising competition policy issues. We conclude by noting some 'restrictions' that are not regarded as harmful because they are clearly ancillary to the licence agreement.

(a) Territorial restrictions—general considerations

7.18 Territorial restrictions take four forms. First, the territory where a licensee is authorised to use the licensed technology or to produce products protected by the licensed technology may be limited. Secondly, the licensor may agree not to license a third party with respect of the territory granted to the licensee (referred to in the TT Guidelines as a 'sole licence'). Thirdly, the licensor may agree itself not to use the licensed technology or to produce products using the technology in the licensee's territory (referred to in the Guidelines as an 'exclusive licence'). Fourthly, the licensee may be limiting in its sale of products made using the licensed technology outside the licensed territory. In EU competition law, this fourth form of territorial restriction is further subdivided into restrictions on *active* sales where the licensee solicits business outside the territory and *passive* sales where customers approach the licensee in its territory for sales outside the territory.

7.19 The grant of a sole or exclusive licence serves to protect the licensee from competition by other licensees and (in the case of an exclusive license) from the licensor itself. Limitations on the licensed territory and restrictions on sales outside the territory serve to protect the licensor or other licensees from competition by the licensee, outside the licensee's territory. Such provisions also indirectly protect the licensee, however, since the licensee may accept such limitations on the explicit or implicit understanding that other licensees will be subject to the same restrictions in its territory. Thus the combination of limits in the territorial grant of intellectual property rights, the grant of exclusive or sole licences regarding specified territories, and territorial restrictions on sale give a licensee or a licensor a degree of protection from competition by other firms using the same technology, and thus potentially a degree of market power. From the perspective of EU competition policy, the creation of market power on this basis raises particular issues because of its effect on integration of the single market, at least where the territories do not cover the entire EU. At the same time, the creation or preservation of some market power on this basis may be desirable or even necessary, because it creates the necessary incentives for investment in technology and removes other hold-up barriers to diffusion of new technology.

The different forms of territorial restriction all contribute to the creation of **7.20** market power. They are nonetheless treated differently under Article 81. The grant of a limited territory is generally not regarded as a restriction of competition under Article 81, since it is inherent in the nature of intellectual property rights that are based on national law. The grant of a sole or exclusive licence may be subject to Article 81, but the Commission has been sympathetic to arguments that such restrictions are either ancillary or pro-competitive, even under the old Article 81 regime. Furthermore, the argument that such licences ought to be regarded as intrinsic to the intellectual property rights themselves has never been fully resolved.[15] Restrictions on sales, in contrast, are regarded as fully within the scope of Article 81[16] and the developed precedent regarding territorial restrictions on sales in distribution agreements provides a model for assessing such restrictions. In the discussion that follows we address first the issues that arise in respect of limited licences, sole licences and exclusive licences, and follow with a discussion of territorial sales restrictions.

[15] The Commission asserted that sole licences would be caught by Article 81 in its *Nungesser* decision [1982] ECR 2015, but this issue was not resolved by the Court in the subsequent appeal.

[16] The doctrine of exhaustion, which holds that IP rights cannot be invoked to prevent sale once a product has been put into free circulation within the Community, has provided a basis for the assertion that restrictions on sales are not essential to the IP right. This argument explains why restrictions on sales by customers of a licensee are clearly covered by Article 81. It does not necessarily follow, however, that sales by a licensee outside the territory are caught on this basis, since the products may be put into circulation for the first time in the sale.

(b) Territorial restrictions—sole and exclusive licences

(i) Sole and exclusive licences are treated the same in agreements between non-competitors

7.21 As already explained above, the TTBER and the Guidelines differentiate between sole licences, where the licensor agrees not to license third parties in a territory, and exclusive licences, where the licensor also agrees not to use the technology or manufacture products covered by the technology in the licensed territory. This distinction is important in the context of licensing between competitors where the grant of an exclusive licence has the effect of removing the competitive constraint previously exerted by the licensor on the licensee. In the case of agreements between non-competitors, in contrast, the substitution of the licensee for the licensor will usually by itself lead to any increase in market power. The economic harm, if any, of an exclusive licence is therefore no different from that of a sole licence and they are therefore analysed in the same way.

(ii) Ancillarity and economic benefits

7.22 Although, historically, the Commission has been reluctant to fully recognise the indispensability of exclusive territorial restrictions in vertical licensing agreements under Article 81(1), the new TT Guidelines acknowledge the case under Article 81(3).[17]

> 'Where the licensee is facing substantial sunk investments, agreements granting an exclusive licence are likely to fulfil the conditions of Article 81(3) except where there are no real alternatives to the licensor's technology on the market or where most of the available technologies have been licensed to the same licensee.'

Under the new paradigm, however, there is also a persuasive case that certain forms of territorial protection are not caught by Article 81(1) in the first place. To understand this case, it is necessary to recall the commercial function of such clauses, i.e. how they figure in the commercial exchange that allows the agreement to be reached.

7.23 The degree of territorial exclusivity that the licensor can offer the licensee lies at the heart of the commercial exchange in a technology licensing agreement. The licensee is initially attracted to the idea of taking on the risks of investing capital in manufacturing premises and plant and a distribution system rather than the costs of engaging in its own R&D programme. However, a licensing agreement often stipulates a minimum royalty provision as well as an upfront fee. Hence, unless there is some protection from competition from the licensor and other licensees in its home territory it will have little incentive to take on the entrepreneurial burden of such risks. The licensor or other prior licensees have

[17] Para 157.

had a head start in learning the most efficient methods of manufacture and distribution of the licensed product. Licensees feel that even with the know-how transferred by the licensor, they need time to tool up, to acquaint themselves with the technology, and to establish themselves as efficient manufacturers with a functioning distribution system before being subjected to competition from the licensor and other licensees within the licensed territory. This need is often so great that without such contractual protection many licensees would be deterred from making the investment.

Both exclusive and sole licences, particularly between non-competitors, are classic examples of clauses that from a purely economic point of view should not be caught by Article 81(1) because the territorial exclusivity offers protection to the licensee for the purpose of inducing it to assume the financial risks of introducing new technology into territory. The force of this point was acknowledged by the ECJ in its analysis of an open exclusive licence in the Maize Seed (*Nungesser*) case.[18] It may nonetheless be difficult to show that in a particular case, the degree of territorial restriction was necessary to induce the licensee to accept the licence. In assessing such clauses, it is therefore necessary to consider the broader Article 81 analysis. **7.24**

If the indispensability of these provisions cannot be shown in a given case, the arguments set out above also provide the basis for assessment of economic benefit under Article 81(3). Even if there is no absolute hold-up problem that makes the restriction indispensable, there is likely to be a relative hold-up problem in terms of the willingness of licensees to invest.[19] By overcoming this problem the sole or exclusive licence contributes to more rapid or effective introduction of the technology and therefore contributes to dynamic efficiency. For all the reasons identified in Chapters 4 and 5, this contribution to dynamic efficiency will inevitably be difficult to quantify. The tendency in the Article 81 analysis will therefore be to shift the assessment to consideration of competitive harm. In the absence of some significant competitive harm, the likely benefits of the provisions in reducing hold-up problems will mean that Article 81 is not infringed. **7.25**

(iii) Sole and exclusive licences—competitive harm and application of Article 81(1)

It is sometimes argued that sole or exclusive licences should not be caught by Article 81 at all because these clauses are simply a transfer of the rights of the licensor to the licensee.[20] They do not reduce rivalry because the licensor has decided not to invest itself in that market. Even if the licensor might have **7.26**

[18] See discussion in Chapter 2. [19] See discussion in Chapter 3.

[20] See Charles River Associates, *Multiparty Licensing*, 22 April 2003.

been a potential entrant in that market, there is no competitive effect on the market because there is still one producer of the technology in question on that market. The licensee did not offer potential competition because it had no intention or incentive to enter the market on any other basis than an exclusive agreement. These arguments have not, however, found favour with the Commission or the Court. This is not surprising, since this argument could also be used to dispute application of Article 81 to any exclusive distribution arrangement where the manufacturer has withdrawn from the wholesale market and the wholesaler was not previously a competitor.

7.27 The principal form of competitive harm identified in the TT Guidelines in relation to exclusive licences is foreclosure. The Guidelines acknowledge that exclusive licences 'between non-competitors—if actually caught by Article 81(1) in the first place, are likely to fulfil the conditions of Article 81(3). It will mainly be in the case of a dominant licensee that intervention may be warranted *up to the level of dominance* because it is only then that real possibility of threat of foreclosure may arise'.[21] This analysis makes some sense if it is recalled that successful foreclosure would require both substantial market power on the downstream market for the licensor, and real dominance for the technology—that rival producers in the downstream market do not have access to viable alternative technology. Even at the level of dominance, however, it is far from clear that intervention is warranted. Where a vertical agreement is *ex ante* a risky proposition and the outcome of the transaction and relationship is a highly successful product, the fact that the acquisition of dominance was by competition on the merits should have some effect on the status of the agreement under Article 81(3). Nevertheless, under Article 81(3) the issue is less one of history and more one of current market position. If the nature of the dominance is well short of monopoly and there are competing technologies in the market, Article 81(3) could still be applicable; assessment must be made of the strength of intertechnology competition in the territorial market. This will require an assessment of the actual and potential intratechnology competition from producers of the licensed technology in other territories. As long as the barriers to entry to the market are not high and there is some actual intertechnology as well as intratechnology competition, there is a case for exemption. It is difficult to imagine that a licensing agreement can suddenly become void in such circumstances.

(iv) Sole or exclusive territorial grants in non-competitor licensing—not generally a competitive risk

7.28 The overall approach to sole or exclusive territorial grants in agreements between non-competitors is generally very positive. The TT Guidelines

[21] TT Guidelines para 166.

acknowledge that the right to grant an exclusive licence is generally necessary in order to induce the licensee to invest in the licensed technology, particularly where the licensee must make large investments in further developing the licensed technology. To intervene against the exclusivity once the licensee has made a commercial success of the licensed technology would deprive the licensee of the fruits of his success and would be detrimental to competition, the dissemination of technology and innovation. The Commission has acknowledged that it will only intervene exceptionally against exclusive licensing in agreements between non-competitors, irrespective of the territorial scope of the licence. So, an exclusive licence for the whole of the internal market or even the whole world is exemptible and exclusive licences may even be non-restrictive of competition in the first place. The Guidelines leave implicit that the same reasoning would support a finding that the restriction is not caught in the first place by Article 81(1).

The exception to this rule is the case where a firm that is already dominant using a rival technology obtains an exclusive licence for a second competing technology. In this case, the Guidelines indicate that where a dominant licensee on a particular technology market obtains an 'exclusive licence to one or more competing technologies,' the agreement is caught by Article 81(1) and will be 'unlikely' to meet the conditions of Article 81(3).[22] This would presuppose that entry to the market is difficult and the licensed technology is a real source of competition on the market because then this would create too great a risk of foreclosure of the market. **7.29**

(c) Territorial sales restrictions in agreements between non-competitors

As already noted above, the competition law treatment of sales restrictions in licence agreements is stricter than for sole or exclusive licensing. The Guidelines nonetheless recognise that sales restrictions, including in limited circumstances restrictions on passive sales, could be necessary or at least useful to avoid hold-up problems. In making this assessment, the Guidelines differentiate between: **7.30**

(i) restrictions on sales by the licensor;
(ii) restrictions on sales by the licensee in territories reserved to the licensor; and
(iii) restrictions on sales by the licensee into territories of other licensees.

As a general matter, restrictions protecting the licensee from the licensor are afforded favourable treatment, restrictions protecting the licensor from the licensee somewhat less favourable treatment, and restrictions protecting **7.31**

[22] Para 166.

licensees from each other are subject to relatively strict regulation. It should also be noted that each of the provisions discussed below involves a restriction on the ability of either the licensor or the licensee to sell products covered by the licensed intellectual property rights. A provision that purports to require territorial restrictions on the customers of the licensee will generally be regarded as restrictive by object and unnecessary to protect the interests of the licensor or the licensee.[23]

(d) Restrictions on sales by licensor into the licensee territory

7.32 The TT Guidelines allows restrictions on both passive and active sales by the licensor below the market share threshold of 30%.[24] Above the market share thresholds, the Guidelines indicate that such restrictions may lead to competitive harm, but that the benefits of such restrictions may outweigh their restrictive effects, particularly if the restriction only applies to active sales.

(i) The rationale for licensor restrictions—avoiding sub-optimal investment by the licensee

7.33 The TT Guidelines base the case for restrictions on licensor sales on the asymmetry between the costs of the licensor and the costs of the licensee.[25] They make the point that the licensee's incentive to invest would be significantly reduced if the licensee faces direct competition from the *licensor* whose production costs are not burdened by royalties and that this could lead to

[23] As regards restrictions on licensees this follows from TTBER Article 4, since such a requirement will almost inevitably accompany a restriction on the initial seller (licensor or licensee) it may be regarded as included in the general limitation of TTBER Article 4(2)(b) which categorises as hard core 'the restriction of the territory into which . . . the licensee may passively sell the contract products' unless it falls within one of the listed exceptions, none of which is applicable. The case of restrictions on the licensor is not, however, dealt with explicitly. It is clear that the licensor may agree (expressly or implicitly) to impose restrictions on other licensees. But can a licensor agree to impose active sales restrictions on its own wholesalers to the benefit of a licensee? Provided that the agreements with the wholesalers are themselves consistent with Article 81, such a commitment would appear justifiable on the same basis and under the same circumstances where such a commitment could be justified if the licensor were itself a distributor.

[24] Since the TTBER market share threshold applies to both the technology market and the downstream product market, this 'exemption' is of limited value. It does indicate, however, that where the licensee's share in the downstream market (which is the market protected by the restriction) remains below 30%, the agreement is unlikely to lead to significant competitive harm.

[25] Para 173. Although the Guidelines focus on the fact that the licensor does not need to pay royalties, other factors such as the association of the licensor with trademarks associated with the technology, the familiarity of the licensor with the technology as a result of the development process, or the links made by the licensor with possible users as part of the development and testing process may provide the licensor with inherent advantages that would make the licensee reluctant to take on the commercial risk of promoting the product without some protection.

sub-optimal levels of investment by the licensee.[26] This asymmetry adds to the more general concern that, in the absence of territorial restrictions, the licensor could 'free ride' on investments by the licensee in developing the market for the product and thus create a further disincentive to such investment.

(ii) Are licensor obligations 'ancillary'?

It is unclear from the discussion in the TT Guidelines whether licensor restrictions are ancillary (and thus outside Article 81) because they are necessary for a license agreement to be fully effective or whether avoiding sub-optimal investment is a benefit to be considered in the context of an Article 81(3) analysis. In a case where a potential licensee would decline a licence in the absence of the agreement restricting sales by the licensor, the restrictions can clearly be characterised as ancillary. In cases where the licensee would accept the licence and either pay a lower royalty or reduce its own efforts to exploit the technology, it may be necessary to apply the full Article 81 analysis. **7.34**

(iii) When are licensor restrictions likely to lead to competitive harm?

The TT Guidelines do not specifically address the circumstances where licensor restrictions can lead to competitive harm. Based on the considerations discussed in Chapters 4 and 5, it is evident that the key issue is the position of the licensee in the territory subject to the sales restriction. Only where the licensee holds a position of significant market power in that territory and competition from the licensor could otherwise constitute a significant constraint on that power should competitive harm be an issue.[27] **7.35**

(iv) Balancing and indispensability

The TT Guidelines indicate that licensor restrictions 'when caught by Article 81(1)' are 'likely to fulfil the conditions of Article 81(3) unless there are no real alternatives to the licensor's technology on the market or such alternatives are licensed by the licensee from third parties.'[28] This cryptic passage seems to assume that other licensees (if any) are subject to licence and sales restrictions that limit any competitive constraint that they would impose and that the licensee's rights under rival technologies are also exclusive in nature. Under these circumstances, the licensee would have a monopoly position. In other words, the Guidelines appear to suggest that where the market power of the licensee does not approach the monopoly level, the balance of harms and **7.36**

---

[26] Ibid.

[27] Even if the licensor is constrained, the licensee may still be subject to competition from other licensees or from firms using rival technologies. Furthermore, if the products manufactured using the technology are routinely resold, arbitrage in the downstream market may limit any consumer harm associated with the restriction.

[28] Para 173.

benefits would favour the restriction. The Guidelines also suggest that this analysis will be easier if the restriction on the licensor is limited to active sales. Although the Guidelines are also cryptic on this point, the implication appears to be that passive sales restrictions reduce the competitive harm and could therefore 'tip the balance' in close cases. It could, however, be argued that where there is only a relative hold-up problem, a ban on active sales would in most cases be sufficient to protect the licensor, so that a passive sales ban would fail the indispensability requirement under Article 81(3).

(e) Restrictions on sales into licensor territory

7.37 The TT Guidelines allows a licensor to bar all sales (active and passive) into territories that the licensor has reserved to itself. Above the 30% market share threshold, the Guidelines indicate that Article 81 may be applicable although some restrictions protecting the licensor will be valid on ancillarity grounds. Where competitive harm is significant, however, restrictions on active sales may be permitted where restrictions on passive sales are unduly restrictive.

(i) Rationale for the restriction

7.38 The primary purpose of a restriction protecting the licensor's territory is to resolve a hold-up problem that might otherwise limit the willingness of the licensor to make its technology available to third parties. The Commission observes in the Guidelines that 'a technology owner cannot normally be expected to create direct competition with himself inside his own technology'. In other words, there will be many cases where a party that controls technology and can exploit that technology itself may view exploitation itself as more rewarding than licensing, if licensing means that its return on its own exploitation efforts will be reduced. Even if the licensor might eventually license the technology in territories that it cannot effectively exploit itself, the licensor may delay such licensing until it has successfully established the technology in its own territories or may impose higher royalties to compensate for the revenues lost due to competition from the licensee.[29]

(ii) Ancillarity

7.39 The Guidelines suggest that restrictions on sales into a licensor territory may fall outside Article 81(1) where 'on the basis of objective factors' the licensor would

[29] A common example of this problem, which is not discussed in the Guidelines, involves the case where the licensee is a firm that has access to more extensive distribution or production resources than are available to the licensor. In this case the asymmetry between the licensor and the licensee means that the licensor may not be able to maintain its own downstream business if the licensee comes on to the market. The licensor will therefore either not license at all or license another firm that is less efficient, leading to a welfare loss.

not have licensed without such a condition. In effect, the Guidelines recognise that a territorial sales restriction may be ancillary in order to avoid the hold-up problem on the part of the licensor.[30] The Guidelines are silent, however, on what objective evidence would be necessary to show that a licensor would not have licensed without protection against active and passive sales by the licensee.[31] The Guidelines also do not consider whether a restriction on sales in a licensor territory might be ancillary for the period necessary to allow the licensor to establish its own distribution in its territory, even if it could be subject to Article 81 thereafter.

(iii) Competitive harm

Where it is evident that the licensor *would* license but subject to a higher royalty or to other restrictive provisions (such as quantity restrictions), the Guidelines indicate that the restriction can lead to competitive harm where the licensor has 'a significant degree of market power'. It should be noted that the 'market power' in this context is not the position of the licensor in the technology market, but rather in the downstream product market in the territory or territories where the restriction protects the licensor. Here again a full analysis of likely effects should consider the extent of actual and potential intertechnology competition as well as the competitiveness of downstream markets before concluding that there is significant competitive harm. **7.40**

(iv) Economic benefit and balancing

The Guidelines do not consider the position where a restriction on licensor sales may not be necessary for the licence arrangements to go forward, but where there are nonetheless economic benefits that flow from the provision (e.g. rapid adoption of new technology or lower royalty rates). The suggestion seems to be that once there is a significant degree of market power, the balance of harms and benefits goes against the restriction. It is not clear, however, that this is the appropriate resolution for this problem. First, from a practical perspective this makes too much turn on the ancillarity analysis—once the licensor has significant market power, only a successful ancillarity argument would justify the restraint. Secondly, even if the rationale for a restriction protecting the licensor **7.41**

---

30 Para 172.

31 From an economic perspective, the issue would be whether it would still be profit-maximising for the licensor to grant the licence in the absence of the restriction. The likely revenue loss for the licensor in the event that licensee started supplying the protected market, compared with the likely royalty stream from the licence, would be relevant in this regard, as well as other benefits to the licensor from increased market penetration for its technology, in the form of network effects. The presence of similar provisions in other licences in the same or related sectors may also provide an 'objective' indication that such provisions are regarded as necessary to avoid hold-up problems.

is arguably less powerful than for one protecting the licensee, it is hard to see how this difference in benefits justifies fundamentally different treatment of two restrictions with essentially similar effects. There seems to be a strong argument for viewing restrictions on sales into licensor territories as valid, except in exceptional circumstances. This argument would be even stronger if the restriction is limited to active sales.

(f) Restrictions on sales between licensees

7.42 The calculus changes when the focus shifts to sales restrictions placed in the licensing agreement to protect other licensees of the same licensor. Under these circumstances the special considerations justifying protection of the licensee from the licensor (asymmetry of competitive advantages) or protection of the licensor from the licensee (otherwise the licensor might not license at all) are absent. Furthermore, where licensees are granted separate territories within the EU, the allocation of territories between licensees creates or perpetuates differences between EU national markets, which are traditionally viewed with suspicion under EU competition law rules. As a result, the TTBER and the Guidelines take a stricter approach to sales restrictions designed to protect licensees from each other. Limits on passive sales by licensees are only allowed in limited circumstances. Limits on active sales, however, are easier to justify and are usually permitted.

(i) The rationale for protecting licensees from each other

7.43 The rationale for territorial restrictions protecting licensees is essentially the same as the rationale for protecting licensees from licensors. The threat of competition from other licensees before the licensee has had time to master the practices that lead to efficient production of the licensed product will deter licensees from entering the licensing agreement. That is the rationale for the two year protection against passive sales for licensing agreements which is not found in the Vertical Agreement BER in respect of distribution.[32] There is also a minor 'free rider' concern, that licensees will have less incentive to invest in developing the market for a product if other licensees profit from that investment by selling in the territory without engaging in their own promotional efforts.

(ii) Competitive harm

7.44 The competitive harm associated with sales restrictions protecting licensees from each other is the same as that associated with sales restrictions protecting

[32] There is also an argument that some market power for licensees is desirable in the licensing context since the additional profit margin gives licensees scope to invest more heavily in exploitation of a new technology.

the licensor; the protected licensee will enjoy enhanced market power due to the limit on competition by other licensees. The Guidelines suggest that these restrictions are likely to lead to competitive harm 'where the individual licensee has a significant degree of market power'. It follows that to assess the impact of this type of clause in individual cases, it is necessary to consider the market position of the protected licensees—not the position of the licensee subject to the restriction.[33] In making this assessment, the key will be the extent to which the protected licensees are subject to effective competition in the downstream market, since it can be assumed that all licensees are subject to similar restrictions.

(iii) Balancing and indispensability

As already noted, the Guidelines differentiate sharply between restrictions on all sales (i.e., restrictions on passive sales) and restrictions that only limit active sales. This distinction reflects an indispensablity assessment; restrictions that include a limit on passive sales cause more competitive harm and are only necessary to achieve the benefits of the licence during a limited start-up period. Where a licensee cannot make any sales in a protected territory, the clause will therefore be treated as a restriction by object that is unlikely to merit exemption, except for a period two years from the date on which the individual licensee benefiting from the restrictions first puts the product incorporating the licensed technology on the market inside its exclusive territory.[34] **7.45**

Restrictions limited to active sales, in contrast, receive relatively favourable treatment. As the Guidelines state, 'such restrictions may fulfil the conditions of Article 81(3) where they are necessary to prevent free riding and *to induce the licensee to make the investment necessary for efficient exploitation of the licensed technology inside his territory and to promote sales of the licensed product.*' The Guidelines further suggest that in these circumstances restrictions can be exempted as long as they are not associated with a market sharing arrangement and as long as they are needed to keep the licensing agreement profitable. This may extend for the entire period of the validity of the licensing agreement. **7.46**

33 The application of the TTBER which depends on the market share of the licensor and the licensee, is somewhat anomalous in this regard.

34 TTBER Article 4(2)(b)(ii). The TT Guidelines do not specifically address the status of this obligation above the market share thresholds, but the policy basis for allowing this restriction would appear equally applicable in applying the general Article 81 test. The allowance of a two year period of protection shows an understanding of the need to encourage investment in IP licensing because it ensures that every licensee gets protection against rival licensees and the licensor itself for a minimum of two years from the time it first markets the product in its territory. The thinking is that the licensee gets an initial period to tool up to match the efficiencies of production of its rivals. This represents a significant advance over the position under the old block exemptions, which allowed a period of five years from the time the product was first put on the market in the EU.

(iv) Exclusive licence and sales restrictions outside the EU

7.47 The TTBER and the Guidelines do not directly differentiate between restrictions applicable within the EU and restrictions applicable to territories elsewhere in the world.[35] The implication appears to be that the assessment remains unchanged as long as the jurisdictional requirements of Article 81 are met, i.e. as long as the agreement affects trade between Member States. Since licence arrangements that treat the EU or an area including the EU as a territory are not unusual, the application of Article 81 to such agreements deserves more attention.

7.48 As a preliminary matter it is important to recall that the rules on free movement do not apply to products introduced to sale outside EU until those products have been lawfully resold within the EU. This means that the holder of an exclusive licence within the EU may lawfully exercise intellectual property rights to exclude the import of products produced by licensees outside the EU. Thus, to the extent that intellectual property rights in question confer such powers, contractual sales restrictions on a licensee or licensor outside the EU may be superfluous.[36] Following the approach set out in paragraph 164 of the Guidelines, where a licensor was not previously a competitor in the EU, neither the exclusive licence nor sales restrictions on the licensor should be regarded as problematic under Article 81.

7.49 As regards restrictions on sales by licensees outside the EU, the analysis should logically depend on where the beneficiary of the restriction is active. Where a licensor or licensee in the EU is the beneficiary of the restriction, then assessment of potential competitive harm requires consideration of the position of the beneficiary on the same basis that restrictions within the EU are assessed. Article 81(1) could be applicable if the beneficiary has significant market power and if the licensee subject to the restriction would impose a significant competitive constraint on the beneficiary in the absence of the restriction. In light of import costs and the possible invocation of intellectual property rights, the competitive constraint imposed by firms outside the EU cannot be assumed.[37] Even where Article 81(1) is applicable, the economic benefits of a restriction on sales in third countries will usually mean that sales restrictions are justifiable.

---

[35] TT Guidelines at para 164 refer to worldwide exclusive agreements in contrast to agreements limited to a particular territory, using an EU Member State as an example of a more limited territory.

[36] This creates a potential difference between technology licensed on the basis of patents or copyright, which confer such protection, and technology licensed in the form of know-how, which does not.

[37] See *Javico International* v *Yves Saint Laurent (Case 306/96)* [1998] ECR I-1983. The Commission Guidelines on the Effective Trade Concept at para 109, following the ECJ judgment in *Javico*, dicate that sales restrictions imposed on resellers outside the Community will only be caught by Article 81 where significant price differences would make imports into the Community likely.

The hold-up problems that justify restrictions on sales by licensees in the EU may be magnified in respect of licensees in third countries where costs and retail prices are significantly lower.[38] This could justify restrictions on passive as well as active sales.[39]

Where the beneficiary of the restriction is operating outside the EU, it is questionable whether Article 81 should be applicable at all. In theory, a restriction on sales to a non-EU territory could affect competition in the EU, if sufficient products would otherwise be re-exported back to the EU to create a competitive constraint on EU licensees with significant market power.[40] This effect would require assessment of relative prices, transport costs, and the ability to assert intellectual property rights to prevent or limit imports. It would also depend on the volume of likely exports (bearing in mind that the EU licensee subject to the restriction will limit its exports if they affect its pricing in its 'home' market). Even if Article 81(1) were deemed applicable, the Article 81(3) analysis would still appear to favour the restriction, since it promotes licensing in the affected third countries. **7.50**

(3) Field of use and customer restrictions

(a) Defining 'fields of use' and 'customer restrictions'

The TT Guidelines define a 'field of use' restriction as a provision by which 'the licence is either limited to one or more technical fields of application or one or more product markets.'[41] This definition provides two alternative ways to identify a 'field of use': (i) limitation to a technical field of application; or (ii) limitation to a specific product market. **7.51**

---

[38] The ECJ in *Javico* (point 19) recognised that sales restrictions in export contracts are 'designed to enable the producer to penetrate a market outside the Community by supplying a sufficient quantity of contractual products to that market'. Particularly where price levels in a market outside the EU are low, an exporter may be reluctant to sell at all if it cannot impose territorial sales restraints in the customer. This could even support an argument for ancillarity of the restriction, although the judgment in *Javico* does not go in that direction. It is interesting to note that the application of Article 81(3) in this context would be excluded by the Commission Guidelines on Article 81(3), as discussed in Chapter 2, because the beneficiaries of the restriction in the third country are not the same 'consumers' as those who have suffered the 'harm' (in the EU). This serves to illustrate the difficulty with the Commission approach to this issue from a policy perspective.

[39] In *Javico International* v *Yves Saint Laurent (Case 306/96)* [1998] ECR I-1983, the Court ruled that an export restriction imposed on a third country distributor was not restrictive 'by object' in contrast with the position where such a prohibition is imposed on a distributor within the EU.

[40] This goes substantially beyond the analysis applied to restrictions within the EU where the focus is on the protected territories and not on potential re-export from those territories to other territories.

[41] Para 179.

(i) The first alternative—a technical field of application

7.52 As regards the first alternative, the Guidelines appear to contemplate that a technical field of application can be defined narrowly, based on use in specific types of products. Thus use of a new engine technology in four cylinder engines would be a different technical field of use from employment of the same technology in six cylinder engines, and use of a technology to make chipsets with up to four CPUs would be a different field of use from employment of the same technology to make chipsets with more than four CPUs.[42] The Guidelines emphasise, in this regard, that the same technology can be used for different fields of use in the same market, referring to the engine technology and chipset technology examples.

(ii) Second alternative—sales in different product markets

7.53 As regards the second alternative, definition of a 'field of use' is based on different product markets, the Guidelines confirm that this test focuses on the downstream markets where the licensee is active. The Guidelines give the example of a new moulding technology that can be used to make plastic bottles and plastic glasses, suggesting that a manufacturer of plastic glasses could be limited to using the technology for that purpose, even if it were also active in the plastic bottle market.

(iii) Contrasting customer restrictions with field of use restrictions

7.54 The Guidelines contrast a field of use restriction with a customer restriction. According to the Guidelines a 'customer restriction presupposes that specific customer groups are identified and that the parties are restricted in selling to such identified groups.'[43] The Guidelines acknowledge that a technical field of use or product market may also comprise a specified group of customers, but suggest that the more favourable treatment accorded a field of use restriction remains applicable as long as the field of use can be 'defined objectively by reference to identified and meaningful technical characteristics of the licensed products.'[44]

(b) Application of article 81—field of use restrictions

7.55 The TT Guidelines accord very favourable treatment to field of use restrictions in agreements between non-competitors. If field of use restrictions are viewed as

[42] Ibid. It will be noted that this approach to defining a technical field of use does not necessarily coincide with the test for identifying a separate application for an innovation under IP law.

[43] Para 180.

[44] Ibid. This puts down a clear challenge for drafters of IP licence agreements.

simply a partial allocation or subdivision of the exclusive intellectual property protection conferred upon the rights holder, then the competition effects of such a restriction in agreements between non-competitors are little different from unilateral action. Even when combined with territorial exclusivity, field of use restrictions can be pro-competitive in effect by encouraging the licensor to license its technology for applications that fall outside its main field of exploitation. Moreover, as the Guidelines point out if licensors could not prevent licensees from operating in fields where they exploit the technology themselves, it would create a disincentive for licensors to license or would lead them to charge a higher royalty. Hence in 'agreements between non-competitors the licensor is normally also entitled to grant sole and exclusive licences to different licensees limited to one or more fields of use'.[45]

(c) Application of Article 81—customer restrictions

The TTBER and Guidelines treat customer restrictions in the same way as they treat territorial sales restrictions. Thus the treatment of a customer restriction depends on whether the restriction limits: **7.56**

(i) sales to a group of customers reserved for the licensor (allowed in most cases, particularly if confined to active sales);
(ii) sales to a group of customers reserved to the licensee (allowed in most cases); or
(iii) restrictions on sales to a group of customers reserved to another licensee (allowed for two years in most cases—after that, active sales only).[46]

(4) Output restrictions

(a) The rationale for output restrictions

An output restriction involves agreement by the licensee to limit the number of units it will sell of products incorporating the licensed technology. The purpose of an output restriction is the same as that of a restriction on sales to a territory or defined customer group—to protect the licensor and other licensees from a **7.57**

45 Para 184.

46 The competitive harms and economic benefits of sales restrictions also provide insight into what is meant by a 'customer group'. Taking the example of a restriction protecting the licensor, the licensor may have invested substantial resources into developing specific customers for use of the products produced by the technology, and sales by a licensee to those customers would 'free-ride' on that investment. This threat could lead the licensor to limit or refuse licensing altogether, which provides the justification for the restraint. On this basis, the key is that the agreement must define the customer group. This could be done either generically (e.g. 'all car manufacturers') or by listing the customers in the group.

producer who might otherwise 'flood the market' with products incorporating the licensed technology.

(b) Competitive harm and ancillarity in output restrictions

7.58 In agreements between non-competitors the competitive harm of output restrictions is in principle the same as for other sales restrictions—a reduction in the competitive constraint on the licensor and other licensees.[47] Because an output restriction lacks the market partitioning aspect of other sales restrictions, such restrictions are accorded more favourable treatment. Indeed the TT Guidelines appear to suggest that output restrictions may be ancillary in many cases.[48] The Guidelines also observe, however, that output restrictions may be combined with other restrictions, including restrictions involving territories or customers, and can increase their effects.[49]

(c) Balancing and indispensability

7.59 Output restrictions are covered by the TTBER subject to the 30% market share limits. Outside the TTBER, the Guidelines signal a relaxed approach to standalone output limitations, 'the licensor should normally be free to determine the output produced with the licensed technology by the licensee'.[50] On this basis, output restrictions are only a real concern in practice where they are combined with the more suspect sales restraints.

(5) Captive use restrictions

7.60 The TT Guidelines define a captive use restriction as 'an obligation on the licensee to limit his production of the licensed product to the quantities required for the production of his products and for the maintenance and repair of his own products'.[51] Thus a firm that makes headlamps for cars might license its technology to a car manufacturer, but limit the licence to the car manufacturer's own production. The Guidelines suggest that this type of restriction is normally justified where the licensor is a supplier of components (as in the car

[47] The TT Guidelines para 176 suggest that this effect will depend on the market position of the licensor and the extent to which the licensee is prevented from meeting customer demand by the restriction. To the extent that there are other licensees, however, they are also protected by the limitation, so their position is also relevant to the assessment of competitive harm.

[48] See para 178. 'If the licensor were not free to determine the output of the licence, a number of license agreements might not come into existence in the first place'.

[49] It would be more correct to say that the output restriction reduces the constraint that the licensee would otherwise exert on the firm (licensor or licensee) that benefits from the restrictions in question—it does not increase the competitive harm of those other restrictions but it does remove a factor that could reduce the competitive harm of those arrangements.

[50] Para 178. [51] Para 186.

headlamps example). In other cases, however, the Guidelines suggest that such restrictions may not satisfy Article 81 outside of the TTBER market share thresholds. The discussion of captive use restrictions is one of the few places where the Guidelines appear to have missed the key issues raised by application of Article 81 to licensing.

(a) Competitive harm

The TT Guidelines identify two types of competitive harm associated with captive use restrictions: (i) restriction of intrabrand competition; and (ii) exclusion of arbitrage between licensees.[52] To assess the first harm requires, of course, identification of the beneficiaries of the restriction and assessment of the full range of constraints applicable to each beneficiary, including producers using rival technologies, passive sales from other licensees and arbitrage effects in downstream markets based on resale of products sold by other licensees, allowing the licensor to discriminate in setting royalties. Competitive harm from a captive sale provision presupposes that the captive seller would otherwise constitute a significant competitive constraint on a seller otherwise not subject to significant competitive constraints. The second 'harm' only applies where the first harm can be shown; limiting arbitrage by the captive seller would protect other licensees and allow the licensor to maintain higher royalties to those sellers. As in the case of territorial or customer restrictions, this 'harm' presupposes limited intratechnology competition and licensees with significant market power. **7.61**

(b) Rationale for the restriction

The TT Guidelines observe that, where the licensor is itself a manufacturer of products using the technology, a captive use restriction serves the same function as a territorial or customer restriction protecting the licensor. The Guidelines fail to recognise, however, that the primary purpose of a captive use restriction is to make an efficiency-enhancing transaction possible that would not take place otherwise, because the incremental revenue from captive sales is limited compared with the incremental revenue from other licensing transactions. To understand this point it is necessary to consider the 'counterfactual'—the position that would apply if the licensor cannot limit the licensee to captive production. **7.62**

The key point in this regard is that a captive licence may not create new demand for the licensed technology (which will often be the case where the licensor grants a licence for a new territory or a new field of use), but rather will shift demand from an existing licensee to the captive producer. If the licensor does have real market power in the technology market (which is a prerequisite for competitive **7.63**

---

52 Para 188.

harm) the potential captive licensee may still need the product, but will need to acquire it from another licensee. From the licensor's perspective this may not make a major difference, since the licensor still obtains its royalty for products purchased from the licensee. For the potential licensee, however, this will lead to production inefficiencies—a direct consumer loss (otherwise the potential licensee would not seek the licence). The licensor may be willing to grant a licence because of the possibility of incremental revenue, but not if doing so will affect the financial incentives of other licensees. The captive use restriction makes this possible. Contrary to the suggestion in the Guidelines, it is hard to see how customer restrictions benefiting the licensor would fully satisfy this need.

(c) Balancing and indispensability

7.64 The TT Guidelines suggest that captive use restrictions will normally either fall outside Article 81 altogether or meet the requirements of Article 81(3) in cases where the licensor is a supplier on the relevant components market.[53] In other cases, the Guidelines suggest that where a captive use restriction does lead to competitive harm, the harm will either outweigh the benefits or less restrictive alternatives are available.[54] This appears to be an unduly harsh approach in light of the considerations outlined above. A better view would be that captive use restrictions are normally allowable regardless of whether the licensor is a components supplier.

(d) The condition—no restriction on sale of spare parts

7.65 The TT Guidelines suggest that that Article 81(3) could only be applicable if 'the licensee is not restricted to selling the licensed product as replacement parts for his own products'.[55] This 'condition'[56] appears to be based on the premise that a captive sales provision that does not allow spare parts sales leads to more extensive competitive harm. This assumes, of course, that captive sales restrictions do lead to competitive harm caught by Article 81. In a real case, the applicability of this condition might be disputed if the spare parts market is in fact competitive.

(6) Royalties

7.66 The obligation on the licensee to pay royalties is a central part of a licence agreement and is therefore not in itself restrictive of competition. As the

[53] Para 189. [54] Para 190. [55] Para 189.

[56] The use of the term 'condition' appears to refer back to practice under Regulation 17/62 or under the block exemptions, where the Commission could impose conditions on applicability of an Article 81(3) exemption. Under the new paradigm, the spare part requirement must be linked directly to one of the specific Article 81 criteria.

TT Guidelines state, the 'parties to a licence agreement are normally free to determine the royalty payable by the licensee and its mode of payment without being caught by Article 81(1). This principle applies both to agreements between competitors and agreements between non-competitors.' As a result, licence agreements can contain a variety of methods for calculating royalties, including lump sum payments, royalties based on a percentage of total sales revenue, or royalties based on the number of customers or locations using the technology (particularly where the technology is covered by copyright).[57]

In general, royalty arrangements only raise significant Article 81 issues in licence agreements between competitors. It is possible, however, to structure an obligation to pay royalties in a way that leads to a restriction of competition subject to Article 81. In this section we address the principal ways in which royalty calculation in agreements between non-competitors *might* restrict competition: **7.67**

(i) disguised price-fixing (primarily through fixing royalties per unit of sales;
(ii) minimum royalty obligations;
(iii) royalty obligations extending beyond the expiration of underlying IP rights; and
(iv) royalties calculated on sale of products not covered by the licensed technology.

As this analysis shows, even in these situations it will be unusual for Article 81 to apply to a royalty clause in an agreement between non-competitors.

(a) Royalties as disguised price-fixing?

There is a sense in which royalty obligations based on a payment for each unit sold impose a floor for the retail price, since the licensee will normally charge some greater amount for the product in order to break even. Where the licensee and licensor are not competitors, however, this effect is not regarded as equivalent to imposition of a minimum retail price. In its economic effects it is no different from the position of a manufacturer that charges distributors for each unit sold. The distributors will normally sell at a mark-up, but this does not constitute setting a minimum retail price. The prohibition on price-fixing only becomes relevant where the licence agreement goes beyond setting the royalty to limit the ability of the licensee to set the selling price itself. Where the licence **7.68**

57 Para 156. 'Royalty obligations may for instance take the form of lump sum payments, a percentage of the selling price or a fixed amount for each product incorporating the licensed technology. In cases where the licensed technology relates to an input, which is incorporated into a final product it is as a general rule not restrictive of competition that royalties are calculated on the basis of the price of the final product, provided that it incorporates the licensed technology. In the case of software licensing royalties based on the number of users and royalties calculated on a per machine basis is as a general rule compatible with Article 81(1).'

does restrict a party's ability to determine its prices when selling to third parties, the provision will be regarded as a hard core restriction.[58] As long as the royalty is established purely by reference to the number of units sold, or as a percentage of total sales income, leaving the actual pricing to the licensee, it is not caught by Article 81.[59] This is true even if the royalty clause is accompanied by a list of recommended prices. The TT Guidelines note that the provision of a list of recommended prices or the imposition of a maximum price on the licensee by the licensor is not considered in itself as leading to fixed or minimum selling prices.[60]

(b) Duration of royalty payments—expiration of patents or disclosure of know-how does not bar continued royalty payments

7.69 Under the old regime, it was sometimes argued that charging royalties in respect of intellectual property rights that had expired or know-how that had entered into the public domain was anticompetitive, because it extended the market power created by intellectual property rights or secret technology beyond the limits created by law or by the nature of the technology itself.[61] Insofar as this argument was based on a concept of competitive harm, it focused on the disadvantage faced by licensees that face a continuing royalty obligation compared with competitors that can use the same technology without payment since it is in the public domain. The Guidelines recognise, however, that this harm for the restricted licensees does not equate to a loss in consumer welfare.[62] The Guidelines therefore indicate that licensors may agree with licensees to continue to pay royalties until the end of the agreement or the regular expiry of the patents, even where the intellectual property rights have completely *lost* their value because the patents have prematurely lost their validity or the know-how has become publicly known other than by the action of the licensor.

7.70 The Guidelines reason that once these rights expire, *third parties* can legally exploit the technology in question and compete with the parties to the agreement. Such actual and potential competition will normally suffice to ensure

---

[58] Article 4(2)(a) of TTBER Recital 21 reminds that the setting of rates of royalty to achieve a hard core restriction puts the agreement as a whole outside of the TTBER.

[59] These could also be combined—a fixed amount per unit or a percentage, whichever is higher.

[60] Para 92.

[61] This was a feature in particular of the old US precedents on 'misuse' of IP rights which carried over to US antitrust analysis and the first generation of EC thinking in this area.

[62] This is a clear example of the way that the new paradigm informs the Article 81 analysis. Under the old approach these clauses were regarded as restrictive, but capable of exemption because they allowed parties to adjust the timing of payments for technology licences to reflect their own assessment of commercial risks, and the opportunity to spread the royalty over a longer period reduces the need for a prohibitive upfront payment.

that the obligation in question does not have appreciable anticompetitive effects.[63]

(c) Minimum royalties—a disguised exclusive dealing requirement?

An obligation to pay a fixed base royalty regardless of the sales of products incorporating licensed technology could theoretically be regarded as leading to the same effects as an agreement not to take a licence from competing technology providers. Since the licensee needs to pay royalties to the licensor anyway, if the licensor needs to pay royalties to another licensor it will 'pay double'. Despite this potential effect, however, the TT Guidelines conclude that obligations to pay minimum royalties 'are generally not restrictive of competition within the meaning of Article 81'.[64] Although the Guidelines do not elaborate on the basis for this conclusion, this conclusion makes sense because the ability to set a base price for *access* to the technology is an essential element of the licensing process. In other words, a licensor is not required by Article 81 to base royalty payments on the extent to which technology is actually used. If a licensee decides to contract with multiple licensors and thereby incur additional costs, that is the licensee's problem.[65] **7.71**

(d) Calculating royalties based on sales of products that may not incorporate the licensed technology—another form of exclusive use requirement

The method of royalty calculation in agreements between non-competitors that raises the greatest practical issues according to the TT Guidelines involves the calculation of royalties based on the total sales of the licensee. As in the case of minimum royalties, the perceived risk is that this method of royalty calculation would lead to *de facto* exclusive use of the licensor's technology, since the licensee would need to pay a double royalty to use a third party technology. In contrast to the position with minimum royalties, this method of calculation cannot be described as inherent in the licence relationship. Accordingly the Guidelines suggest that Article 81 may be applicable under the same **7.72**

---

63 Para 159. 'Notwithstanding the fact that the block exemption only applies as long as the technology is valid and in force, the parties can normally without falling foul of Article 81(1) agree to extend royalty obligations beyond the duration of the licensed intellectual property rights.'

64 Para 155.

65 The same analysis applies to an obligation on the licensee to sell a minimum quantity of products incorporating the licensed technology, which is also mentioned in paragraph 155 of the Guidelines. As long as this provision is linked to an obligation to pay royalties on a per unit or percentage of sales basis, the effect of this provision is to set a minimum royalty. If the licensee breaches the provision, the licensor's remedy will be to obtain damages for lost profits (based on lower royalty payments) and termination of the agreement (which the Guidelines view as lawful where a licensee deals in competing products).

circumstances where an explicit non-compete obligation would be regarded as leading to competitive harm.[66] Calculating royalties based on total sales can also, however, facilitate the licence relationship because of the practical difficulty of determining whether a product sold by the licensee in fact incorporates the licensed technology. On this basis the Guidelines indicate that the conditions of Article 81(3) may be applicable if 'there is no other practical way of calculating and monitoring royalty payments'.[67]

(7) Tying and bundling

7.73 Tying is defined as obligation imposed by the licensor making the licensing of one technology (the tying product) conditional upon the licensee taking a licence for another technology or purchasing a product from the licensor or someone designated by him (the tied product). Bundling occurs where two technologies or a technology and a product are only sold together as a bundle. In both cases, it is a condition that the products and technologies involved are separate products. This is normally not the case where the technologies or products are linked in such a way that the licensed technology cannot be exploited without the tied product or one part of the bundle cannot be exploited without the other. In the following the term 'tying' refers to both tying and bundling. Tying typically involves a tying product (where the licensee has some market power) and a tied product. A licensor may have market power with respect to technology and use that power to induce purchase of other inputs or more rarely may use market power in respect of another input to persuade a customer to use its technology.

[66] This rule is generally consistent with the position taken by the ECJ in *Windsurfing Intl* v *Commission* [1986] ECR 611. In *Windsurfing* the licensing agreement provided a method of calculating royalties for the patented rig based on the net selling price of the entire sailboard (rig and board). The Commission held that this method contravened Article 81(1) because it regarded the net royalty clause as an attempt to prevent the separate sale of rigs to unlicensed third parties. The Court was more understanding of the practicalities in the case and held that once the method of calculation was adjusted to meet some of the Commission's objections, so that the royalty rate for the separate sale of rigs was not lower than the rate under the old agreement, the royalty rate for the rigs based on the product as a whole did not violate Article 81(1). In respect of rigs, its object or effect was not restrictive of competition. Further, the Court indicated that the calculation of royalties for the patented part of a product based on the price of the product as a whole could be justified because it was impracticable to establish its value, either because the number of items manufactured or consumed was difficult to establish or if there was no separate demand for the patented product. On the other hand, the Court held that insofar as the calculation of royalties based on a 'bundling' of patented with unpatented products restricted the sale of the unprotected product, the board, it was not only caught by Article 81(1), it was unexemptible under Article 81(3).

[67] Para 160. The reference to Article 81(3) implies that the provision remains restrictive but is nonetheless indispensable under these specific circumstances. It is implicit in the discussion that the reference to another practical way of calculating royalties is meant to cover alternative ways of calculating running royalites—the discussion in the Guidelines would be redundant if a flat rate royalty were regarded as an alternative, since this possibility will always exist.

Tying may also involve licensing of different technologies together, where market power in respect of one technology is used to induce a licensee to take a licence under a second technology.

(a) The rationale for tying

For the purposes of technology licensing in agreements between non-competitors, there is a difference between quality control led tying and pure tying. Quality control led tying involves the requirement that the licensee use specified inputs, often provided by the licensor, in order to ensure that the products produced by all licensees meet common standards. The TT Guidelines recognise that the licensor has a legitimate interest in ensuring that the quality of the products are such that it does not undermine the value of his technology or his reputation as an 'economic operator', particularly where the licensor's trademark is associated with the product. In other tying cases, the interest of the licensor may simply be to increase sales of the tied product. Tying can also give rise to efficiency gains by performing a metering function[68] In other cases there may be no 'tying' at all, a licensor and a licensee may find it efficient to license a technology and purchase a related input (or even a non-related input) in the same agreement. **7.74**

(b) Competitive harm

For tying to cause competitive harm, three requirements must be satisfied. First, there must be significant market power in respect of one product—the tying product. Secondly, there must be some compulsion exerted on the customer to take the products as a package—if the customer had the option to purchase the products separately there is usually no tie.[69] Thirdly, conditions in the market **7.75**

---

[68] See e.g. *Vaessen/Morris* [1979] 1 CMLR 511 where a manufacturer of sausage-making machines was prepared to offer a free licence to use a sausage-filling machine if the licensee agreed to buy its skins from him. The tie helped him 'meter' the use of the machine as well as to produce income. The Commission found that the tie had the effect of foreclosing other sausage skin makers. In a situation where the licensor offers the licensee a form of payment for the package of patented and non-patented material which in practice is not dramatically different from the overall amounts which would have been payable on the patented material alone, and includes the advantage of avoiding an upfront payment, there seems to be room for exemption. As long as the market share is not so high that there are no substitutes, the tie-in can be viewed as efficiency enhancing and not an aggrandisement of market power by the licensor.

[69] The fact that the package is offered at a discount over its constituent parts should not by itself indicate commercial compulsion, as long as this discount can be attributed to the tied products (and those products are not sold below cost). The Guidelines para 195 discuss a case where a licensee agrees to purchase a catalyst from the licensor that allows it to exploit the technology more efficiently. From the example, it appears that this was a free contractual choice of the parties—buying the catalyst was not a requirement for obtaining the licence. Under these circumstances, it is hard to see why this is treated as a tying arrangement. The fact that

for the second product, the tied product, must be such as to make competitive harm likely in that market. Competitive harm in the second market takes the form of foreclosure—rival suppliers in the second market may be restricted in their access to that market in a way that makes unilateral or coordinated effects likely.[70] In intellectual property licensing, foreclosure effects can occur in a technology market or on the market for the second input required by the licensee.[71] The possible competitive harm in the context of intellectual property licensing depends upon the nature of the alleged tie.

(i) Competitive harm where the tying product is not a technology

7.76 Where the tie involves a licensor with significant market power in respect of a non-technology input, the Guidelines correctly suggest that the analysis should follow that for direct obligations on the licensee not to use competing technology.[72] It is nonetheless essential that there be a high level of market power in respect of the tying product in order to establish the existence of the tie (for the purposes of Article 81).[73]

7.77 In this context, the market share of the tying product may not be a reliable guide to the licensor's ability to coerce the potential licensee into obtaining a licence. Where the tying product is one that the potential licensee would purchase from the licensor in any case, even if only as part of a range of related inputs, then the licensor may have more leverage. It is also essential to show that there is some element of coercion. A grant of a discount in respect of the tied product should not by itself be sufficient to show that the licence incorporates a tie.

(ii) Competitive harm where the tying product is a technology and the tied product is not

7.78 Where the competitive harm involves foreclosure of a market for a non-technology input, the first requirement is that there be real market power in

agreements to purchase two inputs are included in a single contract should not be sufficient to create a tie for Article 81 purposes.

[70] See generally the discussion in Chapters 4 and 5 for a fuller discussion of the appropriate analysis for foreclosure.

[71] The Guidelines para 193 identify a possible impact of the tying market since the tie would force a new entrant in that market to enter 'several markets at the same time'. It will be noted, however, that this effect only occurs if the requirements for effective foreclosure in the markets for the tied products are met. Since, as explained in Chapter 4, foreclosure is a process that leads to economic harm and not an economic harm in itself, the increase in barriers to entry in the tying market (and the resulting unilateral effects in that market) are an economic harm deriving from the foreclosure process.

[72] See discussion at para 7.89.

[73] The Guidelines para 193 refer to 'a significant degree of market power'. It is also essential that there be two distinct products. If the input cannot be used without infringing IP rights belonging to the licensor, requiring the licensee to take a licence is not a tie.

the technology market. As explained previously, market power in technology markets does not depend on market share as much as on whether commercially realistic alternatives exist for the technology in question. The second requirement is that there must also be a real 'tie'—a link of the willingness to license the technology to agreement by the licensee to purchase the input. The fact that the licensee agrees to purchase the input in the same contract should not be sufficient to establish this link. The third requirement is that the market for the tied product should be susceptible to foreclosure. As explained in previous chapters, this means that reducing sales opportunities for rivals is likely to create, maintain, or increase the market power of the licensor in that market with resulting unilateral effects. The fourth requirement is that the tie should contribute materially to that foreclosure effect.[74] Taken together, these requirements mean that tying should only become an issue where (i) the licensor is the leading supplier in the markets for both the technology and the related input (or is likely to become the leading supplier); and (ii) the licensee is an important customer for the related input.

(iii) Competitive harm where technologies are provided in a package

The Guidelines scarcely touch on what is probably the most important tying-related issue in licensing: the application of Article 81 to package licensing. It is common for licensors to offer licences under a package of intellectual property rights. Sometimes the licensor will offer a standard package without offering component parts separately. Licensees sometimes invoke competition law concerns to object to this practice, particularly where they allege that they do not 'need' the full package.[75] **7.79**

The key question in assessing these claims is whether the package really involves separate 'products' for purposes of identifying consumer harm. The fact that a package includes rights to seven (or 107) patents does not mean that it involves a tie of separate products. As explained in Chapter 6, the assessment of competitive harm required under the new Article 81 paradigm requires that technology markets be assessed functionally by reference to the technology—not formally by reference to the various legal rights involved. The fact that a licensee might control or have access to intellectual property rights covering aspects of the technology does not affect this analysis if the licensee is unable to operate successfully in the downstream market without the licensed technology.[76] **7.80**

---

[74] The Guidelines para 193 state that 'the tie must cover a certain proportion of the market'.

[75] *U.S. Philips Corp.* v *ITC* (Fed. Cir. 21 September, 2005) (reversing finding by International Trade Commission that package license including non-essential patents constituted unlawful tying).

[76] In this context the 'tie' does not lead to any consumer welfare loss, because the market power of the licensor derives from the value of the technology for users in the downstream market, if the licence must be limited to specific IP rights the royalty for those will be higher.

Where a package contains genuinely separate technologies, it is also necessary to establish coercion. The fact that royalties for the package are lower than for each of the technologies licensed separately does not necessarily indicate that the licensee had 'no choice' but to take the package, as long as competing technology licensors could match the discount in their pricing for competing technology in the tied market. The fact that additional intellectual property rights are thrown into a package 'for free' or for a small additional royalty also does not create a tie—from an economic perspective these are effectively discounts on the supposedly tied products and nothing more. Where there *is* a tie on this basis, it is still necessary to examine carefully the extent of real foreclosure effects. Effects on a licensee that controls its own technology but does not license it to third parties are unlikely to lead to competitive harm.[77]

(c) Balancing and indispensability in tying cases

7.81 It will be evident from the discussion above that 'tying' should lead to competitive harm for purposes of Article 81(1) only in limited circumstances. Since these circumstances require substantial market power in a number of markets, the TTBER is not relevant to tying allegations. Where the requirements for proving competitive harm are met, the TT Guidelines differentiate sharply between tying for quality control purposes and tying for other purposes. As regards quality control, the Guidelines acknowledge that an obligation on the licensee to observe minimum quality specifications, insofar as they are necessary for a technically proper exploitation of the licensed technology[78] or for ensuring that the product of the licensee conforms to the minimum quality specifications that are applicable to the licensor and other licensees, would normally not be restrictive of competition under Article 81 or alternatively would be exemptible under Article 81(3).[79] In other cases, the implication of the Guidelines is that the competitive harm of the tie will outweigh its benefits (if any).

[77] The TT Guidelines, echoing the similar passage in the Vertical Guidelines, also refer to the possibility that 'tying may allow the licensor to increase royalties, in particular where the tying product and the tied product are partly substitutable and the two products are not used in fixed proportion. Tying prevents the licensee from switching to substitute inputs in the face of increased royalties for the tying product.' This passage assumes the ability of a licensor to increase royalty levels over the course of a licence period, which would be highly unusual. It also assumes an ability to reduce use of the tying technology in favour of using the tied technology, which will also be highly unusual. This supposed competitive harm is therefore not dealt with in the main text.

[78] Para 194. In *Velcro/Aplix* [1989] 4 CMLR 157, the Commission held that only as long as other suppliers are unable to provide the product and therefore there is no technically adequate substitute would a tie-in be viewed as technically necessary.

[79] Para 194.

(8) *No-challenge clauses*

(a) Rationale for the restriction

No-challenge clauses are obligations undertaken by the licensee not to challenge the validity of the licensor's intellectual property rights. Licensors traditionally seek to include such clauses in their licences for two reasons. First, a licensee could use the knowledge that it gained as a result of working the licensed technology as a basis for formulating a challenge to the licensor's rights. Secondly, in cases where the licence relationship requires continuing cooperation between the licensor and licensee, the licensor may be reluctant to cooperate with a partner that at the same time is mounting a legal challenge to those rights.[80] **7.82**

(b) Ancillarity

It is possible that the concern by a licensee regarding use of knowledge gained in the licensing process to challenge its rights could lead the licensee either not to license at all or to license only 'trusted' firms that were unlikely to mount a challenge. In this context it might be possible to argue that a no challenge clause is 'ancillary' to the agreement and therefore outside Article 81. Under the old regime, however, this factor was viewed as completely overshadowed by the issue of public policy that the rules of competition law should not indirectly encourage the weakening of the integrity of the patenting process.[81] The TT Guidelines do not reflect any ancillarity analysis. **7.83**

(c) Competitive harm

The competitive harm associated with a no-challenge clause involves the risk that invalid intellectual property rights give their holders market power that is not justified by the policies underlying those rights. This unjustified market power in turn leads to higher prices or lower output. The existence of no-challenge clauses serves to increase this effect, by reducing the number of successful challenges to intellectual property rights particularly since, as the TT Guidelines point out, a current licensee is also a likely challenger. On this basis, the Guidelines suggest that Article 81(1) will be applicable to a no-challenge clause 'where the licensed technology is valuable and therefore creates a competitive disadvantage for undertakings that are prevented from using it or are only able to use it against the payment of royalties'.[82] It will be noted that **7.84**

[80] In the context of settling disputes over validity of IP rights, a no-challenge clause also serves to resolve the underlying dispute. Since infringement actions usually involve disputes between competitors or potential competitors, the implications of a no-challenge clause in this context are dealt with in Chapter 8.

[81] See e.g. *Vaessen/ Morris* [1979] 1 CMLR 511 at paras 34–35.

[82] Para 112.

this language focuses on harm to competitors and not on genuine competitive harm. It reflects a formalistic concern for the 'the interests of undistorted competition' and a desire to achieve 'conformity with the principles underlying the protection of intellectual property' that is not found in other parts of the Guidelines.[83]

(d) Economic benefit

7.85 The TT Guidelines indicate that no-challenge clauses caught by Article 81 are unlikely to meet the requirements of Article 81(3). No-challenge clauses relating to know-how are, however, treated differently and can be exempted because 'once disclosed it may be impossible to recover the licensed know-how. An obligation on the licensee not to challenge the licensed know-how therefore promotes dissemination of new technology in particular by allowing weaker licensors to license stronger licensees without fear of a challenge once the know-how has been absorbed by the licensee.'[84]

(e) Indispensablity—termination rights as an alternative

7.86 One reason for limiting the application of Article 81(3) to no-challenge clauses is that a licensor can protect its legitimate interests to a substantial extent by terminating the licence in the event of a challenge. The Guidelines state that the licensor should not be forced to continue dealing with a licensee who challenges the very subject matter of the licence agreement. They also state that this implies that, upon termination, any further use by the licensee of the challenged technology is at the challenger's own risk.

(9) Restrictions on development or use of competing technology

(a) The rationale for the restriction

7.87 It is common for a licensor to require that a licensee agree not to use rival technologies as a condition for obtaining a licence. This provision protects the licensor from the risk that the licensee either does not exploit the technology at

---

[83] This approach was probably dictated by the extensive precedents under the old regime applying Article 1(1) to non-challenge clauses. In *Vaessen/Morris* [1979] 1 CMLR 511 the Commission found that the presence of a no-challenge clause was a restriction of competition because it prevented the licensee from removing 'an obstacle to his freedom of action'. In *Davidson Rubber* [1972] CMLR D52 the Commission insisted upon the licensor removing the no-challenge clause as the price of granting the exemption. See also *AOIP* v *Beyrard* [1976] 1 CMLR D14. In *Bayer and Henneke* v *Sullhofer* [1988] ECR 5249, the Court focused primarily on limited areas where a no-challenge clause would fall outside Article 81, but indicated that where the parties were in a strong market position the clause would have a restrictive effect.

[84] Ibid.

all or limits its efforts to exploit the technology. Failure by a licensee to exploit the technology fully can lead to direct revenue loss for the licensor and can also affect the reputation of the technology in other territories or markets.[85] Furthermore, where the nature of the licensed technology requires continued cooperation between the licensor and licensee to promote and improve the technology and products using the technology, parallel promotion and use of rival technologies by the licensee may prejudice the atmosphere of trust required for effective collaboration. There is also a concern that use of a competitive technology in parallel could result in misappropriation of the licensed technology because the licensee could use the licensed technology to assist where it is using the rival technology, particularly where the licensed technology includes know-how.[86]

(b) Types of restriction on competing technologies

The TTBER and the Guidelines differentiate among three types of restrictions on competing technology in agreements between non-competitors: **7.88**

(i) restrictions on use of third party technology;
(ii) restrictions on development or use of technology owned by the licensee (but not currently licensed to third parties); and
(iii) restrictions on independent research and development activities.

Similar issues are also raised by other clauses that arguably have a similar effect—minimum sales requirements, certain royalty obligations, 'best endeavours' clauses, and clauses allowing termination where the licensee sells competing products. These types of provisions are dealt with in turn below.

(c) Restrictions on use of third party technology

(i) Competitive harm

The TT Guidelines describe an obligation not to use third party technologies as a 'non-compete' obligation. (This terminology is borrowed from the terminology used in the context of distribution agreements—in practice this restriction would more usually be described as an 'exclusive use' obligation.) The Guidelines suggest that these provisions can lead to competitive harm through foreclosure of competitive technologies or by facilitating collusion **7.89**

[85] This applies particularly where the licensee has been granted an exclusive territory. See Guidelines para 202.

[86] See Guidelines para 201. The Guidelines also refer to the case where the licensor needs to make client-specific investments such as training or special development of the technology for the needs of the licensee. The Guidelines suggest, however, that less restrictive means are available to deal with the hold-up problems associated with these concerns, such as a lump sum payment by the licensee.

between licensors.[87] The Guidelines provide an extensive theoretical discussion regarding the methodology for assessing foreclosure effects.[88] This discussion is important because it shows that simple calculation of the percentage of the market 'foreclosed' does not suffice to assess foreclosure in technology cases. For practical purposes, however, the issue can be assessed fairly simply. First, are technologies available or likely to become available that compete with the licensed technology? Secondly, do non-compete obligations make it likely that an existing technology will exit the market, that the owner of an existing technology will reduce investment in support of that technology, or that a new technology will not be introduced? Thirdly, will this allow the licensor to increase royalty income or avoid a reduction in royalty income? Fourthly, does the licence agreement under consideration contribute significantly to this effect? All of these questions, which are dependent on the nature of the market which may change over time, must be answered affirmatively for there to be a substantial 'foreclosure' effect.[89]

(ii) Balancing and indispensability

7.90 Although the Guidelines discuss both the competitive harm and the economic benefits associated with non-compete clauses at length, they do not provide precise guidance on how the balance should be struck. The suggestion appears to be that where appreciable foreclosure effects can be shown, non-compete clauses will only be justified where there is a serious risk of misappropriation of the licensor's technology—particularly where that technology is mainly know-how. In other cases, the Guidelines suggest that other provisions, such as minimum royalty or minimum output obligations, have a less restrictive effect.

(d) Restrictions on use or development of technology already controlled by the licensee

7.91 In most cases where a licensee controls technology that is a competitive rival for technology controlled by the licensor, the licensee and the licensor will be viewed as 'competitors'. Under those circumstances, the permissible extent of restrictions on use of competing technology by the licensee must be considered in accordance with the stricter approach described in Chapter 8. Where the licensee is not currently licensing its technology and the licensor is not an actual

[87] Para 198.

[88] Paras 199–200. For a more detailed discussion of foreclosure issues see Chapter 4 above at paras 4, 49–4.55.

[89] As regards the role of non-compete obligations in facilitating collusion, it may be repeated here that this requires that the market in question (the technology market in this case) also present the other features that make collusion likely. For the reasons set out in Chapter 5, this is unlikely in the context of technology markets.

or potential competitor in the downstream product market, the TTBER treats the agreement as one between 'non-competitors'.[90] Likewise if a licensee is not currently an actual or potential competitor in the downstream product yet controls a technology that could be used in that market or could be developed for use in that market, the agreement remains one between non-competitors for purposes of the TTBER. The Guidelines make it clear, however, that even if in these situations the agreement is nominally one between non-competitors, a restriction on the use of technology owned by the licensee or on development by the licensor of its own technology is 'normally considered to be restrictive of competition and not to satisfy the conditions of Article 81(3)'.[91] Thus the Guidelines treat this restriction as something close to a restriction by object.[92]

(e) Restrictions on development of new technology by the licensee

7.92 Where the licensee does not already control a potential competing technology, the TT Guidelines take a cautious view on application of Article 81. On the one hand, the risk of competitive harm is viewed as significant even where the market shares of both licensor and licensee are below 30%, because the licensee may nonetheless be regarded 'as a (potential) source of innovation on the market'.[93] On this basis the TTBER does not cover such restrictions, except in the case where the limitation is necessary to prevent disclosure of secret know-how to third parties.[94] On the other hand, the Guidelines suggest that

[90] This will be the more common case, as it will cover the position where an incumbent on the downstream market seeks to upgrade its product offering. An example would be a pharmaceutical producer that obtains rights to a new drug that offers improved therapeutic effects over a drug that the company currently markets to treat the same condition.

[91] Para 115.

[92] The Guidelines justify this strict treatment on the basis that '[t]his technology constitutes a competitive constraint in the market which should be preserved'. There is merit in this proposition unless there are other alternative technologies in the market that render the competitive constraint of the licensee's efforts insignificant. A similar analysis would be applicable if a licensee comes into possession of a competing technology as a result of a merger or acquisition subsequent to the original licence. An attempt to enforce the restrictions on use of competing technologies in this context would raise the same concerns about elimination of an existing competitive constraint.

[93] Para 116.

[94] TTBER, Article 5(2). Where R&D activities with third parties could lead to disclosure of secret know-how to those third parties, a provision preventing such disclosure would be ancillary for purposes of Article 81. This is viewed as an exception to the general rule of non-interference with R&D and the restrictions must be necessary and proportionate. However, the Guidelines' assessment of necessary and proportionate would include a requirement that employees trained in and responsible for the use of the know-how would not be allowed to work on related R&D with third parties. A mere confidentiality clause in the employee's contract would not be viewed as sufficient protection. This line is justified by the importance of the need to give the licensor a real assurance that the know-how will not only remain confidential, but also be devoted to the exploitation of the licensed technology. Anything less reassuring to the licensor would probably 'chill' know-how licensing.

where there are several technologies available in the market and the licensee does not have special assets and skills that make it a more likely source of innovation than other firms in the market, 'the restriction on research and development is likely to either fall outside Article 81(1) for lack of an appreciable effect or meet the requirements of Article 81(3)'.[95] The key points according to the Guidelines are the competitiveness of the technology market and whether the licensee possesses 'the necessary assets and skills to carry out further research and development'. Where the licensee is a likely source of innovation, and competition for innovation is limited, the Guidelines suggest that that there will be significant competitive harm that will outweigh any economic benefits associated with the restriction.

(f) Provisions with equivalent effect to non-compete obligations—minimum sales requirements, royalty provisions, 'best endeavour' clauses, and termination in the event of use of competing technology

7.93 Some common clauses in licence agreements can be viewed as having similar effects to direct agreements not to use competing technology. Thus a commitment by a licensee to sell a minimum number of products incorporating the licensed technology has the effect of restricting the ability of the licensee to sell products using rival technologies, if the minimum requirement constitutes a substantial proportion of its likely sales. Similarly, royalty provisions providing for payment regardless of actual sales or royalty payments based on sales in the downstream market regardless of whether they use the licensed technology could limit the ability of the licensee to use rival technologies (and pay royalties twice). A clause requiring the licensee to use 'best endeavours' to promote and sell products using the licensed technology will have the practical effect of limiting its 'endeavours' to promote and sell products using rival technologies. The ability of a licensor to terminate a licensee that sells products using a competing technology may be a deterrent on the willingness of a licensee to research new technology or license third party technology.

7.94 The different treatment of these various clauses provides an insight into some of the nuances in application of the new paradigm to intellectual property licensing. Thus minimum royalty payments are considered unproblematic (and, as seen above, a less restrictive alternative to a non-compete obligation). Royalties based on all sales, in contrast, are treated as directly equivalent to a non-compete obligation, even though the economic effects of the two provisions are essentially the same. Minimum sales restrictions are treated like minimum royalty obligations. As regards 'best endeavours' clauses, even under the old

[95] Para 116.

regime it was accepted that this obligation did not in itself restrict competition. Implicit in this assessment is a determination that research and development regarding a competing technology is not inconsistent with using best endeavours to promote the licensed technology. The Guidelines also take the view that a clause allowing the licensor to terminate a licence in the event that the licensee sells competing products does not lead to competitive harm, because it does not reduce the incentive of the licensee to innovate.[96] It may be noted that, even were this not the case, the breakdown in trust between the licensor and licensee once the latter is actively promoting rival products would provide a powerful additional reason for allowing the licensor to dissolve the relationship.

(10) Improvements and grant-backs

The transfer of technology often leads to development by licensees of improvements to the original technology. For the licensor this creates the unpleasant prospect that licensees could through their improvements make the licensor uncompetitive in markets for products that the licensor has itself created. This in turn could create a hold-up problem—a disincentive for the licensor to make its technology available in the first place. It is commonplace for licensors to require that the licensee disclose the existence of any improvements arising while the agreement remains in effect and allow the licensor to use those improvements and license those improvements to third parties. This type of clause is described as a 'grant-back' requirement, since it involves a grant of rights by the licensee back to the licensor. By itself, this obligation does not raise serious competition issues. It addresses the potential hold-up problem, since the licensor will never be at a technological advantage compared with any licensee. More serious issues may be raised, however, where (i) the licensor demands that the licensee continue providing access to improvements on the technology even after the agreement expires (or in the case of patents, after the patents have expired); or (ii) where the licensor demands that the licensee either assign any rights covering the improvements to the licensor or grant the licensor an exclusive licence in respect of those improvements.[97] **7.95**

[96] Para 116. The Guidelines note that 'Article 81(1) only applies where the agreement reduces the licensee's incentive to improve and exploit its own technology.' The Guidelines conclude that a termination right triggered by competing sales cannot have this effect because 'the agreement can only be terminated when a commercially viable technology has been developed and products produced on the basis thereof are ready to be put on the market'.

[97] This second category of restriction is sometimes referred to as a 'grant-back' in contrast to the first type of grant-back, described earlier in this paragraph which is sometimes referred to as a 'license-back'.

(a) Ancillarity

7.96 The Guidelines do not discuss the application of Article 81(1) with respect to non-exclusive grant-backs for severable improvements, other than to note that such obligations are covered by the TTBER as long as the market share requirements are met.[98] There must be a strong argument, however, that non-exclusive grant-back obligations are ancillary as long as they are drafted to cover genuine improvements on the licensor's technology. The ability of a licensor to keep up to date with its own technology and to promise licensees that they can keep up to date, would appear indispensable to effective licensing. This analysis would apply even if the obligation extended beyond the life of relevant intellectual property rights, since the licensees would be in a less favourable position for developing post-expiration improvements if they had not had the advantage of a licence in the previous period and the licensor might not have granted the licence without the benefit of the grant-back provisions.

(b) Competitive harm and grant-back clauses

7.97 Grant-back clauses can arguably lead to competitive harm because they reduce the incentive for licensees to invest in developing improvements on the technology. Where licensees must share any technological advances that they make with the licensor on a royalty-free basis and where the licensor is in turn entitled to share that technology with other licensees, a licensee is less likely to invest in developing the technology.[99] The licensee's incentives diminish further (according to this argument) where the licensee must grant an *exclusive* licence on improvements to the licensor, since this deprives the licensee of the possibility of profiting through licensing third parties. This disincentive is regarded as particularly significant where it applies to *severable* improvements—intellectual property rights covering technology that 'improves' the licensed technology, but can also be practised without infringing the licensed technology.[100] The Commission

[98] Para 109.

[99] We stress that this is the *theory* of competitive harm asserted for grant-back clauses. In practice, the development of improvements will often be a by-product of the licensee's efforts to bring its own products on to the market. This investment will take place regardless of whether the licensee is obliged to share or assign the improvements.

[100] See TTBER Article 1(n). As long as the original technology is protected by valid IP rights, the ability of the licensee to exploit non-severable technology is dependent on the rights granted by the licensor. Likewise the ability of the licensee to profit from licensing that technology to third parties is dependent on those parties obtaining the necessary rights from the licensor. As regards severable improvements, in contrast, in the absence of a contractual restriction the licensee could exploit its innovations independently. See Guidelines: para 109, 'Exclusive grant-backs and obligations to assign non-severable improvements are not restrictive of competition within the meaning of Article 81(1) since non-severable improvements cannot be exploited by the licensee without the licensor's permission.'

considers that exclusive grant-back rights for severable improvements could have significant competitive harm even where the licensed technology has a market share of less than 30% and where the licensee's share of the downstream market is less than 30%.[101] Where a licensor imposes grant-back obligations on all its licensees, the effect may also be to limit product differentiation based on technology thus potentially facilitating collusion between licensees (and the licensor if the licensor is vertically integrated).[102] Where the obligation continues beyond expiration of intellectual property rights, it was once argued that the grant-backs would have the effect of giving the original licensor an extension of its original rights beyond the period granted by the relevant legislation.

(c) Economic benefits of grant-back clauses

A grant-back clause resolves the hold-up problem already identified above. **7.98** Grant-back clauses can have further important benefits, however, by making the original licensor a 'one-stop shop' for all licensees wishing to have the most up-to-date technology. If the original licensor has the right to sub-license the improvements developed by any licensee, then any improvement can benefit all licensees and their customers. This leads to a direct consumer benefit. It also makes a licence more attractive to potential licensees because it allows the licensor to offer not just the existing technology, but access to the top version of the technology on an ongoing basis. This benefit does not end at the point when intellectual property rights covering the original technology expire, but continues as long as different licensees (and the licensor) continue to develop improvements. The licensor may argue that this benefit is enhanced where it controls the intellectual property rights (either due to assignment or to exclusive licensing), since this allows the licensor to police infringement by non-licensees to the benefit of the whole licensee community. (As an exclusive licensee, the original licensor will know which third parties are authorised to use the technology, which individual licensees will not.) The licensor may also argue that centralising the control of the technology allows cost savings in obtaining and preserving the necessary intellectual property protection.

(d) Balancing and indispensability

The speculative nature of competitive harm resulting from grant-back clauses **7.99** means that it will be difficult to evaluate the likelihood of competitive harm at

[101] Under TTBER Articles 5(1)(a) and 5(1)(b) assignments to the licensor and exclusive licences for the licensor are excluded from the TT Guidelines.

[102] See TT Guidelines para 110. 'The negative impact of grant back obligations can also be increased in case of parallel networks effects.' (This potential effect only becomes relevant, of course, where the other prerequisites for successful collusion are present.')

the point where an agreement is made.[103] The TT Guidelines indicate in respect of exclusive grant-back obligations that the competitive harm will depend in part on the position of the licensor in the technology market[104] and in part on whether the licensee is compensated for the improvements that it contributes.[105] The benefits of grant-back clauses are usually evident, but it is not always clear that an exclusive grant-back obligation will be indispensable to obtain those benefits. The inchoate nature of harm and of benefits (at least of the additional benefits of an exclusive grant-back) make this a difficult area to apply the new Article 81 paradigm. The TTBER and the Guidelines appear to recognise this problem and apply a pragmatic solution that should be workable in practice.

(e) A pragmatic approach to non-competitor grant-backs

7.100 The application of Article 81 to grant-backs in agreements between non-competitors appears to include the following elements. First, grant-backs restricted to non-severable improvements are allowed, even if they are exclusive.[106] Secondly, non-exclusive grant-back obligations are allowed, either because they are ancillary or because their benefits generally outweigh their restrictive effects. Thirdly, exclusive grant-back obligations or obligations to assign rights in improvements to the licensor are disfavoured. While such obligations may theoretically be justifiable, in practice the licensor can never be certain

[103] As already noted, the theory that the effort expended in developing improvements is dependent on possible future licensing revenue does not correspond with commercial experience in many industries where the generation of IP rights (the prerequisite for royalty revenue) is a by-product of commercialisation efforts that would happen anyway. Furthermore, as noted in Chapter 4, any competitive harm based on a reduction in dynamic efficiency is by its nature inherently speculative. The position might, however, be different where the licensee has a history of successful innovation in related fields and has been active as a licensor in those fields.

[104] According to the Guidelines para 110, 'the stronger the position of the licensor, the more likely it is that exclusive grant back obligations will have restrictive effects on competition in innovation. The stronger the position of the licensor's technology the more likely it is that the licensee will be a source of innovation and future competition.' This passage is difficult to follow. If the concern is that the licensee might eventually develop rival technologies that compete with the licensor, it is hard to see how the grant-back obligation substantially affects that result. If the concern is that the exclusive grant-back will slow the development of the licensed technology, then perhaps the effect is greater if the technology is more important. It may be noted, however, that the licensor has a strong interest in seeing the technology improve and is unlikely to impose contractual terms that will slow the technology's development. The licensor may take the view that centralising improvements will lead to better and more focused development efforts.

[105] Ibid. The suggestion is that that licensees that are compensated may still have an incentive.

[106] The Guidelines indicate that these restrictions fall outside Article 81 because exploitation would infringe the licensor's rights. This argument would not apply, however, after the expiration of IP rights owned by the licensor. If the grant-back obligation extends beyond the life of the IP rights, then the continued exclusivity obligation should logically be assessed on the same basis as any other exclusivity obligation.

of their enforceability, unless the relevant technology is highly competitive. Fourthly, there is no general bar on grant-back obligations that extend beyond the termination of the original licensed intellectual property. This position is in marked contrast to the position under the old block exemption that allowed such clauses only where reciprocal rights to ongoing improvements were available to the licensee.[107]

(11) Other ancillary restrictions

The TT Guidelines set out a list of examples of obligations in licence agreements that are 'almost always *not* restrictive of competition within the meaning of Article 81(1).' These include: **7.101**

(a) confidentiality obligations;
(b) obligations on licensees not to sub-license;
(c) obligations not to use the licensed technology after the expiry of the agreement, provided that the licensed technology remains valid and in force;
(d) obligations to assist the licensor in enforcing the licensed intellectual property rights;
(e) obligations to pay minimum royalties or to produce a minimum quantity of products incorporating the licensed technology; and
(f) obligations to use the licensor's trademark or indicate the name of the licensor on the product.[108]

Each of these clauses falls into the category of ancillary restrictions,[109] i.e. those restrictions which are indispensable to effectuating the main purpose of the licensing agreement. Each one has first to be assessed in the light of its commercial indispensablilty, i.e. as a restraint that is essential in order to convince one or both of the parties that the licensing agreement is worth entering into in the first place. Then this commercial analysis has to be translated into a case for legal indispensability both for the purpose of assessment under Article 81(1) and, if necessary, for exemption under Article 81(3). **7.102**

(a) Confidentiality obligations

It has long been acknowledged by the Commission that a licensor is entitled to place on the licensee an obligation not to divulge know-how communicated by **7.103**

[107] The old rule appears to have been based more on an idea of fairness to the licensee than on any consideration that a one-way improvements obligation might actually lead to competitive harm.

[108] Para 155.

[109] See Gonzalez Diaz, E, 'Some Reflections on the Notion of Ancillary Restraints under EC Competition Law' [1995] *Fordham Cor Law Ins* 325 at pp 350–4.

the licensor during the contract. Moreover, the licensor is entitled to hold the licensee to that obligation for a period after the expiry of the contract until the know-how enters the public domain because in both instances non-disclosure is an indispensable condition of know-how licensing.[110] As the Commission stated in *Delta Chemie*,[111] the commercial value of the know-how rests very largely in its confidential character and each disclosure brings prejudice to the holder of that know-how. Since there is no time limit on the know-how protection through contract, it is legitimate to impose an obligation as long as the know-how does not enter the public domain. In *Boussois/Interpane*,[112] the Commission cleared a clause that placed a five year limit on the secrecy obligation on the licensee. The parties had agreed that that represented the average life of the know-how and accepted that the obligation remained only as long as the know-how was not obtained by third parties from other sources. The Commission noted in clearing the clause that the ban did not extend beyond the useful life of the technology.

7.104 The Guidelines make it plain that confidentiality obligations are not normally caught by Article 81(1), but leave implicit the point that the licensor may insist that licensee is bound by the confidentiality obligation after the agreement has expired (Article (2)(1)). This provision is an essential minimum standard of protection for the licensor; the only issue is whether it offers sufficient protection for sharing secret know-how with the licensee.

(b) The ban on sub-licensing and assignment

7.105 The obligation not to sublicense the know-how which is communicated under the licence or the patent which is licensed or to assign is not caught by Article 81(1), as long as the know-how has not entered the public domain and the patents are still in force. This is a classic case of an ancillary obligation in a technology transfer. Licensors wish to preserve some control over those individuals who will exploit their licensed technology while it is protected by an intellectual property right. In *Burroughs Delplanque*,[113] this protection was regarded as within the scope of the patent. In *Rich Products/Jus-Rol*,[114] the ban on sub-licensing to any undertaking other than a wholly owned subsidiary of either the licensor or licensee was cleared because it protects the licensor's right to use the know-how as it sees fit as long as the know-how has not entered the public domain. Competition policy would view as restrictive a

---

110 *Burroughs Delplanque* [1972] CMLR D67 at para 48.
111 *Delta Chemie/DDD OJ* [1989] 4 CMLR 535.
112 *Boussois/Interpane* [1988] 4 CMLR 124.
113 *Burroughs Delplanque* [1972] CMLR D67.
114 *Rich Products/Jus-Rol* [1988] 4 CMLR 527.

ban once the patent expires or the know-how loses its secrecy because the contract would unfairly restrain the licensee while its competitors would be free to use the intellectual property. However, as long as the know-how has not entered the public domain and the patents are still in force, it would not be restrictive of competition to require a licensee not to sub-license the know-how that is communicated under a licence even after the contract has expired.

(c) Obligations not to use the licensed technology after the expiry of the agreement

A third indispensable clause in intellectual property licensing is the post-term use ban. The Commission has long accepted that an obligation placed upon the licensee to cease using the licensed technology after the termination of the agreement is essential to a licensing agreement and not a restriction on competition because otherwise the licensor would be forced to transfer the know-how or patents in perpetuity. It is the presence of this obligation in the agreement that distinguishes the licence from the assignment. In *Rich Products/Jus-Rol*[115] the Commission cleared a post-term use ban on know-how that applied for 10 years following termination noting that it was inherent in this type of agreement. If the owner was to lose its exclusive right to use know-how at the expiry of the licensing agreement it would be less willing to grant a licence and this would be harmful to the transfer of technology. **7.106**

A post-term use ban becomes anticompetitive only if it persists after the know-how becomes freely available to others through no fault of the licensee or the licensed technology is no longer valid and in force. In *Delta Chemie*[116] the Commission reiterated that an obligation upon the licensee to cease using the licensed intellectual property right at the end of an agreement is an 'essential condition' insofar as it has not entered the public domain. The limit in time may remove the licensee from the market for the licensed product at the end of the agreement, but it is a necessary condition for the transfer of technical knowledge and should not be prohibited by Article 81(1). This result is reflected in the Guidelines which allow the post-use ban, subject to the proviso 'that the licensed technology remains valid and in force'. **7.107**

The requirement that the technology remain 'in force' makes sense where the licensed technology is protected by intellectual property rights. The position is, however, different where know-how is no longer secret as a result of the actions of the licensee. In such a case, the licensee ought not to profit from that act by escaping the post-use ban. This result is reflected in Article 2 of the TTBER that **7.108**

[115] Ibid. [116] *Delta Chemie/DDD* [1989] 4 CMLR 535.

provides where know-how becomes publicly known as a result of action by the licensee, the post-use ban remains valid.

(d) Quality controls and use of trademarks

7.109 The concern of the licensor to ensure that the licensee preserves quality standards lies at the heart of intellectual property licensing, particularly where the licensor's trademark is associated with the licensed product. This requirement leads to direct consumer benefits in terms of product quality and strengthens the reputation of the technology for the benefit of all licensees. In general the insistence on minimum quality specifications by itself as long as they are agreed upon in advance and based on objectively verifiable criteria has not been viewed as restrictive of competition under Article 81(1).[117] It has also been accepted that the licensor can reserve a right to monitor quality standards by carrying out related checks.[118] In each of these cases, the Commission has applied an ancillarity analysis. There was some question under the old regime, however, regarding the formal status of quality control requirements under Article 81. In some cases, the Commission or the Court indicated that the limitation on the licensee's autonomy inherent in a requirement to adhere to quality standards was sufficient to bring such provisions within Article 81(1).[119]

7.110 It is hard to see that quality standards can lead to any identifiable consumer harm (in the absence of a *de facto* tying requirement, as discussed previously in this chapter). Even if the Article 81(1) precedents were deemed controlling, however, the clear consumer benefits of technical standards and the lack of any serious competitive harm would clear the restrictions under Article 81(3). On this basis, the Guidelines acknowledge that an obligation on the licensee to observe minimum quality specifications, insofar as they are necessary for a technically proper exploitation of the licensed technology[120] or for ensuring that the product of the licensee conforms to the minimum quality specifications

---

[117] See, e.g., *Delta Chemie/DDD* [1989] 4 CMLR 535 (quality standards applied to manufacturing and marketing were held not to be caught by Article 81(1) because they were 'inspired by the legitimate desire of the licensor to ensure the strict conformity of the products'. The Court did not need to address the possible application of Article 81(3) in that case.

[118] Cf. *Raymond Nagoya* [1972] CMLR D45. See Korah, V, *Technology Transfer Agreements and the EC Competition Rules*, 1996, Oxford University Press, at p 175.

[119] Thus in *Windsurfing* [1986] ECR 611, the Court appeared to accept the Commission's assertion that since quality standards come within the scope of the patent they should fall outside the scope of Article 81(1) only where they are intended to ensure 'that the technical instructions as described in the patent and used by the licensee may be carried into effect'. This analysis reflects a form-based approach to Article 81(1) since it focuses on the restriction of competitive autonomy inherent in quality restrictions.

[120] Para 194. In *Velcro/Aplix* [1989] 4 CMLR 157, the Commission held that only as long as other suppliers are unable to provide the product and therefore there was no technically adequate substitute would a tie-in be viewed as technically necessary.

that are applicable to the licensor and other licensees, would normally not be restrictive of competition under Article 81 or alternatively would be exemptible under Article 81(3).[121]

The arguments for quality control also inform assessment of requirements regarding use of the trademark. The use of the trademark by each licensee creates an economic benefit by ensuring that the reputational advantages of the technology are shared by the licensor and all other licensees. There is unlikely to be any competitive harm resulting from this restriction by itself. This reputational benefit is in turn based on the uniform quality of products produced using the technology. Therefore, the guidelines make the point that 'where the licensees use the licensor's trademark or brand name or where it is otherwise obvious to consumers that there is a link between the product incorporating the licensed technology and the licensor, the licensor has a legitimate interest in ensuring that the quality of the products are such that it does not undermine the value of his technology or his reputation as an economic operator'. 7.111

[121] Para 194.

8

LICENSING BETWEEN COMPETITORS

A. Introduction

When firms that are already competitors enter into agreements to license intellectual property, the potential for anticompetitive harm increases enormously. Where competitors wish to create a genuine cartel, intellectual property licensing can provide an effective tool for organising cooperation, coercing participation and policing compliance. It is not surprising that the great cartels prior to the Second World War were organised in part on the basis of networks of intellectual property licences.[1] It is also not surprising, in light of the real possibility of competitive harm resulting from licensing arrangements between competitors, that assessment of intellectual property licensing gives rise to greater scrutiny where agreements between competitors are concerned. **8.01**

The greater scrutiny required for licensing arrangements between competitors is reflected in both the TTBER and the TT Guidelines. The TTBER treats **8.02**

[1] See generally Wells, W, *Antitrust and the Formation of the Postwar World*, 2002, Columbia University Press, New York, pp 125–136.

agreements between competitors differently at two levels: (i) the market share threshold for the TTBER safe harbour is 20%, compared with 30% for other agreements; and (ii) a number of restrictions that could be allowed in agreements between non-competitors are deemed 'hard core' where the agreement involves competitors. The Guidelines reflect the same approach. For many restrictions, the Guidelines suggest that there could be serious concerns in the context of competitor licensing that would not otherwise be present. For some restrictions, principally those considered as hard core under the TTBER, the Guidelines indicate that Article 81(1) will be infringed in most cases and Article 81(3) is unlikely to be applicable.

8.03 While the cautious approach to licensing among competitors set out in the Guidelines is understandable, the level of concern regarding intellectual property arrangements among competitors and the condemnation of some particular restrictions may be misplaced. Licensing of intellectual property rights among competitors is commonplace, but intellectual property based cartels are rare. Application of Article 81 to such arrangements in actual cases requires the same detailed assessment applicable in other Article 81 cases. The concerns expressed in the Guidelines often provide a good starting point for this analysis, but a full assessment may show that the benefits of an agreement do exceed its likely competitive harm, and that apparently hard core restrictions are necessary to achieve those benefits.

8.04 In this chapter we discuss some of the issues relevant to this full assessment. We start by considering what is meant by 'licensing between competitors' and how different competitive relationships prior to the agreement determine the nature of competition policy concerns. We then provide an overview of how the general approach to intellectual property licensing under the new paradigm changes in the context of licensing between competitors. We conclude by assessing the types of restrictions identified in the TTBER and the Guidelines in the context of these different competitive relationships.

B. Preliminary Issues—When is a Licence Agreed between 'Competitors'?—How Do Different Competitive Relationships Affect the Competition Law Analysis?

8.05 The obvious starting point to this analysis is whether the parties to a licence agreement are in fact 'competitors'. This depends on the definition of the relevant markets, as discussed in Chapter 6. As explained in Chapter 6, there will always be at least two relevant markets—a technology market and a downstream product market. Whether firms are competitors in respect of downstream products is a relatively straightforward empirical question.[1A] Determining whether firms are competitors in the technology market, however, raises more complicated issues.

First, as explained in Chapter 5, there is a strong argument that the fact that a firm has 'technology' that it can license does not make it a competitor in the technology market. The key question should be whether that 'technology' constitutes a product that satisfies an identifiable customer need. Secondly, rival technology owners are not 'competitors' where their technologies are blocking, i.e. where either firm could invoke intellectual property rights protecting its technology to prevent use of the rival technology.[2] Whether technologies are in fact blocking depends on the validity and extent of the underlying intellectual property, which will often be uncertain. The Guidelines suggest that the parties need strong evidence in the form of a court judgment or report from an independent expert to make this determination. **8.06**

The strict approach taken in the Guidelines to 'proving' an actual blocking position is hard to reconcile with the general approach to application of Article 81 under the new paradigm. Where a firm has a reasonable basis for belief that use of its own technology could infringe intellectual property rights controlled by a rival, that reasonable belief will often deter the potential infringer from introducing its own technology on the market.[3] The same deterrent effect will affect the willingness of third parties to use the potentially infringing technology. It is the deterrent effect of potential infringement claims and not their 'objective' validity that limits the competitive constraint provided by the potential infringer. On this basis, evidence that at the time of the agreement the party accepting a restraint had a reasonable basis to believe that there was substantial risk of infringement should be sufficient to categorise that agreement as one between 'non-competitors'.[4] **8.07**

The limits on the concept of 'competitors' on the technology market means **8.08**

[1A] The TT Guidelines discuss in this regard the problem of 'breakthrough products', discussed above in Chapter 6. The definition of the geographic market can also impact on whether the parties were competitors as discussed below in connection with territories ourside the EU.

[2] See TT Guidelines point 28.

[3] Under the general Article 81 approach, which is repeated in the Guidelines, a potential competitor must be someone who is likely to enter the market. Where the deterrent effect of infringement action makes potential entry unlikely, the potential infringer is not a potential entrant, even where the deterrent effect is based in part on subjective considerations.

[4] The approach taken in the Guidelines is unrealistic in light of the inherent uncertainties of IP infringement litigation, particularly where patents are involved. It may be possible for an expert to determine whether the claims made in respect of two rival technologies in patent applications conflict, it is less easy to determine whether the application of those claims conflicts with a rival claim, and less easy still to determine whether the claims will be upheld if subject to legal examination. The expense required for an independent expert's report will also be disproportionate in many cases where technology has a limited application.

that it is possible to have cross-licences between non-competitors. To the extent that such licences include restrictive clauses they need to be assessed according to the general principles set out above in Chapter 7. It is possible to have cross-licences among firms that are only active in the technology market. It is also possible, and relatively commonplace, to find licences or cross-licences where the parties are competitors in the downstream market, but not in the technology market. The potentially restrictive impact of the provisions in a licence agreement between competitors will depend in large measure on the nature of these relationships. In particular, the Commission has indicated that the most serious competition risks arise in the case of reciprocal licences—where competitors in a technology market license each other under their rival technologies, or competitors in a downstream product license each other for technology used to provide those products.

C. Applying the New Paradigm to Licensing between Competitors

8.09 Referring back to the 10 step Article 81 analysis set out at the end of Chapter 3, the analysis of licensing between competitors differs from the assessment of pure vertical agreements described in Chapter 7.

8.10 **Differences in analysis of licensing between competitors and assessing vertical agreements** The differences are as follows:

(a) An agreement between competitors is 'by its nature' potentially subject to Article 81—there is no need to identify potentially restrictive clauses (Step Two).

(b) Restrictions that are 'ancillary' to a licence arrangement may nonetheless be relevant to the Article 81 assessment of the agreement as a whole (Step Three).

(c) The range of restrictions deemed to have the *object* of restricting competition is broader (Step Four).

(d) The theories of potential competitive harm are broader and the potential for harm is greater (Step Five).

(e) The benefits for the licensee (and the associated contribution to consumer welfare) may involve creating 'design freedom' rather than the transfer of technology (Step Six).

(f) The legitimate interests of the competitor licensor may justify additional protection in order to avoid hold-up problems (Step Six).

(g) Because of the increased potential for competitive harm—it is more likely that the balance of harms and benefits will lead to prohibition of the arrangements—even if there are no outwardly restrictive provisions (Step Seven).

(h) An agreement is more likely to 'eliminate competition' (Step Ten).

(a) A licence agreement between competitors can 'restrict competition'

Where competitors enter into an agreement that relates to the market where they compete, application of Article 81 does not require identification of a contractual constraint on the conduct of either party. The agreement is potentially restrictive by virtue of the relationship between the parties (although it is still necessary to show restrictive effects, i.e. competitive harm). In practice, royalty-free, non-exclusive licences, with no additional restrictive provisions, will seldom cause competitive harm. Where the existence of the agreement does have significant unilateral or coordinated effects, however, the lack of outwardly? restrictive provisions within the agreement does not render Article 81 inapplicable.[5] **8.11**

(b) Ancillary restraints are not necessarily outside Article 81

In agreements between non-competitors, contractual restrictions that are necessary to achieve the legitimate purpose of the agreement fall outside Article 81. This rule does not apply where direct competitors enter into an agreement because, as discussed in the preceding section, the agreement itself may be restrictive of competition. If competitive restraints are necessary to make the agreement work, those restraints may nonetheless increase the likelihood of competitive harm associated with the agreement. Restrictions that could be classified as ancillary in an agreement between non-competitors will, however, still be indispensable to achieving the benefits of the agreement (Step Nine in the Article 81 analysis). An agreement containing such restrictions may still satisfy the requirements of Article 81(3), even if the restrictions would otherwise fall within the category of 'hard core restraints'. **8.12**

(c) The range of clauses that are 'restrictive by object' is broader in agreements between competitors

Article 4 of the TTBER contains separate lists of 'hard core' restraints depending on whether the parties to the agreement were previously competitors. Royalty calculation and payment, as well as restrictions on downstream pricing, are dealt with differently when the licensor and licensee are competitors. The biggest difference is not, however, between 'competitor' agreements and 'non-competitor' agreements, but rather between 'unilateral agreements' and 'reciprocal agreements'. Where reciprocal agreements are involved, quantity **8.13**

[5] On this basis, the Commission considers that cross-licences between competitors, even if non-exclusive, may be restrictive of competition where they have the effect of creating *de facto* industry standards.

restrictions, restrictions on production for the licensee's own use, and virtually all customer and territorial restrictions (including 'second source' agreements) are condemned as hard core restraints. These distinctions are carried over to the TT Guidelines, which suggest that provisions falling into the hard core category can only exceptionally fulfil the requirements of Article 81(3). This approach makes sense, to the extent that it means that restrictions falling in these categories require some justification—they should be reasonably necessary for achieving significant efficiency benefits. It is important, however, that 'exceptionally' should not be read to mean 'never'. There will clearly be cases where reciprocal agreements including these hard core clauses nonetheless can satisfy Article 81.

(d) Agreements between competitors give rise to both a broader range of potential competitive harms and greater overall potential for harm

8.14 It is intuitively obvious that agreements between competitors can give rise to greater competition concerns than agreements between firms that are not competitors. It is nonetheless important to recall *how* licences of intellectual property can lead to competitive harm. In Chapters 4 and 5 we have discussed in detail the theories of competitive harm under Article 81 and the way in which these theories apply to intellectual property licensing. There are three ways in which agreements between competitors can lead to competitive harm. The first is through coordinated effects taking the form of either direct agreements to increase prices or collusion that leads to increased prices. Aside from direct agreement on the prices to be charged for licensed products, agreements for setting royalties and agreements leading to *de facto* standards could lead to coordinated effects in appropriate circumstances. The second is by unilateral effects, where an agreement reduces or eliminates the competitive constraint previously exerted by one party on the other party. Agreements restricting territories or customers can have such effects. The third is by reducing dynamic efficiency, either by restricting the ability or incentive of third parties to bring innovations to market or by reducing the incentive of the parties to introduce future innovations. The Guidelines suggests that any cross-licence could have this effect, as could provisions governing grant-back of improvements, or royalty arrangements based on total sales.

8.15 As explained previously, with the exception of explicit price-fixing, the economic harm, if any, in an individual case depends on the nature of the markets and the position of the parties in those markets. The likelihood and impact of coordinated effects depends on the importance of the parties in the market and the importance of the licensed technology both as a cost element and as a definition of the downstream product's characteristics. The extent of unilateral effects depends on the market position of the firm benefiting from a restriction and the amount of competitive constraint that the restricted party would otherwise exert. The extent of effects on dynamic efficiency is most problematic, since the counterfactual—the likelihood of substantial innovation in the absence of the agreement—is almost impossible to prove.[6] While the competitive relationship between the parties makes serious negative effects possible, it does not make them certain or even likely.

The real danger associated with agreements between competitors is that competitors can and do use intellectual property licensing as a 'cover' for old-fashioned cartel activity—whether as a device for coordinating prices or as a basis for sharing customers or geographic markets. The best arrangement for achieving this goal would be a reciprocal licence where each competitor agreed to stay away from specified customers or territories. This is the real basis for the extended list of hard core restrictions, which covers the agreements most likely to be used as a pretext for anticompetitive activities. The possibility that a licence is a cover for collusion or market sharing may justify challenging any agreement where there is no obvious efficiency benefit. But as already noted, where there are clear efficiency benefits it remains necessary to consider the actual competitive harm. **8.16**

(e) 'Design freedom' and competitor licensing

The model for intellectual property licensing that underlies most competition law analysis views intellectual property development and licensing as a separate function from production and marketing of downstream products. In this model, a potential licensee identifies a technology that it wishes to exploit and seeks out the owner of the technology in order to obtain the rights to use that technology. Alternatively, a firm that has developed an innovation and is not itself in a position to exploit that innovation can seek out potential licensees to exploit the innovation and pay the licensor royalties in return. **8.17**

In reality, innovation and the protection of innovation through registration of intellectual property rights are integral parts of the development and production of products in many industries.[7] The development of technology that can be licensed is not the principal goal of this exercise and the value of intellectual property rights does not derive from their role as a means for facilitating technology transfer. Firms in these industries have, however, learned that the portfolio of intellectual property rights built up as a result of their development **8.18**

[6] The Guidelines evade this problem by describing innovation as a key parameter of competition, so that a reduction of the incentive to innovate constitutes a restriction of competition. This is a rare example of regression in the Guidelines to the formalistic approach of the old regime, which cannot be readily reconciled with an approach to Article 81 based on economic effects.

[7] Product development for consumer electronics or telecommunication equipment is a good example of this phenomenon.

activities can become a significant source of revenue. Often products already on the market may infringe these rights even though the firms in question have not consciously sought to exploit the technology covered by these rights. Indeed given the vast number of patents registered or applied for, a firm bringing a new product onto the market cannot be sure that its innovations do not infringe existing intellectual property rights. The proliferation of intellectual property rights and the ability of intellectual property rights holders to use their portfolios to tax innovation is sometimes referred to as the 'patent thicket'. The risks that the patent thicket creates for firms creating innovative products has been identified as a real threat to innovation.[8]

8.19 Many firms seek to escape from the patent thicket by obtaining licences in advance from firms that are likely to control relevant intellectual property rights. The goal of the licensee in this case is to clear a window for innovation—an area where it can develop its own technology free of the threat of infringement litigation. From the perspective of the licensee, restrictions on what it can do with the technology covered by the licensed rights outside that window are irrelevant. The licensee will therefore be willing to accept any restrictions in that regard in order to reduce the cost of the licence that it requires.

8.20 Freedom to innovate can be the basis for licensing between non-competitors. It can be particularly important, however, where firms are already competitors, especially in markets characterised by product differentiation. In these markets the most likely possessors of blocking intellectual property rights are existing competitors, yet potential licensors may not regard the proposed innovative products as the most direct substitutes for their own product line. In some cases, differentiated products may be both partial substitutes and complements for products produced by the licensee, creating a further incentive for the licensor to enter into a licence, while increasing the welfare benefits of the transaction.

(f) The interests of the licensor when licensing a competitor

8.21 When a firm licenses another firm to use its technology and the licensee is not a competitor, its interests are usually relatively straightforward. The licensor is interested in maximising royalty income by promoting the maximum use of the technology. Where the licensor is already in the market, the licensor must consider whether the additional revenue derived from licensing (including the benefits of complementary products produced by the licensor), outweighs the loss of sales due to new competition. This balancing problem becomes acute where the licensee is an existing competitor. An existing competitor is more

[8] See e.g. Shapiro, C, 'Navigating the Patent Thicket: Cross-licences, Patent Pools, and Standard-setting, I', *Innovation Policy and the Economy* 119 (2000).

likely to make inroads in selling to the licensor's existing customers and may well be able to invent around the licensor's technology. In short, the potential costs to the licensor of licensing to a competitor are more likely to outweigh any incremental revenue, creating a hold-up problem that can block useful innovation, including licensing to create 'design freedom' as discussed above.

Resolution of this hold-up problem may require restrictions on the use of the licensed technology, including restrictions that fall within the TTBER hard core list. In particular, restrictions on field of use or customer restrictions may be required to make the licensor willing to release its rights. The licensee will often be willing to accept these restrictions if they coincide with use or target customer group for the licensee's product. Under these circumstances, far-ranging restrictions may still give rise to significant efficiency-enhancing benefits. **8.22**

(g) The competitive harm is more likely to outweigh the efficiency benefits in agreements among competitors

As shown in Chapter 7, in the absence of price-fixing or territorial divisions within the EU, for agreements between non-competitors the balance of harms and benefits will almost always lead to a finding that the agreement meets the requirements of Article 81. For agreements among competitors, in contrast, the balance can more easily go the other way. Even where it is possible to identify significant consumer benefits (as in resolving the hold-up problems discussed in the preceding paragraph), those benefits may not save the agreement where there is significant competitive harm. This means that the really hard cases for applying the new paradigm will almost always be agreements involving licensing between competitors. **8.23**

(h) Agreements between competitors are more likely to 'eliminate competition'

Where the parties to an agreement were not competitors prior to that agreement, it is intuitively unlikely that their agreement could lead to the elimination of competition (although in theory it could eliminate competition between their shared customers or suppliers). Where competitors enter into an agreement, in contrast, the potential for elimination of competition becomes greater. In the licensing context, this becomes a significant issue in the context of cross-licensing or pooling, as discussed in Chapter 9. Where otherwise beneficial agreements threaten to freeze the competitive structure of a market, the Commission has signalled that the parties to those agreements may need to make concessions, such as an agreement to license third parties, in order to ensure that competition is not 'eliminated'. **8.24**

D. Analysis of Specific Restraints

8.25 In Chapter 7 we discussed the application of Article 81 to specific contractual obligations commonly found in licensing agreements. In this section we consider how Article 81 applies to each of those clauses where the parties *were* competitors prior to the agreement. The focus in this discussion is on the differences between 'non-competitor' licensing and competitor licensing.

(1) When is a licence agreement without restrictive clauses restrictive of competition?

8.26 One specific type of 'restriction' that can arise in an agreement between competitors that does not occur in agreements between non-competitors is the 'restrictive effect' of an agreement with no restrictive provisions.[9] These effects can arise in two ways: (i) by reducing incentives for future innovation; and (ii) by increasing the likelihood of successful collusion. The possibility for such effects depends on the nature of the pre-agreement competitive relationship between the parties. It cannot be assumed, however, that an open licence agreement between competitors is likely to lead to competitive harm, even where the parties have substantial market power.[10]

8.27 Where the parties to the agreement were previously competitors only in the technology market, a royalty-free agreement that allows each party access to the competing technology without limitations on customers, fields of use, territories, or sub-licensing could still lead to competitive harm in the form of a reduction in future innovation. In theory, since as a result of the agreement both parties become free to market both technologies, the incentive that each party has to invest in better technology is reduced, since it will not bring a competitive advantage in relation to the other party and the other party will share in any benefit. In practice, however, like any theory involving a reduction of dynamic efficiency, this kind of competitive harm requires careful assessment of the commercial factors that drive innovation in the affected markets. First, competitive harm is only likely if the parties to the agreement are real competitors in the technology market; their respective technology packages must be sufficient to meet the same identifiable need on the part of licensees (or end consumers of the products produced using the technology). Secondly,

[9] For these purposes the provisions discussed in the subsequent sections of this chapter are referred to as 'restrictions' even though some may not constitute restrictions in a formal sense (e.g. field of use restrictions that can be categorised as a form of limited licence).

[10] This is particularly important from an administrative perspective. Neither the Commission nor a national authority should be able to make a finding that the parties to an agreement were competitors in a strong market position and use this finding as a basis to shift the burdens to the parties to prove efficiency benefits and ultimately to impose obligations on the parties, particularly in respect of licensing third parties.

competitive harm is only likely if the parties together have genuine market power (remembering that high combined market shares are not necessarily an indicator of market power). Thirdly, the technology needs to be susceptible to significant improvement; where technologies are mature, the loss in dynamic efficiency is unlikely to be substantial. Taken together these factors make competitive harm unlikely where the competitive relationship was restricted to the technology market.[11]

Where the parties were also competitors in the downstream product market prior to the agreement, there may also be an impact on dynamic efficiency, by reducing the scope for competition with innovative products. The factors cited above are equally applicable here, with the caveat that in the product market, the combined market shares of the parties will be relevant to considering market power. Competitive harm will only be likely in the downstream market, however, in markets where differentiation by innovation is an important parameter of competition. **8.28**

A licence between competitors may also lead to competitive harm in the downstream product market by facilitating collusion. If the result of licensing (whether reciprocal or unilateral) is to make products more uniform, it will be easier for the parties to collude. It should be recalled, however, that a single factor that facilitates collusion does not by itself create a likelihood of competitive harm, unless the other market characteristics that make collusion likely are present. In practice, licensing without restrictive provisions or royalties is likely only to feature in coordinated effects cases as part of a broader pattern of relationships among competitors that foster collusion. **8.29**

(2) Retail pricing

In agreements between non-competitors, it is generally permissible for licensors to impose requirements regarding maximum selling prices or recommended selling prices. Such provisions are effectively barred in agreements between competitors. This reflects the potential role that reciprocal agreements between competitors on pricing could have in facilitating collusion. In effect, such arrangements between competitors constitute a form of horizontal price-fixing.[12] **8.30**

(3) Royalties

As explained in Chapter 7, the TTBER and Guidelines accord the parties to an licence agreement substantial discretion regarding the calculation and payment **8.31**

[11] This analysis applies whether the relationship is unilateral or reciprocal.
[12] Compare TTBER Article 4(1)(a) and Article 4(2)(a).

of royalties. This discretion is reduced, however, where the licence agreement is between competitors in the downstream market. This reflects concerns that royalty arrangements between competitors could:

(a) constitute price-fixing;
(b) facilitate collusion; or
(c) create disincentives for the use by the licensee of its own technology.

(a) Price-fixing—royalty payment in 'sham' licences

8.32 The concern regarding price-fixing arises from the fact that where royalties are payable, the sales price will always be higher than the royalty payment. Where each party to an agreement has a royalty obligation to the other, they thus create a floor for retail prices.[13] This concern leads the Guidelines to treat reciprocal royalty obligations as a kind of mild restriction by object. Where such obligations are present, the burden shifts to the parties to identify some efficiency benefits associated with the arrangements. As soon as the parties can show *some* benefits, however, the reciprocal obligations will meet the requirements of Article 81(3), unless the nature of the obligations is such as to facilitate collusion.[14]

(b) Collusion—disproportionate royalties

8.33 The concern about collusion has the same basis as the concern about direct price-fixing. Where the principal competitors in the downstream market (or producers of the closest substitutes) are subject to reciprocal royalty obligations, they can manipulate the level of those royalties to raise the price floor in the downstream market. This implies that the royalty obligations exceed a normal commercial price (because otherwise the final price would not be above the 'competitive level'). It also implies that the parties are not subject to effective competition in the downstream product market (because otherwise the conditions for coordinated effects would not be met). On this basis the Guidelines suggest that reciprocal running royalty arrangements will be regarded as causing serious competitive harm where they are 'clearly disproportionate compared to the market value of the licence and . . . have a significant impact on market prices'.[15]

[13] See TT Guidelines para 80 (explaining TTBER Article 4(1)(a)). The Guidelines suggest that this concern could also apply in the context of a unilateral licence between competitors coupled with an obligation on the licensor to purchase products from the licensee.

[14] Guidelines para 157.

[15] Guidelines para 158. The Guidelines suggest that royalty levels paid by other licensees in the same product market (for the same or competing technologies) may be relevant to assessing whether royalties are disproportionate.

(c) Price fixing and limitation on licensee technology—royalties based on units sold

A further area of distinction between competitor and non-competitor agreements involves royalty calculation that is based on the total number of units sold, irrespective of whether the products include the licensed technology. As explained in Chapter 7, the TTBER (as interpreted in the Guidelines) allows royalties based on total sales. Above the TTBER thresholds, the Guidelines take a generally favourable view of such provisions, subject to assessment of possible foreclosure effects.[16] For agreements between competitors in contrast, the Guidelines treat royalty calculation based on total units sold as a hard core restriction. Royalty calculation based on all units sold is regarded as both a form of price-fixing within the meaning of TTBER Article 4(1)(a)[17] and a restriction of the licensee's ability to use its own technology under TTBER Article 4(1)(d).[18] The Guidelines acknowledge that such obligations may 'exceptionally' fulfil the requirements of Article 81(3) 'where on the basis of objective factors it can be concluded that the restriction is indispensable for pro-competitive licensing to occur'.[19] **8.34**

From an economic perspective, the classification of royalty obligations based on all units sold as a form of price-fixing seems somewhat harsh.[20] Whether such royalty terms could actually affect the final retail price will depend on whether the requirements for successful collusion are satisfied, e.g. whether firms not subject to the restriction impose effective competitive constraints on the parties. It is also not hard to envisage circumstances where this obligation would not limit utilisation by the licensee of its own technology. This could be the case in particular where the licensee obtains a licence in order to have design freedom. **8.35**

(4) Territorial restrictions

(a) Territorial restrictions in unilateral licences between competitors

The TTBER treats territorial restrictions in non-reciprocal licences between competitors in the same way that it treats licences between non-competitors. Above the TTBER thresholds, the position changes. As explained in Chapter 7, Article 81 will normally be inapplicable to exclusive or sole licences between **8.36**

[16] Guidelines para 160. [17] Guidelines para 81. [18] Guidelines para 95.

[19] Guidelines para 81. The Guidelines give the example of the case where it would be impossible to monitor whether in fact the licensee had produced the products using the licensed technology.

[20] The treatment of this restriction as 'hard core' may have been intended to avoid an overt clash with the ECJ ruling in Case 193/83 *Windsurfing* v *Commission* [1986] ECR 611, where similar clauses were ruled to constitute restrictions of competition subject to Article 81(1). See TT Guidelines para 81, n 42.

parties that were not competitors prior to the agreement.[21] Likewise sales restrictions (on active sales and even on passive sales) are accorded relatively relaxed treatment. Where the parties are competitors on the downstream market, however, the scope for both exclusive licences and exclusive sales territories is more limited.[22]

(i) Exclusive licences—competitive harms

8.37 As regards exclusive licences, the TT Guidelines focus on the competitive significance of the licensor, who will effectively either refrain from production in the licensed territory or (in the case of a worldwide exclusive licence) exit the market altogether. The Guidelines suggest that if 'the licensor has a limited market position on the product market or lacks the capacity to effectively exploit the technology in the licensed territory, the agreement is unlikely to be caught by Article 81(1)'.[23] While this statement is undoubtedly correct as far as it goes, the real focus should be on the position of the licensee—the beneficiary of the arrangement—because any competitive harm resulting from the agreement depends on unilateral effects benefiting that licensee. While the competitive constraint imposed by the licensor prior to the licence is significant, the other competitive constraints on the licensee (including the possibility of arbitrage) may be more important in determining whether the licensee will benefit from a significant increase in market power as a result of the agreement.

(ii) Sales restrictions—competitive harms

8.38 As regards sales restrictions, the logical approach is to focus on the market position of the party benefiting from the restriction, i.e. the firm in the territory where active or passive sales are restricted. Where the beneficiary of the restriction has genuine market power, then a restriction on sales into that territory by a firm that previously did sell competing products in that territory is likely to lead to unilateral effects.[24] Thus if the licensee is restricted from active sales in the territory of the licensor, the market position of the licensor would be decisive,

---

[21] As explained in Chapter 7, a sole licence involved agreement by the licensor not to license a third party to exploit the technology in the territory. An exclusive licence involves agreement by the licensor not to exploit the technology in the territory itself. Within the EU, the rules on free movement mean that the exclusive right under the relevant IP does not provide rights to prevent sale of products produced using the technology in the EU outside the territory. These sales may be the subject of separate contractual sales restrictions.

[22] Where the licensor is not an actual or potential competitor on the downstream market, the Guidelines suggest that Article 81 is inapplicable. See Guidelines para 164.

[23] Guidelines at para 164.

[24] It will be recalled that this assessment is ongoing. If, at a future date, the beneficiary has developed substantial market power, restrictions on sales by the other party may give rise to competitive harm. This may in some circumstances change the Article 81(3) analysis, where the benefits to dynamic efficiency may in the end be outweighed by the competitive harm of continuing the restrictions.

while a restriction on sales by the licensor would require assessment of the market position of the licensee.[25] This is very close to the position taken in the TT Guidelines. The Guidelines confuse the position somewhat, however, by suggesting that Article 81 becomes applicable once *one* of the parties has market power, irrespective of whether that party is a beneficiary of the restriction. The Guidelines then suggest that the lack of market power on the part of a beneficiary makes *Article 81(3)* applicable.[26]

Agreements where a licensee agrees not to sell in a territory allocated to a third party raise different issues. There could be a risk that a licensor could use such agreements to craft a market-sharing cartel among competitors in the market. The TTBER allows restrictions on *passive* sales in these circumstances, provided that the beneficiary was not a competitor at the time the agreement was made.[27] This difference in treatment may make sense for defining the scope of the TTBER, since the need to assess the market position of the licensor, the licensee, the third party, and of other parties subject to similar restrictions probably makes a bright line rule approving such arrangements inappropriate. Treating a restriction on sales into a third party territory as a hard core restriction, however, appears equally inappropriate. Where a third party is the beneficiary of the protection, then the focus should be on the market position of that third party. Even if the third party was a competitor of the licensor and licensee prior to the agreements, economic harm is unlikely where the third party does not possess significant market power and where the licensor and other licensees are not the principal constraints on exercise of market power by that third party. **8.39**

(iii) Exclusive licence and sale restrictions—welfare benefits

As in the case of non-competitor licensing discussed in Chapter 7, the primary contribution of territorial restrictions to consumer welfare is to overcome hold-up problems—the reluctance of a licensor to grant a licence at all or the reluctance of a licensee to make the investments necessary to exploit the licensed technology in the absence of territorial protection. The Guidelines suggest, however, that such hold-up problems become less likely where the **8.40**

[25] Guidelines para 170.

[26] Ibid. This may reflect a carry-over from the old Article 81(1) analysis where territorial restrictions were regarded as restrictive by object as long as they were 'appreciable' See e.g. Case 19/77 *Miller* v *Commission* [1978] ECR 131 at para 7. (In this connexion it must be held that, by its very nature, a clause prohibiting exports constitutes a restriction on competition . . . since the agreed purpose of the parties is the endeavour to isolate a part of the market.') Under the old approach, this meant that the analysis shifted to the relevant block exemptions which allowed restrictions on active sales and, to a lesser extent, restrictions on passive sales.

[27] Where the protected party is an existing competitor, the clause is treated as a hard core restriction. See Guidelines para 171.

protected party has significant market power.[28] Where the protected party enjoys a strong market position, the Guidelines imply that the indispensability requirement of Article 81(3) would not be met.[29] Furthermore the Guidelines focus on active sales, implying that any restriction of passive sales would not meet the requirements of Article 81(3) where the TTBER thresholds are exceeded.[30]

(iv) Exclusive licence and sales restrictions—balancing benefits and harms

8.41 The focus of the TT Guidelines on indispensability considerations obscures the biggest problem that arises in applying Article 81 to territorial restrictions where competitors are involved. In some cases the competitive harm that results from the reduction of competition among existing competitors will outweigh the economic benefits of the licence agreement, even where those territorial restrictions are indispensable for achieving the benefits of the agreement. A territorial restriction between competitors *can* give rise to significant competitive harm, where the beneficiary of the restriction has a strong market position and the parties accepting the restriction would otherwise constitute a significant competitive constraint on the beneficiary.

8.42 Whether the agreement is, on balance, beneficial under these circumstances depends on the benefits of disseminating the technology, which depends in turn on the extent to which the technology represents an advance on the technology otherwise available to the potential licensee. Where the technology represents only an incremental improvement, the welfare benefits attributable to the agreement may not justify agreements that include clauses leading to significant anticompetitive effects. Where the technology represents a significant advance, in contrast, restrictions that lead to some unilateral effects may be acceptable, if those restrictions are truly indispensable to rapid dissemination of the technology. Of course, where the technology in question is a genuine 'breakthrough', the parties to the agreement are not 'competitors' and the problem does not arise.

---

[28] Guidelines para 170. The Guidelines do not discuss this point in detail, but the implication appears to be that a licensor with a strong market position will be less concerned about competitive pressure from the licensee and therefore less determined to limit or prevent such competition. As regards licensees, the argument would be that a strong licensee can 'hold its own' against competition from the licensor and will still be in a position to make the necessary investments. These assessments may be correct in specific cases, but as the discussion of hold-up issues in Chapter 4 suggests, there is no real economic basis for this assertion.

[29] The position taken in para 170 of the Guidelines referred to in the main text can usefully be contrasted with the position taken in para 173 regarding the same restrictions on non-competitors. Paragraph 173 explains the hold-up problem that would lead a licensor in an agreement with a new entrant to accept territorial restrictions in order to induce the licensee to invest in exploiting the licensor's technology.

[30] Article 4(1)(c)(iv) TTBER allows restrictions on both active and passive sales that benefit either the licensor or the licensee.

(v) Exclusive licence and sales restrictions outside the EU

The TTBER and the TT Guidelines do not directly differentiate between restrictions applicable within the EU and restrictions applicable to territories elsewhere in the world.[31] The implication appears to be that the assessment remains unchanged as long as the jurisdictional requirements of Article 81 are met, i.e. as long as the agreement affects trade between Member States. Since licence arrangements that treat the EU or an area including the EU as a territory are not unusual, the application of Article 81 to such agreements deserves more attention. **8.43**

As a preliminary matter it is important to recall that the rules on free movement do not apply to products introduced to sale outside EU until those products have been lawfully resold within the EU. This means that the holder of an exclusive licence within the EU may lawfully exercise intellectual property rights to exclude the import of products produced by licensees outside the EU. Thus to the extent that intellectual property rights in question confer such powers, contractual sales restrictions on a licensee or licensor outside the EU may be superfluous.[32] Following the approach set out in paragraph 164 of the Guidelines, where a licensor previously had a limited presence in the EU, neither the exclusive licence nor sales restrictions on the licensor should be regarded as problematic under Article 81. **8.44**

For other cases, the first issue to consider is definition of the geographic market. Where the EU or parts of the EU are separate geographic markets, the parties to the agreement may not have been actual or potential competitors prior to the agreement, even if they were active in respect to the same products in markets outside the EU. Thus even if the TTBER and Guidelines are fully applicable, the less restrictive rules for pure vertical relationships will be applicable. **8.45**

Where the geographical market is worldwide or where the parties were competitors in the EU prior to the agreement, the analysis should logically depend on where the beneficiary of the restriction is active. Where a licensor or licensee in the EU is the beneficiary of the restriction, then the analysis should in principle be the same as for any other agreement between competitors, although the competitive constraint imposed by firms outside the EU cannot be assumed and **8.46**

[31] The Guidelines at para 164 refer to worldwide exclusive agreements in contrast to agreements limited to a particular territory, using an EU Member State as an example of a more limited territory.

[32] This creates a potential difference between technology licensed on the basis of patents or copyright, which confer such protection, and technology licensed in the form of know-how, which does not.

therefore requires careful assessment.[33] Where the beneficiary of the restriction is operating outside the EU, it is questionable whether Article 81 should be applicable at all. However, if the restriction on sales in a territory outside the EU is accompanied by 'unilateral' withdrawal of the beneficiary from the EU markets, the agreement could be treated as an implicit creation of an exclusive territory with the EU and assessed on that basis.

(b) Territorial restrictions—reciprocal agreements

8.47 The application of the TTBER and TT Guidelines to reciprocal agreements is relatively straightforward. Reciprocal exclusive licences (where the licensor withdraws from the protected territory) are regarded as hard core restrictions. Reciprocal sole licences (where each licensor agrees not to license third parties) are allowed under the TTBER, but may be restrictive where the parties have a 'significant degree of market power'.[34] Restrictions on either active or passive sales are regarded as hard core restrictions that are 'unlikely to fulfil the conditions for application of Article 81(3)'.[35]

8.48 As noted earlier in this chapter, the general concern regarding reciprocal agreements is justified where it involves withdrawal from territories where the licensor and licensee previously operated.[36] Such agreements clearly can serve as a tool for dividing markets and lead to significant unilateral effects. Furthermore, the acceptance of the reciprocal licence suggests that neither technology is clearly superior to the other. This suggests in turn that a reciprocal licence is less likely to involve the diffusion of new technology and therefore less likely to lead to efficiency gains. Nonetheless the blanket condemnation of these arrangements in the Guidelines may not be justified.

8.49 The principal context in which reciprocal restrictions may be justifiable involves the case where the products sold by the parties are only partial substitutes. In these circumstances the products produced by the parties (and the underlying

[33] The Commission Guidelines on the Effective Trade Concept (at para. 109), following the ECJ judgment in Case C-306/96, *Javico International* v *Yves Saint Laurent Parfums* [1998] ECR I-1983, dictate that sales restrictions imposed on resellers outside the Community will only be caught by Article 81 where significant price differences would make imports into the Community likely.

[34] Guidelines at para 163. The basis for this assertion is that such arrangements 'may facilitate collusion by ensuring that the parties are the only source of output based on the licensed technologies'. Whether reciprocal sole licensing will in fact facilitate collusion on this basis will depend, of course, on the extent to which competitors using rival technologies impose a limit on the ability of the parties to raise prices or restrict output.

[35] TTBER Article 4(1)(c). Guidelines para 169.

[36] These concerns are less clear where the prior competitive relationship was only on the technology market. There is a strong argument that such arrangements should be dealt with as agreements between non-competitors, since the relevant source of competitive harm is the reduction in previously existing competition on the downstream product market.

technology) may be both competing products and complementary products at the same time. Since a properly defined market can still include products that only compete with each other to a limited extent,[37] there will sometimes be cases where providing competing products as a package or designing them so that they work better together can lead to genuine efficiency gains.[38] There can, however, be hold-up problems associated with this relationship—neither party may be willing to create the package if it will then be sold on a competing basis. While the parties could evade this problem by creating a joint venture to market the package in each territory, there is no obvious policy reason why they should not be permitted to reach the same result through their licence arrangements.[39] This type of arrangement may be particularly important where the entire EU forms part of a single territory. Under these circumstances there is no 'partitioning effect' in the EU. There is no difference in the competitive harm within the EU for a reciprocal arrangement compared with a non-reciprocal arrangement.

(5) Output restrictions

As explained in Chapter 7, 'output restrictions' describe any clause that has the effect of limiting the number of units that the licensor may produce with the licensed technology. These include agreements explicitly limiting production, as well as royalty arrangements where royalties increase with volume,[40] or the obligation to make other payments if output targets are exceeded.[41] TTBER Article 4(1)(b) treats 'reciprocal output restrictions' as a hard core restriction. A reciprocal output restriction is *not* an output restriction in a reciprocal licence agreement. It is rather a restriction on a licensee that mirrors a similar obligation on the licensor in respect of the licensee's technology. In the Commission's view, 'the object and likely effect of the agreement is to reduce output in the market'.[42] **8.50**

---

[37] Consider in this regard the competitive relationship between the various suppliers of server software in the *Microsoft* case (Commission Decision of 18 March 2004, not yet officially reported).

[38] Another situation where this type of benefit could arise is the case where the parties license each other on packages of technologies used for a variety of related product markets. If there are 10 product markets involved and only one involves an overlap between the activities of the parties, then the reciprocal restrictions in the tenth market may resolve hold-up problems affecting the other nine.

[39] Such an arrangement would still need to meet the requirements of Article 81, including the balance of harms and benefits discussed in the previous section.

[40] Guidelines para 158. [41] Guidelines para 82.

[42] Ibid. In effect, since a reduction of output by the market as a whole is a form of welfare loss, an agreement to reduce total output is a restriction by object. This conclusion may not be problematic where there is a genuine reciprocal agreement to reduce output (although such agreements may not have an effect on output where the parties can increase production using their own technology). A greater methodological problem is posed by the assumption that agreements providing for increases in royalties or other payments are equivalent to quantity limitations. A licensor may be willing to grant the licensee a lower royalty for the initial tranche of units produced, to help the licensee cover its fixed costs of production. Where the licensee's unit costs

8.51 Non-reciprocal quantity restrictions, in contrast, benefit from relatively favourable treatment even where they are imposed on one party to a reciprocal agreement. The Guidelines suggest that such restrictions may be caught by Article 81(1) where the parties have a 'significant' degree of market power. The Guidelines further suggest, however, two scenarios where output restrictions may lead to significant efficiency benefits. The first arises where (i) the licensor's technology is substantially better than that controlled by the licensee; and (ii) the output restriction is higher than the output of the licensee before the agreement. In this case there is no real restriction of output and a definite welfare benefit. The second involves a licence that 'leads to a real integration of complementary assets'. Where this is the case, the Guidelines suggest that quantity limitations may be justified to overcome hold-up issues on the part of the licensor. These benefits will not save the arrangements, however, if the parties have 'substantial market power'.[43]

(6) Customer restrictions

8.52 As explained in Chapter 7, an agreement by which a licensor agrees not to sell (or not to make active sales) to a group of customers reserved to the licensee can lead to competitive harm where the licensee benefits from unilateral effects; freedom from competition by the licensor could allow the licensee to raise prices for the relevant consumer group. Similarly, a restriction on sales by the licensee to customers reserved to the licensor may allow the licensor to raise prices. These effects are a particular concern where the parties were previously competitors in respect of sales to that customer group. For this analysis to apply, the reserved customer group must constitute part of a broader product market. Where a licensor is barred from using technology in one or more product markets—this limitation is treated under the TT Guidelines as a field of use restriction—not a customer restriction.[44]

---

decrease with increased volume (as is usually the case), the licensor may legitimately seek to share in the increased profit. In principle, this royalty structure should be viewed as output-enhancing, since it facilitates initial investment by the licensee. It will only function as an output limitation where the increase makes increased sales by the licensee unprofitable.

[43] Ibid. The analysis of the second Article 81(3) scenario in the Guidelines illustrates two important points regarding the application of the 81(3) paradigm. The use of quantity restrictions to resolve hold-up problems shows how restrictions that would be ancillary (and therefore non-restrictive) in an agreement among non-competitors can require assessment under Article 81(3) in an agreement between competitors. The difference in treatment between agreements where the parties have significant market power (where the benefits may outweigh the harm of the quantity limitation) and agreements where the parties have 'substantial market power' (where the harm outweighs the benefit) illustrates the way that the balancing test in Article 81(3) is meant to function.

[44] Guidelines para 179.

The TTBER and the Guidelines treat customer restrictions in non-reciprocal agreements between competitors in the same way as they treat territorial restrictions. Below the TTBER threshold, parties to a licence agreement may bar sales by the licensee to a customer group reserved for the licensor, and may bar sales by the licensor to a customer group reserved to the licensee. An agreement may also bar active sales by a licensee to customers in a group reserved to another licensee. Outside the 20% market share threshold of the TTBER, the restrictive effects of customer allocation among competitors are regarded as likely to outweigh the benefits where the beneficiary of the restriction has a 'significant degree of market power'.[45] **8.53**

In assessing the application of Article 81 to customer restrictions, market definition plays a key role. If a group of customers constitute a separate market (because of the ability of sellers to discriminate) then the likelihood that the beneficiary of a customer restriction will have significant market power will increase. On the other hand, if there is a separate market for the customer group, the likelihood increases that the parties to the agreement were not actual or potential competitors in that market.[46] **8.54**

The TTBER and Guidelines treat customer restrictions in reciprocal agreements as hard core restrictions that are unlikely to satisfy Article 81(3). As was the case with territorial restrictions, this classification appears harsh. Firms may be able to use complementary technologies to specialise in meeting the specific needs of different consumer groups. In the absence of significant market power, the benefits of these differentiated products could exceed any likely unilateral effects.[47] **8.55**

(7) Field of use restrictions

As explained in Chapter 7, the TT Guidelines define a field of use restriction as a provision limiting a licence to one or more technical fields of application or one or more product markets. In effect, by limiting the scope of the licence the **8.56**

[45] TT Guidelines para 170. See discussion of competitive harm in territorial sales restrictions at paras 8.37–8.39 above. In the Vertical Guidelines the Commission indicated that distribution agreements including customer restrictions were unlikely to satisfy the requirements of Article 81(3) where the beneficiary market share exceeded the 30% threshold for application of the Vertical Agreement Block Exemption.

[46] Where the customer group forms part of a broader market, the possibility of passive sales may constitute a greater constraint on the beneficiary of the restriction than is usually the case where territorial restrictions are involved. This results from the fact that passive sales to a customer group do not necessarily involve additional transport costs or the other transaction costs associated with import into a reserved territory.

[47] Where territorial restraints were concerned, the hard core approach could in part be justified on the basis of the emphasis in EU law on territorial integration. There is no similar policy imperative applicable to customer integration.

licensor protects itself from competition in respect of the fields of use that have not been licensed. Where the licensee is an existing competitor, adoption of the new technology could make continued use of the prior technology in the fields reserved to the licensor commercially unattractive.[48] On this basis using the new technology pursuant to the field of use restriction may have the effect of limiting the competitive constraint asserted by the licensee in the reserved fields of use. Thus field of use limitations in agreements between competitors can function as *de facto* customer restrictions and have the same effects.[49]

8.57 There are, however, two important differences between agreements that expressly allocate customers within a market and limitations on fields of use. First, associating competitive harm with a field of use limitation requires a more complex analysis. Secondly, the benefits associated with field of use limitations are arguably more extensive than those associated with customer restrictions that are not technically based.[50] It is noteworthy that field of use restrictions do not feature in the list of hard core restrictions in the TTBER. This reflects in part the traditionally favourable treatment of field of use limitations.[51] It reflects even more, however, the fact that the Commission was persuaded that even reciprocal field of use restrictions do not necessarily involve significant competitive harm and often give rise to substantial economic benefits.

(a) Competitive harm in field of use limitations

8.58 A non-reciprocal field of use limitation in an agreement between firms that were previously competitors in a downstream market[52] can lead to competitive

---

[48] As explained in Chapter 7, there is an argument that a field of use 'restriction' in an agreement between non-competitors is not a 'restriction' for purposes of Article 81 at all, but is rather a limited licence. Where the limited licence is part of an agreement between competitors, however, it is the potential change in the competitive relationship that constitutes the 'restriction' and not the limitation on the licensee's autonomy.

[49] See Guidelines para 183. The language of the Guidelines is somewhat confusing in that they appear to suggest that the competitive harm is a restriction of output by the licensee. A field of use restriction cannot readily be categorised as an output restriction (and thus presumptively restrictive by object) since in general the function of a field of use restriction is to increase output. Even if a restriction has the effect of restricting output by the licensee outside the field of use there is likely to be an increase in output within the licensed field of use.

[50] The Guidelines state that a field of use must be defined by reference to 'identified and meaningful technical characteristics of the licensed product'. Where this is the case, 'the fact that a technical field of use restriction may correspond to certain groups of customers within a product market does not imply that the restraint is to be classified as a customer restiction', para 180.

[51] As explained in Chapter 7, field of use restrictions have traditionally been viewed as a form of limited licence that is not by itself 'restrictive'.

[52] Although the Guidelines refer without distinction to 'competitors', the competitive harm associated with field of use limitations only arises where the limitation affects competition in the downstream market.

harm where the effect of the agreement is to reduce the competitive constraint previously exerted on the licensor. The fact that the licensee reduces or even eliminates production in the non-licensed field or fields of use should not, however, be sufficient to show competitive harm. The relevant question is whether removing the competitive constraint previously provided by the licensee creates or strengthens market power on the part of the licensor in the market or markets covered by the non-licensed fields of use.

The Guidelines refer to two situations where field of use restrictions are unlikely to be restrictive of competition for purposes of Article 81(1), even in the case of reciprocal agreements between competitors.[53] The first example provided by the Guidelines of an agreement normally falling outside Article 81 involves a 'symmetrical' field of use restriction. This describes the case where the parties license each other in the *same* field of use. The Guidelines suggest that this kind of arrangement is 'unlikely to restrict competition that existed in the absence of the agreement'. This is presumably based on the idea that the competitive constraint exerted on the parties within the licensed field of use is unchanged (in contrast to the position in respect of asymmetrical restrictions described below). **8.59**

The second example of a field of use provision falling outside Article 81 according to the Guidelines involves the situation where one party wishes to gain design freedom through a licence. In order to limit the royalties payable (or induce the licensor to grant the licence at all), the licensee will typically seek a licence covering the specific application that it intends to develop. The Guidelines note that the continuing risk of infringement action in respect of the unlicensed fields of use does not represent a change from the position without the licence. If in return the licensor demands a licence from the licensee (which is not uncommon) and that licence is also limited to specific fields of use, the second licence needs to be assessed as a unilateral licence agreement. **8.60**

The most problematic form of agreement identified in the Guidelines involves an 'asymmetrical' field of use restriction between competitors—a reciprocal agreement by which the parties license each other for *different* fields of use. The impact of this kind of agreement depends on the nature of the prior competitive relationship and the relationship of the various licensed fields of use. Where the field of use reserved to either licensor covers the use of the technology in a specific product market where the other party was previously a competitor, unilateral effects may be possible in that market. This analysis would need to be made for each such market. The reciprocal nature of the arrangement does not affect the likelihood of competitive harm, but may call for more careful **8.61**

[53] Guidelines para 183.

consideration of alleged efficiencies. Where the fields of use cover different technical applications within the same market, the analysis is essentially the same as for customer restrictions (provided that the field of use restrictions in each case are likely to lead a licensee to reduce or eliminate sales of products incorporation the reserved technical applications). In making this assessment the key issue is not whether the beneficiary has or will obtain market power, but the extent to which removing the previous constraint will change that position.

(b) The benefits of field of use restrictions

8.62 The TT Guidelines recognise three benefits attributable to field of use limitations. The first is to resolve a hold-up problem. The Guidelines observe in this regard that 'if the licensor could not prevent licensees from operating in fields where he exploits the technology himself or in fields where the value of the technology is not well established, it would be likely to create a disincentive for the licensor to license'.[54] This hold-up problem may be different from that dealt with by territorial or customer restraints because of the role of 'fields where the value of the technology itself is not established'. Where the licensor is developing its technology for other uses, but its development efforts are not yet complete, the role of the field of use restriction may be to protect the licensor's development efforts. If the licensor cannot protect those efforts, it may forgo further investment in that area. In other words, the licensor should not need to choose between allowing broader use of the technology (a consumer benefit) and investing in its own exploitation in secondary fields of use (also a consumer benefit).

8.63 The second benefit is that if the licensor cannot protect the reserved fields of use 'it would lead him to charge a higher royalty'.[55] Defining the field of use to cover the actual needs of the licensee allows the parties to calculate a royalty based on the revenue expectations for that business, which may either make entry by the licensee feasible or allow it to expand its own output. This reflects the more general point that the ability to discriminate between customers in different markets can be output-enhancing even where the seller has market power.

8.64 The third benefit identified in the Guidelines refers to the role of field of use limitations in facilitating design freedom. Since the Guidelines have already taken the view that field of use restrictions in this context are not caught by Article 81, it is not clear where this benefit may be relevant.

---

[54] Guidelines at para 182. [55] Ibid.

(c) Balancing competitive harm and consumer benefits

The TT Guidelines do not address the appropriate balance between harms and benefits in assessing field of use restrictions. It is evident, however, that consumer harm may not be substantial even where the licensor has substantial market power. It is also evident that the benefits of a field of use limitation can persist even where the licensor has a market power and the licensee was previously a competitor or potential competitor. This means that even cases of asymmetrical reciprocal restrictions should not be automatically prohibited under Article 81. The difficult issues could arise if a licensor has substantial market power in a market and the licensee was also a major player in that market. If the licensee withdraws from the market or terminates relationships with important customers, the licence arrangements may come under scrutiny. If parties in an established market position withdraw from related markets or customer groups at the same time, scrutiny will be more intense. In these circumstances it will be necessary to identify and document with some precision the hold-up problems that justify the arrangements. If effective market partitioning is the result of the arrangements (as could be the case if the two major players in a market agreed to reciprocal licences on these terms) then even a strong showing of efficiency benefits may not suffice to avoid condemnation under Article 81. **8.65**

(8) Captive use restrictions

Where a licensee obtains the right to use the licensed technology only for production of the licensee's own products or maintenance and repair of its own products, such a licence limitation is described as a captive use restriction.[56] In addition to the competitive harms identified in respect of agreements between non-competitors, discussed in Chapter 7, the Guidelines identify a specific problem that arises only in connection with agreements between competitors. This involves a captive use provision in respect of manufactured components. Where the licensee previously was an actual or likely supplier of components to third parties, the licence may change that position if use of the new technology makes it impractical for the licensee to provide components to third parties in the future. By removing the competitive constraint exerted by the licensee, the agreement could lead to unilateral effects benefiting the licensor. As in the examples of unilateral effects previously considered in this chapter, the extent of these effects here depend on both the extent of the market power of the licensor and the extent to which the licensee was or could become a significant constraint on the licensor. **8.66**

[56] Guidelines para 186.

9

COOPERATIVE DEVELOPMENT AND EXPLOITATION OF TECHNOLOGY—STANDARD-SETTING, TECHNOLOGY POOLS AND JOINT RESEARCH AND DEVELOPMENT

A. Introduction

9.01 In this chapter we address certain issues that arise where firms that may be competitors or potential competitors in technology markets or downstream products collaborate in the development and licensing of technology. There is some overlap between this discussion and the topics discussed in Chapter 8, since licensing between competitors involves cooperative elements, particularly where licensing is reciprocal. Technology-pooling involves an important form of licensing activity. Joint research and development or standard-setting, in contrast, involve technology-related cooperation that may not involve exploitation of intellectual property rights at all, although intellectual property rights, including creation of technology pools, can play an important role in achieving the benefits of joint R&D or standardisation. We deal with these topics in roughly chronological order. Research and development comes first because it deals with the creation of technology. Standard-setting logically follows research and development, since R&D can generate new technology and thus the need for new standards. Patent-pools follows standard-setting because the formation of a pool is often the consequence of a standard-setting process.

B. Joint R&D and IP Licensing

9.02 The assessment of joint research and development activities is discussed in the Horizontal Guidelines[1] and is the subject of a block exemption regulation.[2] Joint research and development may generate technology benefiting from intellectual property rights. Such rights may be assigned to participants and/or licensed to participants and third parties. The purpose of this section is not to discuss the conditions for application of Article 81 to joint research and development in general, but rather to focus on the specific issues that arise regarding the licensing of intellectual property rights generated in the course of joint research and development.

(1) The general approach to R&D cooperation under Article 81

9.03 The block exemption and the Horizontal Guidelines take a generally permissive approach to joint R&D activities. Pure R&D agreements that do not involve joint exploitation of results are regarded as unlikely to cause competitive harm.[3] Likewise, agreements by which a research and development is subcontracted to a third party are unlikely to cause competitive harm.[4] The competitive harm of R&D cooperation only becomes a serious issue where the parties were previously competitors or potential competitors and their cooperation extends to 'joint exploitation'. The block exemption provides a safe harbour for such agreements (subject to certain blacklisted clauses) as long as the market share of the parties in affected markets does not exceed 25%. Above that level the Guidelines indicate that Article 81 may still be satisfied in a significant proportion of cases. Of particular importance for this discussion, the Guidelines suggest that treatment is likely to be favourable where joint exploitation is limited to licensing.

[1] Paras 39–77.

[2] Regulation 2658/2000 on the application of Article 81(3) to categories of research and development agreements ('R&D Block Exemption').

[3] Para 58. Article 81 may be applicable where competition in innovation is significantly reduced—where the number of 'innovation poles' was limited and the cooperation reduces incentives for independent research.

[4] Para 57.

(2) Licensing and R&D

The Guidelines and Block Exemption deal with three different licensing scenarios: **9.04**

(i) licensing to participants in the R&D exercise;
(ii) licensing where R&D involves improvements or refinements in existing products or technology; and
(iii) licensing where R&D cooperation leads to creation of an entirely new product or technology.

In the first scenario, the starting point for the Commission assessment is that all parties must have full access to the results of the cooperation. This means that all parties should obtain an unrestricted licence on any technology generated through the R&D process, whether the resulting intellectual property rights are held by participants or by a joint venture entity. Failure to allow full access to results is a condition for application of the block exemption.[5] The Guidelines indicate, however, that 'exclusive access rights' could meet the requirements of Article 81(3) 'in particular' where they 'are economically indispensable in view of the market, risks and scale of investment, required to exploit the results of the research and development'.[6] Thus the Guidelines identify a particular kind of 'hold-up' problem that could justify limiting and dividing the rights to exploit joint R&D.

9.05 The Guidelines do not elaborate on the nature of 'exclusive access rights'. These rights will normally, however, involve a commitment to limit intellectual property licences to third parties in respect of defined fields of use or geographic application. In addition to the need to identify a hold-up requirement making these rights indispensable, the grant of such rights must be subject to the limitations imposed more generally on territorial, field of use or customer restrictions in agreements between competitors, as discussed in Chapter 8.[7]

9.06 Where R&D efforts involve improvements or refinements of existing products or technology, the Guidelines indicate that there joint exploitation is unlikely to raise issues where it is conducted 'only by way of licensing'.[8] Thus agreements between the participants regarding licensing policy, including decisions to restrict the availability of licences to third parties, are not apparently regarded as

[5] The general principle stated in the text does not apply to academic bodies, research institutes, or specialised companies that provide R&D services and are not vertically integrated in downstream markets, where the block exemption allows restrictions on access to results.

[6] Para 67.

[7] Indeed it is not clear whether there is a principled policy basis for treating such restrictions differently at all because they are the result of joint R&D efforts.

[8] Para 64.

restrictive. Where R&D efforts involve development of entirely new products or technology, the Guidelines identify a risk of 'foreclosure from key technology' for third parties.[9] The Guidelines indicate that such risks do not arise where the parties are willing to license their technology to third parties. They do not, however, identify the circumstances where such an obligation would arise or whether there is a legal obligation to grant a licence under such circumstances.[10]

C. Standard-Setting

(1) Background

9.07 The efficiency benefits of standard-setting are clear. Standards facilitate use of common inputs.[11] Standards facilitate consumer choice by allowing product comparison.[12] Standards can reduce the transaction costs of common commercial transactions.[13] Of particular importance in the context of technology licensing, standards allow products from different producers to operate together. Standards for mobile phones, airline baggage tags, or DVDs allow consumers using products or services obtained from different providers to utilise the same facilities or equipment. While some standards are imposed by law, most modern standard-setting involves discussions and agreement among participants in the affected industries.[14] Even where the resulting standard has legal sanction, agreement among some at least of the affected competitors will usually be the basis for the standard.[15]

---

[9] Para 65.

[10] Arguably an obligation to grant a licence on 'FRAND' terms could arise under the same circumstances where participants in a standard-setting are required to grant such licences. See discussion below at 9.21. In effect the new technology resulting from the R&D exercise will be treated as a *de facto* standard,

[11] For example, standardised beer bottle sizes in some markets have allowed producers to simplify collection and re-use of bottles.

[12] For example, the traditional purity standards for precious metals (e.g. sterling silver) allow a purchaser to compare the quality of different items and of similar items from different sellers. Modern consumer standards, such as 'quality marks' or approval by an industry body serve a similar role.

[13] For example, the use of an industry standard like 'Brent crude' in the oil industry may both allow the parties to the contract to describe the product covered by the contract and simplify establishing the price for future orders.

[14] The Horizontal Guidelines para 159 define 'standardisation agreements' as agreements that 'have as their primary objective the definition of technical or quality requirements with which current or future products, production processes or methods may comply'. The accompanying footnote (n 47) indicates that the conclusions regarding standardisation in the Guidelines were meant to apply not only to formal standard-setting which is mandated or encouraged by EU law, but also to any agreements that led to standardisation, including bilateral agreements.

[15] The fact that a standard is required by a national authority and made legally binding does not mean that competition law is not applicable. The fact that an agreement among firms is made

The clear benefits of standard-setting have led to generally favourable treatment under Community law. Under the old regime, standard-setting agreements belonged to the limited class of agreements for which notification was not a necessary condition for subsequent application of Article 81(3). As a result, there were very few cases in which the standard-setting process was assessed under Article 81. The primary concern in most such cases was the potential barrier to trade between Member States resulting from standards set at a national level.[16] The Commission dealt with two cases in the 1970s where manufacturers attempted to use standard-setting and intellectual property rights together to create barriers to entry for rival manufacturers. In those cases, the Commission established the general principle that the combination of standards and cross-licences or patent pools was not restrictive for purposes of Article 81, as long as licences were available to competitors on 'fair, reasonable, and non-discriminatory terms'.[17] **9.08**

The interface of standardisation and intellectual property rights drew renewed interest in the 1990s. The US Department of Justice issued two important business review letters that focused on the relationship between standardisation efforts and licensing. In the EU, attempts by the European Telecommunications Standards Institute (ETSI) to alter its rules on intellectual property disclosure and licensing became the focus of a major Commission investigation in 1994–95. The general shift away from mandatory notification starting in 1999 has also has meant that standard-setting lost its former special status under Article 81. This provided the context for discussion of standard-setting in the 1999 Horizontal Guidelines. **9.09**

(2) Application of Article 81—the approach of the Horizontal Guidelines

The Horizontal Guidelines devote two and a half pages to a discussion of standardisation. This discussion signals a generally positive treatment of standardisation agreements, which is sensible from a policy perspective. The Guidelines fail, however, to provide a framework for resolving standard-setting issues outside the context of formal standard-setting bodies. The Guidelines also do not consider the relevance of the dual roles played by many participants in standard-setting as both providers and users of technology that forms the basis **9.10**

---

legally binding by administrative action does not mean that that agreement is not caught by Article 81, (see e.g. Case 123/83 *BNIC* v *Clair* [1985] ECR 391: government approval of agreement does not make Article 81 applicable), although subsequent compliance with that standard would not be unlawful. The government action itself may be subject to direct challenge by the Commission, *inter alia* under Article 86.

[16] See e.g. Case 96/82 *IAZ* v *Commission* [1983] ECR 3369.

[17] IGR, Fourteenth Report on Competition Policy, 1984, point 94; *IGR Stereo Television*, Eleventh Report on Competition Policy, 1981, point 92.

of a standard. In consequence, the Guidelines fail to address a key issue regarding the interface between intellectual property rights and standard-setting—the ability of standard-setting bodies to require disclosure of intellectual property rights and licensing terms before setting the terms for a standard.

(a) Standard-setting in recognised standard-setting bodies—a safe harbour

9.11 The Horizontal Guidelines indicate that standard-setting 'by the recognised standards bodies which are based on non-discriminatory, open and transparent procedures' will not 'normally' restrict competition for purposes of Article 81.[18] This 'quasi-safe harbour' is not limited to standard-setting in recognised industry bodies. In principle, any standard-setting exercise that is open and transparent and is either 'part of a wider agreement to ensure compatibility of products' or does not impose an obligation on participants to comply with the standard, broad approval should not be caught by Article 81(1). In practice, however, the effect of this pronouncement is to provide cover for the activities of recognised standard-setting organisations or of *ad hoc* bodies that act in a similar way to recognised standard-setting organisations. This safe harbour applies to the standard-setting exercise itself—additional restrictions imposed by standard-setting bodies or participants in organised standard-setting may still be subject to Article 81.

(b) Standard-setting outside formal standard-setting organisations—application of Article 81(1)

9.12 The Horizontal Guidelines suggest that Article 81(1) may be applicable to standardisation agreements 'insofar as they grant the parties joint control over production and/or innovation thereby restricting their ability to compete on product characteristics, while affecting third parties like suppliers or purchasers of the standardised products'.[19] This formulation suggests a three part test. Under this test, a standardisation agreement would be restrictive of competition for purposes of Article 81 if it:

(i) grants joint control to the parties over production and/or innovation;
(ii) restricts the ability of the parties to compete on product characteristics; and
(iii) affects third parties, in particular by discrimination, foreclosure, or geographic market segmentation.[20]

This assessment would apply to the standardisation agreement itself, taking

---

[18] Para 163. [19] Para 166.

[20] As regards third party effects, para 168 of the Guidelines state that 'standards that are not accessible to third parties may discriminate or foreclose third parties or segment markets according to their geographic scope of application'.

'into account the nature of the standard and its likely effects on the market concerned.'[21]

In addition to the assessment of the standardisation agreement itself, any restrictive provisions in the agreement may be caught by Article 81(1) if they 'go beyond the primary object of standardisation'.[22] As regards such restrictions, the Horizontal Guidelines focus in particular on agreements limiting 'the ability of the parties to develop alternative standards or products that do not comply with the agreed standard'.[23] The Guidelines also identify agreements granting exclusive rights to one or more testing bodies to assess compliance with a standard and restrictions on conformity markings as provisions that might restrict competition. **9.13**

(c) Standard-setting outside formal standard-setting organisations—application of Article 81(3)

The Horizontal Guidelines become somewhat incoherent in their discussion of the possible application of Article 81(3) to standard-setting. First, as regards the economic benefits of standard-setting, the Guidelines suggest that to 'materialise these economic benefits, the necessary information to apply the standard must be available to those wishing to enter the market, and an appreciable proportion of the industry must be involved in the setting of the standard in a transparent manner.'[24] Yet it is unclear how access for third parties is a necessary precondition for benefits, although it can affect the *extent* of benefits. Secondly, as regards indispensability, the Guidelines focus on: **9.14**

(i) whether standards are 'neutral';
(ii) the desirability that standards be set on a non-discriminatory basis; and

---

[21] Para 166. In para 168, the Guidelines suggest that even high market shares do not necessarily indicate that a standard-setting agreement is restrictive, in the absence of barriers to entry by firms outside the standardisation project.

[22] Para 166.

[23] Para 167. The Guidelines state: 'The existence of a restriction of competition in standardisation agreements depends upon the extent to which the parties remain free to develop alternative standards or products that do not comply with the agreed standard. Standardisation agreements may restrict competition where they prevent the parties from either developing alternative standards or commercialising products that do not comply with the standard.' The passage raises two important questions. First, is it really correct that the application of Article 81(1) depends on the ability of the parties to apply alternative standards? In other words, does this statement restate on a more restrictive basis the test set out in para 166 and discussed in the text above? As the analysis in the text suggests, we do not regard this statement as an exhaustive test. Secondly, does this statement apply only to explicit limits on exploitation outside the agreed standard or also to *de facto* limitations, where participants in standard-setting have no commercial interest in exploitation outside the standard?

[24] Para 169.

(iii) the need to justify choice of one technology over other technologies as the basis for the standard.[25]

9.15 The Guidelines also assert that '[a]ll competitors in the market(s) affected by the standards should have the possibility of being involved in the discussions', suggesting that any limitation of the parties involved in the standard-setting must be shown to be indispensable.[26] Thirdly, the Guidelines apply the requirement that competition should not be eliminated as the basis for imposing the obligation that third parties have access to the standard on fair, reasonable, and non-discriminatory terms, even where the 'standard' is a *de facto* standard resulting from an agreement between private firms.[27]

(d) Problems with the Horizontal Guidelines on standard-setting

9.16 Overall, the Horizontal Guidelines do not provide much useful guidance regarding application of Article 81 outside the context of organised industry-wide standard-setting. Where the Guidelines discuss Article 81(1), they provide limited concrete assistance in identifying the circumstances in which standard-setting could lead to competitive harm. Although there is a strong suggestion that Article 81(1) is only applicable in exceptional circumstances, the wording of the Guidelines is sufficiently vague to allow a competitor that feels excluded to invoke Article 81. Where the Guidelines discuss Article 81(3), there is a strong indication that the analysis started with the requirements regarding the standard-setting process that the drafters of the Guidelines wished to impose, rather than with an assessment of the requirements of Article 81.[28] This does not mean that these requirements—access to third parties, neutrality, and non-discrimination—are not justified. Indeed the discussion in the Guidelines can be characterised as an attempt to require agreements between firms to adhere to the same norms as institutional standard-setting where they operate to set *de facto* standards for a whole industry.

---

[25] Para 171. The discussion in the Guidelines does not identify the restriction that is being assessed for indispensability. On its face, the concern is not that there is a 'restriction' in the form of a restrictive provision but rather that the conduct of the parties in setting the standard was itself a separate restriction that somehow infects the resulting agreement. This is a novel application of Article 81, which is not found elsewhere.

[26] Para 172. Here again this discussion does not identify what 'restriction' is being assessed for indispensability, as required by Article 81(3). Presumably there is meant to be an implicit agreement to limit the number of participants, which is itself a restriction of competition. This concept is explored in the next section of the main text.

[27] Para 174.

[28] Thus as noted above, it is hard to fit the Commission concerns about participation into an 'indispensability' framework, and the points about access to the results of standard-setting and transparency are at best indirectly related to benefits.

(3) *Applying the new Article 81 paradigm to standard-setting*

A detailed consideration of standard-setting under Article 81 is beyond the scope of this work. In light of the lack of rigour in the discussion in the Horizontal Guidelines, however, a few comments regarding this assessment are appropriate in order to provide a basis for discussing the issues specific to technology transfer in the context of standard-setting.[28A] **9.17**

The first comment involves the need to focus on consumer welfare considerations in applying Article 81 in the standard-setting process. There appears to be a tendency to invoke competition law as a basis for resolving competing claims between stakeholders on the profits of innovation. Often, however, these disputes involve neither significant effects on price or output for consumers nor significant reductions in incentive for future innovation. The only real question in these cases is, instead, which set of rights holders should be able to benefit from rents attributable to an innovation. In the absence of likely harms to consumer welfare, Article 81 does not provide a good platform for resolving these disputes. Application of competition law can simply add another layer of transaction costs. It is therefore important that claims of competitive harm receive critical examination, particularly where they concern either structural change or reductions in incentive to innovation. **9.18**

The second comment involves the competitive harm that can arise from standard-setting. This falls into two categories. First there is possible competitive harm in the markets for the products produced according to the standard, primarily through facilitating collusion.[29] Second there is the possible competitive harm in the technology market—foreclosure of disfavoured technologies from the licensing market and inhibition of future development efforts for excluded technologies.[30] These effects will, however, be rare, even in cases where the firms involved in the standard-setting exercise have a large aggregate market share. In many cases there will be no effects in downstream markets because the conditions for collusion are either absent or unchanged by the standard-setting exercise.[31] **9.19**

[28A] A good introduction to the legal issues raised by application of Article 81 to standard-setting is Dolmans, M, 'Standards for Standards' [2002] 26 *Fordham Int LJ* 163.

[29] This effect can result from increased transparency in the market that makes price comparison easier, or from an increase in concentration (because the standard excludes some existing or potential competitors).

[30] This can only occur if four conditions are met. First, there must be genuine alternative technologies. Secondly, the number of firms committed to the standard must be sufficient to make competing technologies unviable. Thirdly, the commitment of the firms to the standard must be such as to make use of alternative technologies by those firms unlikely (whether as a result of contractual commitments or because of commercial considerations). Fourthly, the firms setting the standard cannot have the ability to limit the prices charged for access to the technology in the context of setting the standard.

[31] Where third parties do have effective access to the standard, conditions will be unchanged unless increased pricing transparency really does play a substantial role in the market.

Effects on technology markets will often be absent because there are not genuine competing technologies or because there is only room for one technology in the market.[32] Thus application of Article 81 to '*de facto*' standard-setting should only be justified in genuinely exceptional circumstances. This is particularly important in the licensing context in light of the suggestion in the Technology Guidelines that cross-licences may lead to creation of *de facto* standards.

9.20 The third comment involves the use of Article 81 as a basis for imposing controls on the process of standard-setting. The Horizontal Guidelines suggest that Article 81 may mandate extensive requirements regarding participation by interested participants, transparency in decision-making, and 'access' for third parties to the standard. We have already noted that imposing such conditions in the hope of making standard-setting 'fairer' may not be consistent with a consumer welfare-oriented approach to Article 81. There may nonetheless be circumstances where ensuring broader access to the standard both reduces any likely competitive harm and increases economic benefits.[33] The proper legal basis for imposing these conditions on the standard-setting process is that, in their absence, the economic benefits of the exercise do not outweigh the competitive harm.[34] Particularly where there is significant potential competitive harm in the downstream market, a more inclusive process may reduce the likelihood of collusion and increase potential benefits. The suggestion that these conditions derive from either the 'indispensability' requirement[35] or from the 'no elimination of competition' requirement of Article 81(3)[36] cannot be sustained.

[32] The Horizontal Guidelines can be read as suggesting that the process of standard-setting could be regarded as restrictive where it is not designed to facilitate choice of the best technical alternative. This argument is expanded in Dolmans, n 28A above, at section 4.4. There are, however, considerable theoretical and practical problems with applying Article 81 on this basis. Theoretically where one technology is clearly 'better' than another technology, the choice of the inferior technology may lead to consumer harm by reducing future product quality. In practice, however, the better technology may be the one that has the best chance of succeeding in the market. It is hard to see how a court or competition authority could usefully assess whether one technology is 'better' than another on this basis. It is also difficult to see how the failure of participants in standard-setting to incorporate procedures designed ensure full participation and transparency would make it 'likely' that an inferior technology would be chosen. The interests of the parties in a successful downstream product will normally ensure choice of the best technology from a commercial perspective. Finally, it should be recalled that for purposes of applying Article 81(1), the 'counterfactual' is the position where standards are set in the marketplace without agreement, not the result of a hypothetical 'fair' standard-setting exercise.

[33] Participation by all interested players and transparency can both be viewed in this context as ways of making access by third parties more effective.

[34] Step Seven in the analysis set out at the end of Chapter 3. This means that a showing of significant competitive harm is a necessary prerequisite for imposing access conditions.

[35] As noted in the previous section, the Horizontal Guidelines paras 171–172 attempt to base these requirements on indispensability. This implies, however, that there is some restrictive agreement to limit participation or make the process non-transparent.

[36] See Horizontal Guidelines para 174. This presupposes, that the standard-setting agreement eliminates effective competition between the firms already adhering to the standard, which will seldom be the case.

(4) Standard-setting and IP rights—'FRAND' terms and disclosure or licensing obligations on participants in standard-setting

In this section we address the two most important areas where standard-setting overlaps with intellectual property licensing: **9.21**

(i) requirements that participants in standard-setting arrangements make available any IP rights required to implement the standard on 'FRAND' terms; and
(ii) obligations imposed within in a standard-setting process that oblige participants to disclose the existence of intellectual property rights that might cover implementation of the standard or the terms on which they will agree to license under the standard.

(a) Providing 'access' to the standard—licensing under FRAND terms

Where Article 81 is applicable to a standard-setting exercise, Commission precedent requires that third parties have 'access' to the standard. This is intended to remove the 'barrier to entry' that the creation of the standard would otherwise impose on third parties. Where application of the standard would infringe intellectual property rights belonging to parties involved in the standard-setting process, the Commission has indicated that effective access requires licensing on 'fair, reasonable, and non-discriminatory' ('FRAND') terms. The requirement for licensing on a FRAND basis has been a central element of Commission practice regarding standard-setting and patent pools since the 1970s, but there is little precedent regarding what FRAND actually means.[37] Some commentators have attempted to deduce the meaning of FRAND terms by reference to Article 82(a) (which requires that dominant firms refrain from imposing 'unfair' prices or trading conditions) or to Article 82(c) (which bars dominant firms from applying 'dissimilar conditions to equivalent transactions').[38] These provisions are clearly relevant to assessing the meaning of FRAND; it may not make **9.22**

[37] The FRAND concept appears to follow US precedents. The US cases, however, impose only 'reasonable and non-discriminatory' ('RAND') terms. This leaves open the possibility of an intentional distinction between the concepts. Perhaps it is possible to be fair but not reasonable (or vice versa). As the discussion in the main text indicates, however, not too much should be made of this textual distinction. Subtle textual analysis is not appropriate for Commission documents. These texts need to be translated into the various official languages. Often an English text or French text will have been drafted by someone working in their second or third language. The translation services will not always be able to convey a textual nuance from one language to another. In part for this reason the Community courts favour a 'teleological approach' interpreting legal requirements based on their intended purpose. It is interesting to note that where the TT Guidelines deal with licensing terms (in the context of Article 82) they refer to 'fair and non-discriminatory terms' leaving out the additional reference to 'reasonableness'. See TT Guidelines at para 226.

[38] See e.g. Dolmans, n 28A above, at section 4.7(d).

sense to impose higher obligations regarding the terms of a licence under Article 81 than would be imposed on a dominant firm under Article 82. The legal context of the FRAND obligation does not, however, suggest either an intent to apply an Article 82 based standard or a need to lift standards from Article 82. The discussion below attempts therefore to assess the meaning of FRAND on a principled basis based on the underlying policy goals.

9.23 The requirements that the terms be 'fair' or 'reasonable' must be understood in the context of the reason for licensing on FRAND terms—to allow third parties to enter the market where the standards are applicable. If licensing terms for the technology required to apply the standard constitute a substantial barrier to entry, then they may not be fair or reasonable.[39] It should be recalled, however, that from an economic perspective entry costs only impose a barrier to entry where they exceed the costs already incurred by incumbents in the market. Where technology is involved, the incumbents have already borne the cost of research and development as well as the risk that technology may not have a useful application. Under these circumstances even substantial royalty obligations may not constitute a real barrier to entry in the economic sense.[40]

9.24 The requirement that terms be 'non-discriminatory' must be considered on the same basis. A non-discrimination requirement makes sense as a further guarantee that access will be effective. Where conditions of access are comparable for all competitors, access costs should not create a barrier to entry. In this context, 'non-discriminatory' terms do not necessarily mean identical terms. The key issue should be whether differences between the terms available to two firms or groups of firms in fact create a barrier to entry for one firm or group of firms in the economic sense described above. Comparable treatment satisfies this goal.[41] Furthermore, where the circumstances of a potential entrant create an objective justification for treating that firm differently, such different treatment does not

[39] It will be noted that the barrier to entry imposed by the need to obtain licences under relevant IP rights applies to the cost of the entire package of rights required for successful entry. By implication, this means that the licensor needs to take account of the reasonable expectation of other licensors in setting the terms for a licence.

[40] The developing law under Article 82(a), discussed below in Chapter 10, may be relevant to this assessment. Under Article 82(a) prices, including royalty terms, may be 'unfair' if they are not consistent with pricing in comparable markets, or if they are intrinsically unfair in light of their inherent value.

[41] An important example of comparable treatment involves the case where participants in standard-setting cross-license each other on a royalty-free basis under IP rights necessary to apply the standard. In effect, each licensee is obtaining the rights under the cross-licence in lieu of a royalty payment from the licensee. This does not mean that a licensor is obliged to offer royalty-free licences to third parties who do not control IP rights that interest that licensor. The benefits under the cross-licences should be sufficiently comparable to royalty payments from the other parties to satisfy the non-discrimination requirement. This reflects the practice of the ECJ in applying Article 82(c).

necessarily constitute discrimination.[42] This will be the case particularly where the firm seeking access possesses advantages in terms of reputation, distribution assets, experience in related markets, or possible economies of scale or scope that will assure it successful entry into the market. For a potential entrant in these circumstances, minor differences in licence terms will not affect its ability to enter the market.[43]

The role of the FRAND obligation as a means for allowing access to the market for products covered by the standard also determines the firms that are required to provide a licence on a FRAND basis. The Commission practice is to require a licence under *essential* intellectual property rights. In the context of standard-setting, an intellectual property right becomes essential where it is not possible to apply the standard without infringing that right. Where intellectual property rights are not essential, i.e. where third parties could obtain the necessary rights from different parties, each with an interest in obtaining the royalty revenues, then a FRAND requirement is not necessary.[44] **9.25**

(b) Conditions on participation in the standard-setting process

The requirement that participants in the standardisation process license third parties on FRAND terms is a direct requirement of Article 81. Where Article 81 is otherwise applicable, the failure of a licensor subject to the FRAND obligation to grant such a licence means that the agreement to establish the standard infringes Article 81.[45] A standard-setting body or the parties to an *ad hoc* standard-setting arrangement may, however, impose a private FRAND **9.26**

[42] Thus where a firm seeking a FRAND licence itself controls IP rights for technology relevant to applying the standard, it would not be discriminatory for a licensor to demand a licence for that technology even though other firms that do not control such IP rights are not subject to a comparable requirement.

[43] It is important to recall that FRAND terms are not an independent requirement under Article 81. They are only justified as a condition to ensure that standard-setting does not exclude third parties and thereby lead to competitive harm. Where an entrant would not otherwise be excluded, the basis for the FRAND requirement disappears. The position may be different, however, where Article 82 is applicable, as discussed in Chapter 10. It is also worth noting that the position may be different where FRAND terms are imposed as a remedy, following a finding that the practices of a standard-setting group have led to exclusionary effects on a market. Under these circumstances a stricter non-discrimination requirement may be justified to ensure that the restrictive effects are brought to an end.

[44] It could be argued that the logic of allowing third party access could also mandate FRAND terms in other circumstances. If producing a product that conforms to the standard can be done in two ways, and different participants in the standard-setting process control IP that covers the two different methods for compliance, then a licence from one of the participants may be a condition for effective market access. In this case, however, competition between the two licensors would ensure a fair price in the absence of collusion.

[45] In theory this means that the agreement establishing the standard is unenforceable and that all parties to the agreement have committed an infringement. In practice, the firm that has refused the licence is likely to be treated as the culpable party. Both the party refused the licence and the other parties to the standard-setting process may have valid claims under

obligation on participation.[46] Given the uncertain application of Article 81, this obligation may be in place even if the standard-setting exercise is not otherwise subject to Article 81. Furthermore, a standard-setting body or the parties to an *ad hoc* standardisation may impose other requirements on participation by owners of intellectual property rights that are not mandated by Article 81. The question then arises whether these obligations are themselves restrictive of competition and barred by Article 81.

9.27 The need for such obligations results from the nature of the modern intellectual property licensing environment. As already noted in Chapter 8, the ubiquity of intellectual property rights in high-tech industries means that no firm can be sure that an innovative product will not infringe some previously unknown intellectual property right. The problem of negotiating the 'patent thicket' can be particularly acute in the standard-setting context, because of the risk that firms could manipulate the process to ensure that intellectual property rights belonging to them are essential for application of the standard. The subsequent assertion of intellectual property rights is then described in the literature as a 'patent ambush' and the previously unsuspected rights that are used to implement the 'ambush' are referred to as 'submarine' patents.

9.28 One way to limit the impact of a potential patent ambush is to require participants to disclose the existence of relevant intellectual property rights before the standard is agreed. This obligation, which allows the participants to 'design around' those intellectual property rights in constructing the standard, is generally regarded as unproblematic under Article 81.[47] A more effective obligation might be to require that participants license essential intellectual property rights on FRAND terms. This obligation would also appear unproblematic from an Article 81 perspective, particularly since FRAND licences are required anyway if Article 81 is applicable to the standard-setting agreement itself.[48] In the 1995 ETSI investigation, however, the Commission considered claims that

---

Article 81 against the refusing firm. Furthermore, a court or competition authority may mandate the grant of a licence under FRAND terms as an appropriate way to bring the infringement to an end.

[46] There is no reason why a private licensing commitment would need to follow the Article 81 FRAND requirements set out above, as long as neither Article 81 nor Article 82 is infringed.

[47] See Commission Press Release IP/05/1565, 12 December 2005, 'Commission Welcomes Changes in ETSI IPR Rules to Prevent Patent Ambush'.

[48] In general, the ability of firms to determine whether the benefits to them of participation in the standard-setting process justify the loss of their flexibility in licensing ensures that there is no competitive harm. It is possible that competitive harm could result from this requirement, however, in the limited number of cases where the relevance of an IP right to a standard was not readily ascertainable in advance. Even in this case, however, the competitive harm (as opposed to the harm to the competitor) relates to a reduction in the incentives for innovation. Given the unusual nature of this scenario, this harm cannot be regarded as likely. Dolmans, n 28A above, argues that a rule requiring agreement to license on specified terms could constitute a 'collective

mandatory licensing on FRAND terms was anticompetitive where the standard-setting body had special legal responsibilities in respect to standards for the relevant product.[49] As a general matter, if there are concerns about mandatory licensing in this context, they would appear more suitable for resolution under Article 82 than under Article 81.

Mandatory licensing on FRAND terms may not fully resolve concerns about patent ambushes and submarine patents. The uncertainty about what would constitute FRAND terms in an individual case may leave room for unpleasant surprises (or require expensive dispute resolution). These concerns can be addressed by requiring that participants in the standard-setting process disclose in advance the terms under which they will license intellectual property rights for inclusion in the standard.[50] Where alternative technologies are available, this allows participants in the standard-setting process to take account of their relative cost in determining which technology to incorporate in the standard. In other cases, advance disclosure of licensing terms allows participants either to design around expensive intellectual property rights or to terminate the standard-setting exercise, if the royalty costs will be too high. **9.29**

Advance disclosure of licensing terms has raised concerns on the basis that it could facilitate collusion on royalty levels between participants in the standard-setting process. Whatever the theoretical merits of this objection,[51] it is evident that the results of this obligation will normally be to force overall royalty pricing down.[52] In the Technology Transfer Guidelines the Commission signalled a cautious approval for this requirement, at least where it is implemented as prelude to subsequent establishment of a technology pool.[53] **9.30**

boycott' by participants in the standard-setting body. It is unclear that an agreement on these terms does constitute collective boycott where participants remain free to obtain licences on an individual basis outside the standard-setting body. Classification as a collective boycott does not, in any case, mean that the conduct necessarily leads to competitive harm.

49 See general discussion in Dolmans, n 28A above, at section 4.6(a). In 1994 ETSI made a formal notification under the old Article 81 procedure of changes in its practice for standard-setting. These included obligations on participants to declare the existence of essential IP rights and agree to license on FRAND terms. Following objections from potential licensees, ETSI submitted a revised policy which removed the mandatory licence requirement in favour of a solution that involved disclosure and a commitment to design around any IP rights where the licensee was unwilling to license.

50 See generally Hansen, M, Shah, O, and Ohana, G, 'Disclosure and Negotiation of Licensing Terms Prior to Adoption of Industry Standards: Preventing Another Patent Ambush' [2003] ECLR 644.

51 The difficulty of collusion between licensors has already been discussed in Chapter 5. In this situation it is further complicated by the different relationships between competing, complementary, and essential technologies, as discussed in greater detail below under patent pools.

52 This is thus not a classic example of 'price-fixing' that would be suitable for treatment as a restriction by 'object'.

53 Paragraph 225 in the TT Guidelines states that '[i]n certain circumstances, it may be more efficient if the royalties are agreed before the standard is chosen and not after the standard has

D. Technology Pools

9.31 The Technology Transfer Guidelines define technology pools as 'arrangements whereby two or more parties assemble a package of technology which is licensed not only to contributors to the pool but also to third parties'.[54] As the Guidelines recognise, technology pools can generate significant benefits by reducing transaction costs and reducing total royalties. Commission practice under the old regime nonetheless displayed a tendency in some cases to use Article 81 (or Article 82) as a basis for detailed intervention in the structure and operation of pooling arrangements. While there will be circumstances where such intervention is justified to deal with genuine competitive harm, application of the new Article 81 paradigm requires more attention to the nature of competitive harm in the context of technology pools and to the likelihood of such harm in an individual case.

(1) Competitive harm in patent pools

9.32 The Technology Transfer Guidelines identify three ways in which patent pools can lead to competitive harm.[55] First, the Guidelines indicate that where technologies that are substitutes are included in a pool, setting the pool price amounts to price fixing between direct competitors on the technology market.[56] Second, the Guidelines state that a technology pool can 'result in a reduction of innovation by foreclosing alternative technologies'.[57] The Guidelines suggest that this effect may be particularly important where the pool supports a standard, stating that the 'existence of the standard and the related technology pool may make it more difficult for new and improved technologies to enter the market'. Third, the Guidelines suggest that a technology pool can result in 'collective bundling'—forcing licensees to pay for technology that they do not want or need.[58]

---

been embarked upon, to avoid that the choice of the standard conveys a significant degree of market power on one or more essential technologies'.

[54] Para 210.

[55] The Guidelines also note at para 224 the possibility that patent pools could provide a channel for exchange of confidential information among participants that could facilitate collusion in the downstream market. Whether this is the case will depend on whether the other conditions for effective collusion in that market are met and the nature of the information exchanged. See discussion regarding general requirements for effective collusion in Chapter 4 at paras 4.27–4.30 above.

[56] Para 213. [57] Ibid.

[58] Para 221. The Guidelines describe this effect under the heading of 'foreclosure'. Since, however, it describes a different kind of potential harm, it is described separately in the main text.

(a) Direct competitive harm—pooling of substitute technologies

Where two firms that each control competing technologies form a pool to license those technologies on a joint basis, there is a clear risk of competitive harm. Even where the two firms each remain free to license their own technology independently, setting the joint price for the pooled product is likely to lead to higher royalty levels.[59] On this basis, the Technology Transfer Guidelines state a 'general rule' that 'inclusion of substitute technologies in the pool constitutes a violation of Article 81(1)'.[60] **9.33**

This rule makes sense as long as the term 'technology' is defined in an appropriate way. As explained in Chapter 6, 'technology' needs to be defined on a functional basis to describe the input that a licensee requires to perform a specific function in its own economic activity.[61] In other words, a 'technology' is not necessarily synonymous with an 'innovation' or with a 'solution' to a technical requirement. Furthermore, the intellectual property rights held by a particular licensor do not necessarily define a 'technology'.[62] **9.34**

The functional approach to defining technology accords with the purpose of market definition, to provide a framework for assessing competitive harm. For example, where two firms are each engaged in licensing eye surgeons for use of an intellectual property protected technique for performing laser eye surgery, **9.35**

[59] Lerner, J and Tirole, J, 'Efficient Patent Pools' [2004] 94 *American Economic Review* 691, suggest that even a pool between full substitutes is unlikely to lead to higher royalties as long as the participants are free to license on their own. Their basic insight is that in the absence of the pooling arrangement non-collusive coordination between the parties will lead to royalty rates at the same level. The Tirole/Lerner insight is important because it focuses attention on the counterfactual—the level of interaction that is inevitable in highly transparent concentrated markets, suggesting that the impact on consumer welfare of licensor price arrangements may not be as significant as the Commission Guidelines would suggest. The overall conclusion is more questionable, because (i) the pooling arrangement will lead to better information for each participant regarding the market perceptions of the partner, making collusion outside the pool more effective; and (ii) even in the absence of explicit obligations to license only through the pool, a party may be reluctant to do so if this could lead to termination of the pooling arrangement and thereby to lower overall royalty revenues.

[60] TT Guidelines para 219.

[61] Paras 6.53–6.59. This results from the necessary approach of identifying a 'product' (which must meet a definable separate consumer need) in a market where a sole supplier would enjoy market power.

[62] As noted in Chapter 6, the IP Guidelines do not explicitly address the definition of 'technology' as a product under Article 81 or Article 82. The discussion on technology pools does, however, tend towards an approach that equates 'technology' with IP rights. At para 216, n 69 for example, the Guidelines state that the 'term technology is not limited to patents. It covers also patent applications and intellectual property rights other than patents'. In para 216 itself, the Guidelines refer to 'the package of technologies for the purposes of producing the product(s) or carrying out the process(es) to which the pool relates'. These statements suggest that the Guidelines view a 'technology' as either an IP right or group of IP rights or as the rights attributable to a specific technical solution.

formation of a patent pool covering their respective rights could pool genuine substitute technologies, and lead to higher prices and/or lower output.[63] Where two firms each control intellectual property rights governing alternative solutions for implementation of a broader 'technology' the position may be different. This might be the case, for example, if two firms offered alternative solutions for ordering data on a DVD. From the perspective of potential licensees, the rights to these alternative solutions only have value as part of a package with the other rights needed to produce either DVDs or equipment that can run DVDs. If essential intellectual property rights exist covering other aspects of the DVD specifications, any combination of rights for alternative data-ordering solutions do not increase the market power of intellectual property rights holders in relation to potential licensees. The leverage of the firms holding rights on data-ordering may increase in relation to other licensors, but a resulting reallocation of rents between licensors does not in itself impact consumer welfare.[64]

(b) Foreclosure of complementary technologies

9.36 According to the TT Guidelines, '[o]nce a technology is included in the pool, and licensed as part of the package, licensees are likely to have little incentive to license a competing technology when the royalty already covers a substitute technology'.[65] The Guidelines conclude the assessment of this effect (along with 'bundling', discussed in the next section) with the observation that '[w]hen a pool encompasses non-essential technologies, the agreement is likely to be caught by Article 81(1) where the pool has a significant position on any relevant market'. The sweeping assertion that any pool in a 'significant position' containing 'non-essential technologies' (by which the Commission appears to mean non-essential intellectual property rights) is likely to cause competitive harm requires a careful consideration of the basis for these claims of competitive harm.

9.37 It will be recalled that 'foreclosure' by itself is not a description of competitive harm. Foreclosure describes instead a process by which competitive harm can take place where market conditions are appropriate, where limiting a rival's access to a customer or group of customers would remove a competitive constraint on the firm or firms that execute the foreclosure.[66] For foreclosure to lead to competitive harm, it is necessary to identify the nature of the constraint that would be exercised by the foreclosed firm in the absence of the foreclosure.

[63] The US Federal Trade Commission dealt with a case with those facts in *Summit Technology/VISX*, Consent Decree issued 23 February 1999 in Docket No 9286. Although it has been suggested that even these technologies may have been complementary rather than direct competitors.

[64] This raises again the 'double-marginalisation' problem. If the licensors cannot agree a joint profit-maximising price, then the creation of groups with additional essential IP blocks may lead to a total price for the technology that is higher than the profit-maximising price.

[65] Para 221.

[66] See Chapter 4 above at paras 4.49–4.55.

If that constraint would not be significant, the foreclosure is irrelevant for purposes of applying Article 81(1).

Where a technology pool, by itself or in combination with a standardisation agreement, leads to removal from the market of a genuine alternative technology, then there can be a real risk of competitive harm. Competitive harm requires, however, that a number of conditions be met in addition to the existence of a genuine alternative technology. First, the market conditions must be such that there is room for two (or more) technologies to co-exist. In the not uncommon situation where the market will support only a single solution, incorporation of one alternative in a standard and subsequent incorporation of the relevant intellectual property rights in a pool will not eliminate any competitive constraint. Secondly, the standard-setting process (or market conditions where a *de facto* standard has developed) must, in fact, lead to exclusion or marginalisation of the rival technology.[67] Thirdly, the firm or firms that control the technology benefiting from the foreclosure effect must be in the position to exercise some market power as a result. The ability of potential licensees to control a standard-setting process may inhibit the ability of a licensor included in the standard to charge super-competitive royalties.[68] It should be evident that a 'significant position' by the pool in 'any relevant market' is not sufficient to reach a conclusion on the application of Article 81, even where a genuine alternative technology exists. **9.38**

Where the foreclosure involves firms controlling intellectual property rights that do not protect technology with a 'stand-alone' value, the possible competitive harm is difficult to identify. If the pool has market power in relation to potential licensees because it includes intellectual property rights that are essential for applying a formal or *de facto* standard, the presence of absence of additional firms offering alternative solutions for technical problems within the standard will not change the ability of the pool to charge a monopoly rent. Since the foreclosed firms do not exert competitive constraint, their exclusion should not lead to competitive harm.[69] **9.39**

---

67 As explained in Chapter 5, market shares are not reliable guides to market power in technology markets. For a foreclosure strategy to be successful in removing the competitive constraint exerted by a rival technology, the proportion of the licensee market effectively committed to the standard must be sufficient to make the rival technology non-viable.

68 Where the 'foreclosure' results from a standardisation process where potential licensors were required to disclose their licensing terms, the disclosure requirement (together with FRAND obligations) will limit the ability of the successful licensor to benefit from any supposed foreclosure effect.

69 There could theoretically be circumstances where the exclusion of firms in this position could lead to a reduction in dynamic efficiency, by inhibiting the development of new technology by the excluded firms. As noted in Chapter 5, however, for this long-term effect to be a basis for applying Article 81(1) should require some showing that the firms involved would be likely to produce such technological advances and that the impact of 'foreclosure' is to create a significant change in the likelihood of this result. The effect on dynamic efficiency is too speculative to be the basis for a presumption that Article 81(1) is infringed.

(c) 'Collective bundling'

9.40 The TT Guidelines suggest that the 'inclusion of technologies which are not necessary for purposes of producing the product(s) or carrying out the process(es) to which the technology pool relates also forces licensees to pay for technology that they may not need'. The Guidelines therefore conclude that the 'inclusion of complementary patents thus amounts to collective bundling'. The Guidelines subsequently suggest that this competitive harm may be limited by allowing licensees to decline a licence under some intellectual property rights included in the pool in return for a reduction in royalties.[70]

9.41 There will indeed be circumstances where a technology pool leads to higher prices and/or lower output because the range of licensed technologies is overbroad. These circumstances do not, however, arise in every case where a pool includes complementary patents or even in most cases where a pool includes complementary patents. This effect will only arise where the pool includes genuine separate technologies.[71] Where a pool only includes the intellectual property rights needed to implement a single technology and includes essential intellectual property rights, the market power of the pool is based on the essential rights. The value of those rights to a potential licensee does not change because a particular licensee does not 'need' all the complementary patents included in the pool.[72] Under these circumstances, the inclusion of complementary patents in the pool may shift monopoly rents among licensors or potential licensors, but does not necessarily lead to any consumer welfare loss.[73]

---

70 Para 222, ss (c) and (d).

71 This analysis is consistent with the general legal and economic approach to bundling issues, which require that there be two separate products before there can be a bundling case. It should be noted that a separate technology for this purpose would include functions that are not necessary for downstream exploitation of the basic technology. A technological function that is not essential for exploitation of the technology can give rise to bundling—but not simply an IP right that is not essential.

72 For example, a potential licensee may itself control patents that cover a alternative technical solution to that covered by complementary patents in the pool. The fact that this licensee does not need those complementary patents does not change the value to the licensee of the remaining essential patents which make exploitation of the technology in the downstream market possible. There is no effect on consumer welfare because the licensee pays no more than it would pay in the absence of the pool, if it had to negotiate a royalty with the licensor on an individual basis.

73 Firms that control essential rights may agree to include complementary IP rights in a pool (and thus share the monopoly rent attributable to their essential rights) for a variety of reasons that are not by themselves anticompetitive. First, it is not always evident which IP rights are essential—over-inclusion reduces transaction costs. Secondly, including complementary patents may make adoption of the technology in the marketplace more likely by creating a 'one-stop shop' for licensing. Most important, licensors of essential rights may include licensors of complementary rights in the pool in order to give those licensors a stake in adoption and diffusion of the technology. The inclusion or exclusion of complementary IP rights on this basis can of course have a negative impact on other licensors, but consumer welfare is only implicated in the situations described in the text.

(2) *Economic benefits and balancing under Article 81(3)*

The TT Guidelines recognise that technology pools can lead to significant economic benefits by offering a 'one-stop shop' to licensees and by resolving double marginalisation issues. They recognise that these benefits can arise where pools include complementary patents, and even where pools contain substitute patents, at least where those patents might be both complements and substitutes, or the pool is not primarily composed of substitute patents. The Guidelines do not, however, provide much guidance in determining how these benefits should be weighed against competitive harms in an individual case.[74] Instead the principal concern of the Guidelines is to outline ways in which a pool could be structured to limit competitive harm. The implication is that in many cases these conditions will need to be met, in order to ensure that the balance of benefits under Article 81(3) favours the creation or maintenance of the pool.[75] **9.42**

Instead of creating a model for balancing benefits and harms, the Guidelines create what are effectively two safe harbours for pooling arrangements. The first safe harbour involves limitation of a pool to essential intellectual property rights. The second safe harbour involves both provisions designed to limit foreclosure and implementation of an institutional framework to ensure transparency and fairness. In either case, the pool cannot include additional obligations on parties or licensees that could restrict competition. **9.43**

---

[74] The Guidelines do indicate at para 222(a) that assessment will be influenced by consideration of 'whether there are any pro-competitive reasons for including non-essential technology in the pool.' The Guidelines also indicate at para 224 that the following 'main principles' govern assessment of pools under Article 81:

'(1) The stronger the market position of the pool the greater the risk of anticompetitive effects;

(2) Pools that hold a strong position on the market should be open and non-discriminatory;

(3) Pools should not unduly foreclose third party technologies or limit the creation of alternative pools.'

[75] The practice of the Commission, as reflected in cases where investigations of pooling arrangements have been concluded by settlement, has been to require substantial commitments along the lines outlined in the main text as a prerequisite for clearance of pooling arrangements under Article 81(3) (See e.g. '*Philips/Sony*', Thirty-third Report on Competition Policy, 2004, point 65: joint licensing limited to essential patents; 'DVD Licensing Programme', Thirtieth Report on Competition Policy, 2001, point 147.) The Guidelines themselves do not commit the Commission to this policy, however. In para 222 the Guidelines state only that the assessment will 'take into account' factors tending to reduce foreclosure, and in para 230 the Guidelines state further that the way in which a pool is 'created, organised, and operated can reduce the risk of it having the object or effect of restricting competition and provide assurances to the effect that the agreement is pro-competitive'. Thus the various measures proposed in the Guidelines are described as helpful for ensuring that a pool is consistent with Article 81, but are not proposed as necessary for that purpose.

(a) Safe harbour number one—restrict pool to essential IP rights

9.44 The discussion in the Guidelines of Article 81(1) clearly indicates that Article 81 is not applicable to a pool that only includes essential intellectual property rights. The Guidelines suggest, however, that some rights that are 'essential' today could lose their essential status because of subsequent development of substitute technology. To remain in this safe harbour it is therefore necessary for the pool to incorporate a mechanism for review of the essential status of intellectual property rights in the pool. The Guidelines also suggest that it would be useful to employ an independent expert to evaluate whether the intellectual property rights are indeed essential or remain essential.[76]

(b) Safe harbour number two—measures to limit foreclosure and an institutional framework for the pool

9.45 The second safe harbour applies to pools that include complementary intellectual property rights and even some substitute intellectual property rights, as long as the pool does not comprise substitutes 'to a significant extent'.[77] If such a pool incorporates in full the contractual and structural measures suggested in the Guidelines, the risk that a court or competition authority will find that the pooling agreement infringes Article 81 as a whole must be limited.

9.46 To achieve this status, the pooling arrangements would need to include the following features designed to limit the possibility for foreclosure:[78]

- Licensors must be free to license their technology independently.
- Where the pooled technologies have different applications, different licence packages should be available for different applications.
- Licensees should be able to obtain a licence for only part of a package 'with a corresponding reduction in royalties'.
- Licensees should be able to terminate licence obligations at reasonable notice and 'obtain a corresponding reduction in royalties'.

9.47 The second safe harbour also requires the following measures concerning the institutional framework for the pool:

- Licences should be available on FRAND terms.[79]

---

[76] See Guidelines paras 232–33.

[77] The Guidelines state at para 219 that the 'Commission considers that it is unlikely that the conditions of Article 81(3) will be fulfilled in the case of pools comprising to a significant extent substitute technologies'.

[78] See generally para 222 in the Guidelines.

[79] The Guidelines para 226 discuss this requirement in the context of Article 82. Of course a safe harbour must make the pool secure from challenge under Articles 81 and 82.

- Pool participants should pay the same royalties as third parties.[80]
- Participation in 'a standard and pool creation process' should be 'open to all interested parties.[81]
- Where possible, independent experts should be involved in any relevant technical assessments.[82]
- Appropriate safeguards should be in place for exchange of confidential information, possibly including use of an independent third party to collect and distribute data.[83]
- The pool should have a mechanism for resolving disputes that is independent of the pool and its members.[84]

(c) Restrictive provisions in pooling arrangements

The TT Guidelines address both restrictions on parties to a pooling agreement and restrictions in agreements between pools and third parties. As regards agreements by licensors, the Guidelines indicate that agreements not to license outside the pool or to participate in alternative pools may lead to competitive harm because they 'limit innovation and preclude the development of competing technological solutions'.[85] The Guidelines recognise, however, that discussions between licensors and potential licensors regarding the royalties charged by the pool and the distribution of such royalties are not in themselves normally restrictive of competition.[86] **9.48**

As regards agreements with licensees, the general rules applicable to licence arrangements remain applicable.[87] The Guidelines identify three areas where pool licences require a more restrictive approach: **9.49**

(i) exclusivity requirements;

---

[80] Ibid. As noted in the discussion of FRAND terms earlier in this chapter, this interpretation of FRAND obligations appears inconsistent with policy rationale for a FRAND requirement and with analogous case law under Article 82.

[81] Para 231. It will be noted that this builds on the analysis of standard-setting earlier in this chapter—the safe harbour for institutional standard setting also applies *mutatis mutandis* to the creation of the subsequent pool.

[82] Guidelines paras 232–233. [83] Para 234.

[84] Para 235. It is unclear how the presence or absence of a neutral dispute resolution mechanism affects the tendency of the pool to be restrictive under Article 81 or affects the likely benefits of the pooling arrangements. There may, however, be an argument that some dispute resolution mechanism is required under Article 82, where a pool controls the technology for an industry.

[85] Para 227. As a general principle this may be true, even where the licensors do not control stand-alone technologies. The restriction on the ability of IP licensors to license third parties or to license for other technologies could impede the development of such technologies. Whether such an effect is likely in a specific market will nonetheless remain an empirical question.

[86] Para 225. This includes a requirement of advance disclosure of licensing terms, as discussed earlier in this chapter in the standard-setting context.

[87] See Chapters 7 and 8 above. Since some of the licensors in a pool will also be competitors in the downstream market, these licences need to be assessed as agreements between competitors.

(ii) grant-backs; and
(iii) no-challenge clauses.

9.50 In respect of exclusivity, the Guidelines suggest that any limitation on obtaining licences outside the pool could be restrictive of competition on the same basis that restrictions on licensing outside the pool could restrict competition.[88] As regards no-challenge clauses, the Guidelines suggest that the right of the licensor to terminate a licence where the licensee challenges the validity of the licensed rights should be limited to the rights of the licensor under challenge—the pool licence as a whole should not be terminated.[89] The Guidelines suggest that grant-backs should be non-exclusive and limited to 'developments that are essential or important to the use of the pooled technology'.[90] The discussion in the Guidelines suggests that a 'grant-back' in this context includes a cross-licence at the point where an original licence is concluded.[91] It is thus not restrictive for the pool to require potential licensees to license their own existing essential or important rights as a condition for obtaining a licence under the pool.

[88] Para 227. [89] TT Guidelines para 229.

[90] Para 228. The Guidelines state that this 'allows the pool to feed from or benefit from improvements to the pooled technology', commenting that it 'is legitimate for the parties to ensure that the exploitation of the pooled technology cannot be held up by licensees that hold or obtain essential patents'. It is not clear from this discussion why the obligation to grant back a non-exclusive licence could not apply to all improvements rather than just those that are 'important'. Determining whether an improvement is 'important' raises an additional uncertainty that is not substantially reduced by involving a third party expert.

[91] '[E]xploitation cannot be held up by licensees that *hold* . . . essential patents.' 'Hold' is included in addition to 'obtain'—the usual context for obtaining a grant-back. See Chapter 7 above at paras 7.96–7.97.

10

TECHNOLOGY LICENSING AND ARTICLE 82

A. Introduction

The discussion above has focused on the application of Article 81 of the EC Treaty—the prohibition on agreements or concerted practices that restrict competition. The economics-based 'new paradigm' for application of competition law that is the focus for this work has thus far primarily concerned application of Article 81 and the Merger Regulation. Article 82—the provision in the EC Treaty that covers abuse of dominant position—has so far remained an outpost for the 'Ordoliberal' form-based approach to competition policy that characterised the old Article 81 regime. Even recent cases applying Article 82 emphasise ideal market structures or unarticulated notions of fairness, rather than a focus on actual or likely consumer harm. Enforcement policy under Article 82 is now under review and there is a strong suggestion that the Commission will endorse an economics-based approach. The judicial precedents under Article 82 may, however, leave less room for a coherent economics-based approach. **10.01**

10.02 It is beyond the scope of this work to intervene in the ongoing discussions of whether and how Article 82 enforcement can be adapted to be consistent with the modern approach to Article 81 and the Merger Regulation. Article 82 can, however, play an important role in defining both when and how an owner of intellectual property may be required to license third parties. In particular, an important line of cases, including the *Microsoft* case currently under review in the Community courts, address the circumstances where a dominant firm may be required to license third parties under intellectual property rights. In the discussion below, we consider these cases and other aspects of current Article 82 practice that directly impact on intellectual property licensing. Because this discussion is intended to complement the discussion in the previous chapters of licensing under Article 81, it does not attempt to review the range of issues where intellectual property impacts on the application of Article 82 that do not involve the decision to license intellectual property rights or the terms on which intellectual property licences are granted.

B. The Structure of Article 82 Analysis

10.03 Article 82 states that:

> 'Any abuse by one or more undertakings of a dominant position within the common market or in a substantial part of it shall be prohibited as incompatible with the common market.
>
> Such abuse in particular may consist in:
>
> (a) directly or indirectly imposing unfair purchase or selling prices or unfair trading conditions;
> (b) limiting production, markets or technical development to the prejudice of consumers;
> (c) applying dissimilar conditions to equivalent transactions with other trading parties, thereby placing them at a competitive disadvantage;
> (d) making the conclusion of contracts subject to acceptance by the other parties of supplementary obligations which by their nature or according to commercial usage, have no connection with the subject of such contracts.'

10.04 Where Article 81 focuses on agreements or understandings between firms, Article 82 controls *conduct*. It is based on the premise that a firm or firms that hold a 'dominant position' have a 'special responsibility' to act in a manner that does not harm competitors, customers, and ultimately consumers.[1] Dominant firms are in principle allowed to compete by using normal methods of performance-based competition, but other forms of conduct that harm

[1] *Michelin* v *Commission* [1983] ECR 3461.

competitors or are unfair to customers may be regarded as an 'abuse' of that dominant position.[2]

As currently applied by the Commission and the Community courts, Article 82 involves a three step analysis: **10.05**

(i) define the relevant market or markets;
(ii) determine whether one or more firms hold a 'dominant position' in one or more markets; and
(iii) determine whether the conduct complained of constitutes an abuse.

Each of these steps raises special issues where intellectual property licensing is at issue.

(1) Market definition in licensing cases under Article 82

In principle, the points discussed in Chapter 6 regarding market definition in cases involving technology licensing are relevant where Article 82 applies.[3] The distinction between technology markets and product markets is applicable under Article 82. Most licensing cases will involve dominance by the licensor in a technology market, although dominance by a licensee in the market where the rights will be used may also be caught by Article 82.[4] The effects of the conduct may be felt in a different technology market or in a downstream product market. Since market definition is often determinative of dominance in technology cases, and since dominance carries with it special conduct requirements, it is particularly important to define technology markets on a functional basis, as described in Chapter 6. Otherwise Article 82 will be applied in cases where there is no real economic power, even under the relaxed standards of the present Article 82 case law. **10.06**

(2) Dominance in licensing cases

Article 82 apples both to single firm dominance and joint dominance. Single firm dominance arises where a company 'holds a position of economic strength that enables it to act independently in relation to competitors, customers, and ultimately consumers'.[5] Joint dominance applies the same definition to a variety of situations where a group of companies hold a position of economic strength **10.07**

---

[2] *Hoffman-La Roche* v *Commission* [1979] ECR 461.

[3] The possibility that markets may be defined on the basis of an input for a specific customer or a specific service may be important in technology cases.

[4] See, e.g., *BRT* v *SABAM* (Case 127/73) [1974] ECR 313 (copyright society was dominant in providing performance management services in Belgium on basis of assigned rights).

[5] *United Brands* v *Commission* [1978] ECR 207.

because of the relationships between them. It will be noted that although the application of Article 82 requires a position of economic strength, neither Article 82 nor the case law links that position directly with a consideration of market power as previously discussed in this work.

(a) Single firm dominance

10.08 The traditional approach to assessing dominance involves consideration of a variety of structural factors drawn from the industrial organisation literature of the 1960s and earlier. Thus the most important factor has been market share—50% plus creates a presumption of dominance.[6] Below the 50% threshold other factors become significant, including the relative size of rivals (if a firm is double the size of the nearest rival, that is a strong indication of dominance), and the size and nature of barriers to entry, including access to distribution, access to necessary inputs, and access to intellectual property.[7] The traditional structural approach does not emphasise the competitiveness of the market, although the Court has indicated that where the largest firm in the market has needed to respond to recent price decreases by rivals by reducing its own prices, this can be inconsistent with dominance.[8] The dominance threshold clearly does not equate with a finding of monopoly power, since it allows for some significant competition to remain (sometimes referred to as 'residual competition'). The threshold is sometimes referred to as requiring 'significant market power',[9] although this specific link to a market-power-based standard has not yet been endorsed by the Community courts.

10.09 In the very few cases where Article 82 has been applied in technology markets, the dominance test has not raised issues since the allegedly dominant firm has had a monopoly or near monopoly share in the market as defined. Where this is not the case, however, the traditional Article 82 approach may not provide a useful standard for determining whether a licensor holds the kind of position of economic strength that should make Article 82 applicable. As discussed in Chapter 6, market share does not function well as a proxy for market power in technology markets. There should therefore be a good argument for modifying the Article 82 test where dominance in technology markets is alleged.

---

6 *AKZO* [1993] CMLR 215; [1991] ECR I-3359.

7 *United Brands* v *Commission* [1978] ECR 207.

8 See e.g. *Hoffman-La Roche* v *Commission* [1979] ECR 461. However, even consistent losses by an allegedly dominant company are consistent with a dominant position, see *United Brands* v *Commission* [1978] ECR 207.

9 Commission Guidelines on market analysis and the assessment of significant market power 2002 OJ C 165/6.

(b) Joint dominance

Under the approach applied by the Commission and endorsed by the Court of First Instance (CFI) in *TACA*, joint dominance involves a two tiered assessment. The first level of analysis involves 'jointness'; whether the firms in question should be considered as a group in assessing dominance. The second level of analysis, which only applies if the first is resolved affirmatively, involves assessment of the group to see if it holds a dominant position, following the assessment applicable in single firm dominance cases, as described above. The key question is whether the firms in question can be treated as a collective group for Article 82 purposes. This question has been complicated by the fact that until May 2005 the test for applying the Merger Regulation also involved a finding of dominance and the test for multiple firm dominance was extended with judicial approval to cover oligopolistic relationships. **10.10**

There are four established ways in which 'jointness' may be established for purposes of joint dominance. The first involves links between the firms (whether ownership or contractual links) that give each of them a significant competitive advantage over firms outside the group.[10] The second, related to the first, involves the joint control by otherwise independent firms of an essential facility that gives each of them an advantage in the downstream market.[11] The third arises where the firms present themselves on the market as a group, as was the case of the liner shipping conferences.[12] The final category involves 'oligopolistic dominance' where the firms in a market are interdependent as a result of the market structure.[13] In this situation it is necessary to satisfy the stringent requirements for showing that oligopolistic effects are likely.[14] **10.11**

In the technology licensing context, the concept of joint dominance becomes important in two ways. First, there may be situations where a network of cross-licences among a group of competitors may make them jointly dominant in the downstream product market on the basis of developed contractual links.[15] This may give rise to obligations on each of the firms in the group, including obligations to license new entrants where the legal requirements are met. Secondly, the participants in a technology pool might be regarded as jointly dominant in a technology market on the basis that the pool members are appearing on the market as a single entity. Where there are several complementary pools in a **10.12**

---

10 *Italian Flat Glass* [1992] ECR II-1403.

11 *Almelo* [1994] ECR I-1477.

12 *CEWAL* [1993] OJ L 34/20.

13 *Gencor Ltd* v *Commission* [1999] ECR II-743. Oligopolistic dominance has not yet been asserted as a basis for applying Article 82 in a decided case, but Gencor was cited as a controlling precedent by the CFI in [1995] ECR II-595.

14 See discussion in Chapter 4.

15 The possibility of joint dominance based on cross-licenses was specifically raised by the CFI in the *Italian Flat Glass* appeal, see n 10 above.

market, the pools (and their constituent members) might be regarded as jointly dominant on the basis of oligopolistic dominance. In either case, it will still be necessary to show that the pool or pools have a dominant position in a properly defined technology market.

(3) *Abuse in licensing cases*

10.13 Abuse under Article 82 is generally understood to consist of a general test, first articulated by the Court of Justice in the 1979 *Hoffmann-La Roche* case, and four specific tests based on the four specific examples provided in Article 82. The general test focuses on 'exclusionary abuse'—conduct designed to drive a rival from the market. The four examples focus on specific categories of conduct that could be exclusionary (as in the case of bundling/tying or discrimination) or that could be exploitative (involving direct harm to customers that may be passed on to consumers).

10.14 The general Article 82 test could cover a broad range of conduct in the licensing context. Thus requiring a licensee not to license third party technology, or progressive volume-based royalty terms having that effect, could infringe Article 82 because of their exclusionary effects in the technology market. Because the conditions for applying Article 82 in this context overlap with the conditions for applying Article 81 as set out in Chapter 7 above, these cases are not addressed further here. The cases involving refusal to license, which are discussed in detail below, are also sometimes considered as examples of exclusionary abuse, since they involve leveraging by a dominant firm to exclude a rival downstream market. These cases have been analysed with implicit or explicit reference to Article 82(b) (limiting markets to the prejudice of consumers). The major areas of concern for licensing under Article 82 arise in the context of the examples: unfair pricing or selling conditions; refusals to license; discrimination; and bundling or tying. We address these issues briefly below, followed by an extensive discussion of the most significant issue—application of Article 82 to refusals to license.

C. Unfair Licence Terms (Article 82(a))

10.15 Article 82(a) covers two types of behaviour (i) 'unfair purchase or selling prices' and (ii) 'unfair trading conditions'. Each of these categories is potentially applicable in a licensing context.

(1) *Unfair pricing or royalty terms*

10.16 Unfair pricing under Article 82(a) has generally involved excessive pricing (although unfairly low pricing by a dominant purchaser could also fall under

Article 82(a)).[16] The fairness of the price may be determined by reference to the costs of providing the product, by reference to the prices in comparable markets, or by relation to the intrinsic value of the product.[17] The first criterion—cost plus a reasonable margin—is a filter. If the price does not exceed costs by an unreasonable margin, the price cannot be unfair.[18] The second and third criteria are alternatives. If the price is significantly higher than the prices charged in a comparable market (usually a comparable *geographic* market), then the price may be deemed unfair.[19] A price may also be unfair if the price exceeds the 'intrinsic value' of the product. This concept is not fully developed in the case law, but it is clear that in the Commission's view it includes the intangible value that the product contributes to the products provided by customers of the dominant firm.[20] It is also clear that the existence of valuable intellectual property rights itself usually creates a higher intrinsic value.[21]

It will be very difficult to apply these principles to find that a royalty in an intellectual property licence constitutes an unfair price. Comparisons of royalty rates between markets requires identification of separate geographic markets, yet in most cases the relevant geographic market for technology licensing will be EU or worldwide.[22] The calculation of an intrinsic value faces almost insuperable difficulties in identifying relevant costs. In many cases the costs of developing a technology cannot be allocated between the profits of exploitation by the licensor and the revenues from licensing, which will make identification of a separate 'intrinsic value' for a license difficult. Even where expenses of research and development are exclusively recovered from licensing revenue (as in the case of pharmaceutical product development by non-integrated research groups) the costs of failed development projects must also be covered by the revenue of successful products. Finally where a technology is dominant, the intangible **10.17**

---

16 Thus in principle a licensor could have a claim against a licensee dominant in its downstream market for imposing an unfairly low royalty level.

17 *United Brands* v *Commission* [1978] ECR 207; *SACEM* v *Lucazeau* [1989] ECR 2811.

18 See n 17 above.

19 *SACEM* v *Lucazeau* 1989 ECR 2811. This of course raises difficult issues as to whether the market is comparable in terms of the nature of the products sold and the nature of demand. Thus in *Swedish Ports*, the Commission rejected a comparison of prices at the port of Helsingborg with prices at the Danish port of Elsinore, since the Swedish port had a much more developed infrastructure and costs at the two ports were different.

20 In the *Swedish Ports* cases, the Commission considered that the location of Helsingborg at the narrowest part of the Oresund channel was a factor that contributed to the value of the services provided at the port.

21 *Parke Davis* v *Proebel* [1968] ECR 55; *United Brands*, see n 17 above.

22 See Chapter 6. If the comparison market is outside the EU the likelihood of objective distinguishing factors is far greater.

benefits must be taken into consideration in determining whether a royalty is excessive. In most cases the value of an indispensable technology in the downstream market should justify a high royalty.

(2) Unfair licence conditions

10.18 Unfair trading conditions generally cover contractual terms that are not reasonably required to protect the commercial interests of the dominant licensor and are onerous for the other contracting party.[23] Because terms can be burdensome for a customer without necessarily leading to competitive harm, it is possible for contractual terms to be 'unfair' even though the agreement in question does not infringe Article 81.[24]

10.19 There is not enough experience with the application of Article 82(a) in the licensing context to provide more than anecdotal consideration of what constitutes an unfair licence term. It is clear that in the licensing context a broad range of terms will be regarded as necessary for achieving the objectives of the licence and therefore clearly 'fair'. The various provisions discussed in Chapters 7 and 8, to the extent that they are compatible with Article 81, should usually also be compatible with Article 82(a) to the extent that they generate identifiable economic benefits. There is a suggestion, however, that even some provisions clearly beneficial under Article 81 may infringe Article 82(a) if they are over-broad in their application. For example, obligations on licensors to provide information on sales may need to be restricted to sales of products incorporating the licensed technology.[25] Likewise a license-back obligation, for example, may need to be restricted to the licensed technology.[26] Clauses allowing a dominant licensor to terminate a licence in the event of litigation may create a undue burden on the licensee,[27] as may licence agreements that are insufficiently transparent in

---

[23] *Tetra Pak II Rausing SA* v *Commission* [1990] ECR II-309.

[24] This reflects the fact that 'fairness' is in it a goal under Article 82 as traditionally applied, in addition to economic efficiency and consumer welfare.

[25] See *Philips/Sony* Report on Competition Policy 2003, 197. This case involved undertakings accepted by Commission to resolve Article 82(a) complaints. Since the licensors did not contest application of Article 82(a) and the Commission generally wishes to close the file by reaching a solution satisfactory to the claimant, they should not necessarily be read as defining the law in this area.

[26] *Philips/Sony (2003)*. See also *Elonex/Microsoft*, Report on Competition Policy 2000, 152 (Commission rejected complaint that Microsoft grant-back provisions covered products produced by other Microsoft licensees) and *BRT/SABAM*, n 4 above (Court indicated that terms in agreement between copyright society and copyright owner could be abusive because they required assignment of future rights after termination and required assignment of *all* types of exploitation).

[27] In *AllVoice/IBM*, Report on Competition Policy 2002, 205, the Commission challenged a clause in IBM's licence to AllVoice that allowed IBM to terminate if AllVoice brought

describing the licensed technology and the terms under which royalties are payable.[28]

D. Discrimination

Article 82(c) prohibits **10.20**

> 'applying dissimilar conditions to equivalent transactions with other trading parties, thereby placing them in a competitive disadvantage'.

This provision on its face incorporates a three step test. First, conditions should be 'dissimilar'. Secondly, the transactions must be equivalent, and thirdly, the 'victim' of discrimination must be placed at a competitive disadvantage.

(1) Dissimilar conditions

The wording of the first requirement is important because Article 82(c) does not **10.21** require identical treatment. As long as the conditions offered by the dominant firm are 'similar', Article 82(c) is satisfied. This means that contractual terms between customers can vary as long as those terms do not significantly affect the costs imposed on the customer. It also means that minor price differences should not be caught by Article 82(c), nor should differences in methodologies for price calculation, as long as the resulting price is 'similar'. The European Court has also ruled that a dominant firm can, consistent with Article 82(c), charge one customer for a service, while providing the same service 'for free' to another customer, in return for non-cash benefits provided by the second customer, as long as the value of those benefits approximates or exceeds the price charged to the first party.[29]

(2) Equivalent transactions

The Commission and the Community courts have traditionally regarded trans- **10.22** actions as 'equivalent' unless there are significant differences in the costs of the dominant firm in dealing with different customers.[30] Even volume discounts, which are usually not policed under Article 82, may be suspect where the dominant firm provides an essential service to its customers.[31] In the licensing context, however, the Court has indicated that licensors are entitled to vary their

---

infringement proceedings under its own patents against IBM or any of its customers. The Commission accepted an agreement by IBM to limit the clause to cases where AllVoice sued IBM (excluding customers from its protection).

[28] *Philips/Sony (2003)*, see n 25 above.

[29] *GT Link* v *A/S DSB C42/95* [1997] ECR 14449.

[30] E.g. *Commission* v *Portugal* [2002] 4 CMLR 319.

[31] Ibid.

pricing in accordance with the intensity of use by the licensee.[32] Thus in *Coditel v Ciné-Vog* (*Coditel I*),[33] the Court ruled that 'the right of the owner of the copyright in a film and his assigns to require fees for *any* showing of that film is part of the essential function of copyright'.[34] The Court observed that this right was essential to the intellectual property right because of the nature of a performance right, exploitation of which can be 'infinitely repeated'. It follows from this ruling that variations in royalty rates according to intensity of use cannot be banned under Article 82(c), since to do so would be to limit the essential function. Although this ruling on its face only applies to copyright, the logic applies to any right where exploitation can be infinitely repeated (and where intellectual property law would allow a licence limiting the number of times that the technology is used).

10.23 Article 82(c) also does not apply where customers are in different geographic or product markets.[35] In this situation, even though the price of providing the product for the dominant firm is the same, the different market conditions mean that the transaction is not equivalent. Article 82(c) can apply to price differences between customers in different geographic areas in the same market.[36]

(3) Competitive disadvantage

10.24 The case law under Article 82(c) has nearly read the requirement that dissimilar treatment leads to a 'competitive disadvantage' out of the Treaty.[37] Perhaps the most extreme example of this tendency was the in *GVL* decision where the Commission ruled that a difference in treatment by a German performing rights society that cost the individuals affected the equivalent of 15–20 euros annually put each performer at a competitive disadvantage compared with the performers that received the payment. Thus, although reactivation of this requirement is a valid subject in the ongoing Article 82 review, this requirement does not play a practical role in application of Article 82(c) today.

(4) Application to IP licensing—single firm dominance

10.25 There is little precedent on the application of Article 82(c) to differences in royalty rates or other conditions. While, as a general matter, a dominant licensor may be required to provide similar conditions to similarly situated licensees, that requirement is limited by the following principles:

- Where a licensor is entitled to limit a licensee to a defined field of use, a

---

32 [1980] ECR 881. 33 Ibid. 34 Ibid. (emphasis added).

35 *Tetra Pak II*, see n 23 above. 36 *United Brands*, see n 5 above.

37 *Hoffmann-La Roche*, see n 2 above.

licence granted to a different licensee for a different field of use should not be regarded as a comparable transaction. This follows from the same considerations that mean that Article 82(c) does not apply to different markets. Just as differences in the commercial considerations applicable to different product or geographic markets exclude application of Article 82(c), the different commercial considerations applicable to licensing for different applications should justify different treatment.

- On the same basis, licences for different packages of fields of use or licences including different rights should not be regarded as equivalent. Article 82(c) does not require a seller to assign specific prices to different elements of a product sold as a package.
- Where a licensee will not use the licensed technology at all, as in the case of a licence to allow design freedom, the licensee is in a different position from a licensee that needs the licence in order produce products or services using the licensed technology. Article 82(c) should not bar a licence to the 'design freedom' licensee at a lower rate or even on a royalty-free basis, because the transactions are not equivalent.[38]
- The facts that cross-licences are made on a low royalty or royalty-free basis and that third party licensees are made to pay higher royalties do not infringe Article 82(c), as long as the value provided to the licensor in return for the licence is comparable to that provided by the licensees paying cash royalties.[39]

(5) Joint dominance

The application of the abuse rules in cases of joint dominance has not been a subject for extensive discussion by either the Commission or the Community courts. The indication so far has been that, once joint dominance is found, each firm that forms part of the joint group is subject to Article 82 in respect of its individual conduct, in exactly the same manner as if it possessed single firm dominance.[40] There is, however, a strong argument that Article 82(c) cannot be applied to individual firms that are part of a group holding a joint dominant position, where those firms are in a horizontal relationship. If the **10.26**

[38] Logically this position should be assessed in respect of each licensee at the time that the licence agreement is made, since this is the 'transaction' referred to in Article 82(c).

[39] See, n 29 above. In the cross-licence context, this value may logically consist of licences for technology in different markets or the benefits of commercial cooperation in other fields.

[40] Thus in *CEWAL* [1993] OJ L 34/20, predatory pricing by individual shipping lines were deemed to be abusive. Likewise in *Irish Sugar* [2001] 5 CMLR 1082, where the Commission found 'vertical joint dominance' involving Irish Sugar and a distributor, discrimination by Irish Sugar was deemed to infringe Article 82(c), although it may be noted that in that case the economic position of Irish Sugar in the sugar market was the same as it would have been if the Commission had proceeded on a single dominance theory.

jointly dominant firms are in an oligopolistic market (which will almost always be the case where there is joint dominance), application of Article 82(c) will significantly increase the likelihood of collusion. The possibility of 'cheating' (by offering a lower price to selected customers) is often a key factor that prevents informal collusion and undermines formal cartels. If, however, offering a lower price is prohibited by Article 82(c), then this discipline disappears. This result makes no sense from a policy perspective and is not required by the wording of Article 82.[41]

10.27 The application of Article 82(c) to pricing by firms in a joint dominant position is particularly troubling where the joint dominance involves intellectual property licensors involved in standard setting or pooling arrangements. As explained in Chapter 9, a central feature in the application of Article 81 to such arrangements has been the requirement that licensors remain free to license to third parties. The discipline provided by this requirement will, however, disappear if the licensors are required to offer the same terms to all licensees. Here again, the result makes no sense from a policy perspective and is not required by the wording of Article 82.

10.28 If Article 82(c) is deemed to apply to licensing by individual firms forming part of a jointly dominant group, it is still necessary to consider carefully whether conditions are actually 'dissimilar' or transactions 'equivalent'. It is necessary in this context to distinguish between three groups of licensees:

(A) other licensors participating in a pool (who will often license each other under cross-licences rather than under a pool licence);
(B) licensees from the pool; and
(C) third party licensees outside the pool. This leads to five potential discrimination scenarios:
 (i) discrimination between licensees in Group A;
 (ii) discrimination between licensees in Group B;
 (iii) discrimination between licensees in Group A and Group B;
 (iv) discrimination between licensees in Group A and Group C; and
 (v) discrimination between licensees in Group B and Group C.[42]

10.29 The application of Article 82(c) to scenarios (i) and (ii) appears relatively

[41] Article 3(g) of the Treaty, which mandates the creation of a system ensuring that competition is not distorted, requires an interpretation of Article 82 that does not increase the likelihood of consumer harm. In this case, it would not be difficult to interpret Article 82(c) as applying only to joint products offered by the jointly dominant firms, including cases where those joint products are offered separately.

[42] Discrimination by jointly dominant firms against each other (discrimination *within* Group C) is not considered because it is unlikely to lead to complaints and because the 'transactions' predate the dominance.

straightforward. Subject to different requirements of licensees (who may require different packages of rights) and subject to the ability of the licensor to differentiate according to intensity of use, Article 81(c) could be applied to require similar treatment of third parties for licences outside the pool. Likewise the offer of pool licences on a non-discriminatory basis is pretty clearly covered by Article 81(c) and is also probably mandated by FRAND requirements under Article 81.

The complex questions involve alleged discrimination between the various groups of licensees. In this context, the 'transactions' involved in licensing by pool licensees appear fundamentally different from those with either pool participants or third parties—the rights licensed are different and the licence is granted at a point where the technology is defined and its value capable of assessment. Furthermore, the cross-licences with other pool members do not necessarily involve dissimilar conditions, since even royalty-free licences fall outside Article 82(c) where the licensor receives a similar (or greater) level of benefits than is received from other licensees.[43] This suggests that Article 81(c) should not usually be applicable in scenarios (iii) and (v). On the same basis, the licensing transaction with a third party (scenario iv) may not be equivalent to a cross-licence with a pool participant, and the terms in the end may not be dissimilar. **10.30**

E. Tying or Bundling (Article 82(d))

As explained in Chapter 7, 'tying' describes the practice where the sale of one product (the tying product) is conditional on the purchase by the customer of a second product (the tied product). 'Bundling' describes the practice by which the seller provides the two products as a package, so that the contractual obligation is created by the structure of the product offering rather than as an obligation on the customer to purchase the tied product. Since the application of Article 81 to these practices is now linked to significant market power in the market for the tying product, the analysis under Article 81 and Article 82 is now essentially similar. Reference is therefore made to the discussion in Chapter 7, which is relevant to Article 82 as well. **10.31**

This section focuses on the special features applicable to a tying bundling **10.32**

[43] In many cases pool participants may have cross-licensing arrangements with each other that predate the formation of a particular pool or agreement on a standard for a particular technology. In many cases these arrangements may have been established to allow design freedom. The benefits accruing to each party from the full range of arrangements between them must be taken into account in comparing the value received from cross-licensors to that paid by pool licensees or third parties.

analysis under Article 82. The first is the treatment of bundling. While there may be a question as to whether Article 81 is applicable in the absence of a contractual 'tie', the potential application of Article 82 is not questioned since Article 82 also applies to purely unilateral activity. This principle is also applicable where the 'bundling' occurs as a result of the design of the product. The Commission has taken the view that product design that integrates two separate products and does not allow for purchasing either on its own, constitutes a form of bundling caught by Article 82.[44] The second special feature applicable to tying analysis under Article 82 involves proving competitive harm. Application of Article 81 normally requires a showing of competitive harm in respect of the tied product, while Article 82(d) on its face does not require a showing of competitive harm in that market. A third special feature of Article 82 tying analysis is that provision of an additional product feature at no increased price can constitute tying, whereas under Article 81 it is normally assumed that sale of the tied or bundled product will increase revenues.[45]

F. Mandatory Licensing (Article 82(b))

10.33 The most contentious issue in the application of Article 82 to licensing involves the circumstances where a dominant licensor may be obliged to grant a licence. It has long been understood that the essential feature of an intellectual property right is the ability to exclude rivals from exploiting the protected innovation. Limiting this right in order to preserve or create 'competition' can be viewed as a direct conflict between the goals of the intellectual property system in encouraging innovation and the goals of the competition law system.

10.34 The current position is that a licensor may be obliged to grant a licence, but only under 'exceptional circumstances'. The Court of Justice established in *IMS Health* that such circumstances exist in the case where the potential licensee wishes to exploit the licence to produce a product for which there is an unmet consumer need. The Commission has asserted in its *Microsoft* decision, however, that the criteria applied in *IMS Health* are not exhaustive. The Commission suggested in that case that other criteria, particularly the withdrawal of cooperation in respect of a previous generation of products, could establish the existence of exceptional circumstances. To understand the basis

[44] *Microsoft* COMP/37.792 2004.

[45] Thus in *Microsoft* the Commission applied Article 82(d) to the bundling of the Windows Media Player ('WMP') with the Windows operating system, even though WMP was offered at no charge (as were media players provided by rivals of Microsoft). The Commission argued that there was nonetheless competitive harm because of the impact that dominance for WMP would have on further related markets, such as the market for licensing content providers.

for the current position and the issues that are on the table, it is necessary to consider the way that these rules have evolved in the very particular context of these cases.

(1) *Distinguishing between mandatory licensing and licensing as a remedy*

Where a firm infringes Article 82 (or indeed Article 81) the most effective way to bring the infringement to an end may be to require the infringing firm to grant a licence under the intellectual property rights that it controls.[46] Indeed the imposition under Article 81 of requirements to license under FRAND terms can be viewed as an *ex ante* remedy, as it imposes the condition that would otherwise be required to cure the Article 81 infringement caused by a standard-setting or pooling arrangement. **10.35**

The application of this principle in the Article 82 context was established by the Court of Justice in *Volvo* v *Veng* 1988 ECR 6211. That case arose out of a suit alleging infringement of Volvo's UK design rights for its front wing panels. Veng defended the case on the basis that Article 82 required Volvo to supply panels or license him to make panels. The ECJ ruled that Volvo was under no obligation to license Veng under Article 82, because this would deprive Volvo of the substance of its intellectual property right.[47] The Court also indicated, however, that an obligation to license (or in the circumstances of the case a limitation on Volvo's ability to assert its rights) could arise where Volvo had engaged in other conduct that constituted an abuse of its dominant position in respect of the products protected by the intellectual property rights: **10.36**

> 'It must however be noted that the exercise of an exclusive right by the proprietor of a registered design in respect of car body panels may be prohibited by Article [82] if it involves, on the part of an undertaking holding a dominant position, certain abusive conduct such as the arbitrary refusal to supply spare parts to independent repairers, the fixing of prices for spare parts at an unfair level or a decision no longer to produce spare parts for a particular model even though many cars of that model are still in circulation, provided that such conduct

---

[46] For an example of a decision imposing a licensing requirement as part of a remedy see *Gruner Punkt* (German recyling organisation required to authorise use of its trademark by third parties offering recyling services). The principle of proportionality limits the application of this remedy. Mandatory licensing should only be required where less restrictive remedies would not be sufficient to bring the infringement to an end.

[47] 'It must also be emphasized that the right of the proprietor of a protected design to prevent third parties from manufacturing and selling or importing, without its consent, products incorporating the design constitutes the very subject-matter of his exclusive right. It follows that an obligation imposed upon the proprietor of a protected design to grant to third parties, even in return for a reasonable royalty, a licence for the supply of products incorporating the design would lead to the proprietor thereof being deprived of the substance of his exclusive right, and that a refusal to grant such a licence cannot in itself constitute an abuse of a dominant position.' Point 8.

is liable to affect trade between Member States.' There is no attribution for this quotation.

Thus where abuse in the supply of spare parts could be shown (in this case, excessive prices, refusals to supply independent repairers, or termination of production), the remedy for that abuse could be to prevent the dominant firm from enforcing its intellectual property rights, a form of mandatory licence.

(2) Mandatory licensing and the general law of refusals to supply

10.37 Article 82 has not been interpreted to create a general obligation on dominant firms to deal with potential customers. Instead, Article 82 has been interpreted to create obligations on dominant firms in specific situations. Thus dominant firms are limited in their ability to terminate deliveries to existing customers.[48] Dominant firms may be required to cooperate with rivals where such co-operation is a normal practice in the industry.[49] Most importantly, dominant firms may be required to deal where failure to do so will affect competition in a downstream market where the dominant firm is active.[50] Refusals to deal with customers in downstream markets is condemned as a form of 'leveraging' use of market power to obtain a competitive advantage in a second, related market.[51]

10.38 The circumstances where a dominant firm is required to allow its downstream competitors access to inputs that it controls are now defined in the context of what is referred to as the 'essential facilities doctrine'. In *Oscar Bronner* v *Mediaprint*[52] the ECJ established strict conditions for application of this rule, ruling that a claimant must show:

- The facility was indispensable to access to a market.[53]
- Without access to the incumbent's service, it would be completely excluded from the market.[54]
- Even if its resources were comparable to the incumbent in the primary market, it would still not be able to create a facility that would allow it entry to the market for providing the facility.[55]

---

48 *United Brands*, see n 5 above. 49 *Aer Lingus/British Midland* [1992] OJ L 96/34.

50 See *ICI and Commercial Solvents* v *Commission* [1974] ECR 223.

51 Ibid. See also *Tetra Pak International* v *Commission* [1994] ECR II-755.

52 *Oscar Bronner GmbH & Co.KG* v *Mediaprint* [1998] ECR 1-7791.

53 Ibid. para 41, 'refusing a third party in an after market access to an essential facility would only amount to an abuse when ... "the service in itself is indispensable to carrying on that person's business, inasmuch as there is no actual or potential substitute in existence".'

54 Ibid. para 41. See too *European Night Services* v *Commission* [1998] ECR II-3141.

55 Cf. Forester, I, 'Compulsory licensing of intellectual property rights in Europe: A rare case of aberrant intellectual property rights'. Paper presented to DoJ/FTC Hearings on Competition and Intellectual Property Law and Policy in the Knowledge Based Economy, *Comparative Law Topics, 22 May, 2002.*

- 'For such access to be capable of being regarded as indispensable, it would be necessary at the very least to establish . . . that it is not economically viable to create a second home delivery scheme for the distribution of daily newspapers with a circulation comparable to that of the daily newspapers distributed by the existing scheme.' (para 46)

The obligation in cases that do not involve inputs protected by intellectual property rights on dominant firms to deal with downstream rivals under defined circumstances provides the necessary background for consideration of the cases where intellectual property rights are involved: the Court judgments in *Magill* and *IMS* and the Commission decision in *Microsoft*. **10.39**

(a) Magill

The first case to find that a refusal to license intellectual property rights could constitute an abuse under Article 82 was the 1989 Commission decision in *Magill*.[56] This case involved the attempt by the publisher of a weekly news magazine (Magill) to introduce a comprehensive weekly TV guide on the Irish market that would compete with the channel-specific weekly TV guides sold separately by the respective TV companies. The broadcasters each brought successful actions against Magill for breach of their copyright in the programme information.[57] Following a complaint from Magill, the Commission ruled that the refusal to allow use of the full programme information constituted an abuse under Article 82. In its decision, the Commission found that this practice prevented the meeting of a substantial potential demand existing on the market for comprehensive TV guides.[58] Following a lengthy judgment from the CFI in 1991,[59] the case was finally resolved by the Court of Justice in 1995, nearly six years after the initial Commission decision.[60] **10.40**

The proceedings before the CFI judgment and the ECJ prompted considerable debate on the appropriate balance between the protection of intellectual property rights and the need to promote competition. One feature of that debate was discussion of derivative nature of the market for TV guides. It was suggested **10.41**

[56] *Magill TV Guide/ITP, BBC, and RTE* [1989] OJ L 78/43 (decision dated 21 December, 1988).

[57] The programme information had been provided to Magill subject to a contractual limitation on its use. (Magill was allowed to print weekly highlights in its news magazine but not complete listings.)

[58] The Commission observed that 'this demand results from the advantages offered by comprehensive TV guides, i.e. making advance weekly listings for a great number of programmes available to the consumer in a reasonably practical way and without having to pay a considerable amount of money'.

[59] Case T-69/89, *RTE* v *Commission* [1991] ECR II-485.

[60] *RTE, BBC and ITV* v *Commission* [1995] ECR I-743.

that intellectual property rights were less worthy of protection where used to 'extend the monopoly' to further markets.[61] The ECJ, however, did not directly address these arguments. Instead, after acknowledging the general interest in preserving the integrity of intellectual property rights, the Court ruled that such rights could nonetheless be put aside in 'exceptional circumstances'. The Court upheld the Commission decision finding that exceptional circumstances existed where there was consumer demand for the new product, where the TV companies had a *de facto* monopoly over the listings by virtue of their scheduling of TV programmes, where a licence of the listings was an indispensable input for the comprehensive TV guide, and where they were not themselves supplying the product to consumers. The Court emphasised the significance of the leveraging aspect of the case, observing that 'the appellants, by their conduct, reserved to themselves the secondary market of weekly television guides by excluding all competition on that market . . . since they denied access to the basic information which is the raw material indispensable for the compilation of such a guide'.[62]

(b) IMS Health

10.42 The limits of the *Magill* exceptional circumstances rule were tested in the complex litigation involving IMS Health. IMS is a company that collects and collates data on sales of pharmaceutical products to pharmaceutical companies. To provide data in a manner useful to the customers while preserving anonymity for specific outlets, data is aggregated on the basis of geographic units. In Germany, IMS developed a system based on 1860 geographic units—the 1860 Block System—for this purpose. (IMS has more recently introduced a new structure with 2847 blocks). An ex-director of IMS founded a rival firm—PII—to provide pharmaceutical data collection services in competition with IMS. When PII began using the 1860 Block System for reporting data, IMS brought an action in German courts to block use of its copyrighted system. On 27 October 2000, the German trial court issued an interim order barring use of the 1860 Block System 'or any derivative thereof' by PII.

10.43 On 17 September 2002, a German appellate court affirmed the interim order against PII, but indicated that the scope of the order should be defined narrowly (i.e. use of standard geographic units similar to or overlapping with those used

[61] Advocate General Gulmann argued against this position in his Opinion.

[62] Citing *Commercial Solvents*, see n 50 above. The link between the *Magill* rule and the general rule requiring dominant suppliers to deal with downstream competitors in limited circumstances was reaffirmed in *Oscar Bronner* which cited *Magill* as a precedent for the essential facilities rule. In effect, any obligation to supply involves exceptional circumstances; a more intrusive obligation (i.e. mandatory licensing) imposes higher requirements of 'exceptionality'.

in the 1860 Block System should not necessarily be regarded as derivative of the 1860 Block System). The trial on the merits was suspended, however, because the trial court had referred to the Court of Justice the question of whether the refusal of IMS to license the 1860 Block System could constitute an abuse of a dominant position under Article 82.

In the meantime, NDC (which had acquired PII) complained to the Commission that the refusal by IMS to allow use of the 1860 Block System breached Article 82. On 7 March 2001, the Commission issued a decision imposing interim measures on IMS, which the Commission found to be the dominant provider of statistical services to the pharmaceutical industry in Germany.[63] The Commission decision required IMS to allow NDC to use the 1860 Block System, finding that access to the block system was necessary for effective entry into the market for providing statistical services and refusal was not objectively justified. In October 2001, the President of the Court of First Instance issued an order suspending the effectiveness of the Commission decision.[64] On 13 August 2003, the Commission issued a decision withdrawing the interim measures, ostensibly on finding that the *Camera Care*[64A] grounds for interim measures are no longer satisfied because NDC no longer has an urgent need for protection.[65] **10.44**

In the proceedings before the Court of Justice on the reference from Germany, the Commission and IMS repeated the arguments that they had rehearsed before the Court of First Instance in the appeal from the Commission decision. IMS argued, in effect, that a dominant firm may be obliged to license its intellectual property rights in only two circumstances: (i) where it engages in additional abusive behaviour alongside the refusal to license; or (ii) where it controls an 'essential facility'. The second situation could only arise, according to IMS, in the circumstances identified in *Magill*—the refusal should involve a firm dominant in an upstream market refusing to license in a manner that impedes development of a related downstream market. Where the refusal involves access to the primary market where the dominant firm is active, according to this argument, compulsory licensing would constrain the existence of the intellectual property right itself and not merely its exercise. **10.45**

The Commission for its part, argued that intellectual property is a property right like any other, so that its exercise can be constrained in the interest of protecting competition. According to the Commission, where access to an 'input' that constitutes an essential facility is protected by an intellectual property right, **10.46**

[63] *NDC Health/IMS Health* [2002] OJ L 59/18.
[64] Case T-184/01 *IMS Health Inc/Commission* [2001] ECR II-3193.
[64A] [1980] ECR 119.
[65] *NDC Heath/IMS Health: Interim Measures* [2003] OJ L 268/69.

the right holder can be required to grant access, just as the owner of a physical facility can be required to grant access to that facility. The Commission asserted that the obligation to license in this context does not infringe on the 'essential function' of the right, because the refusal to give access to the 'input' extends the power of the intellectual property holder into a derivative activity.

10.47 The Court of Justice held that the three main conditions of *Magill* 'exceptional circumstances' test were cumulative, i.e. that in order for a refusal to license new entrants to a market dependent upon an indispensable intellectual property protected input, to be abusive, the refusal must meet three conditions:

(i) the undertaking which requests the licence intends to offer, on the market for the supply of data in question, new products or services not offered by the copyright owner and for which there is potential consumer demand;
(ii) the refusal is not justified by objective considerations;
(iii) the refusal is such as to reserve to the copyright owner the market for the supply of data on sales of pharmaceutical products in the Member State concerned by eliminating all competition on that market.

Thus the Court rejected the argument, which was the basis for the Commission's own interim measures decision in *IMS Health*, that the need to foster competition on the downstream market was by itself an essential circumstance that justified a mandatory licence. The Court's interpretation of the 'exceptional circumstances' test to exclude 'clones' or 'me too' products at least attempts to reconcile competition law and intellectual property rights based on their mutual interest in innovation.

10.48 In *IMS*, however, the Court of Justice also seemed to make it a point to indicate that although the *Magill* conditions were cumulative, they did not offer an exhaustive definition of the test of 'exceptional circumstances'. It carefully referred to *Magill* as a case in which the Court held that 'such exceptional circumstances were present in the case'. The Court also held 'that it is clear from the case law . . . that "it is sufficient" to satisfy the three *Magill* criteria in order to show an abusive refusal to license'.[66] That choice of words hints at the possibility that the three Magill criteria were not essential criteria for all categories of exceptional circumstances. That proposition would seem to follow from the purpose of Article 82(b), which is to prohibit as abusive conduct by dominant firms where such conduct limits *technical development of markets* to the detriment of consumers. Limiting technical development is a wider concept than the particular factual circumstances and conditions of the *Magill* case. Hence, both the language of Article 82(b) and the ECJ judgment in *IMS*

[66] See n 64 above. Para 38.

suggest that other types of abuse can also fall within the category of 'exceptional circumstances'.

(c) Microsoft

The Commission decision in *Microsoft*,[67] which is currently under review in the CFI, directly raises the question of whether the exceptional circumstances test for refusals to license can be extended beyond the specific criteria identified in *Magill*. The EU investigation of Microsoft commenced in 2000 following a complaint by Sun Microsystems, one of Microsoft's most important competitors in the workgroup server market. Sun complained that Microsoft was leveraging the Windows 2000 and Microsoft's Office Suite monopoly to obtain a further monopoly for Microsoft's workgroup server operating system (OS) in the workgroup OS market. A key part of this leveraging strategy, according to Sun, was Microsoft's refusal to provide adequate information about the interface codes that would enable Sun servers to interoperate smoothly with Microsoft's 'integrated' package of Windows 2000, Office Suite and workgroup server operating system. The refusal of Microsoft to disclose how the integration between Windows and Office Suite and its server operating system worked limited Sun's ability to offer certain services to Windows-based users of its own non-Microsoft workgroup server. **10.49**

In a decision issued on 24 March 2004, the Commission found that Microsoft had abused its near monopoly in the Windows operating system by deliberately restricting interoperability between the Windows OS and non-Microsoft workgroup servers such as those operated by Sun Microsystems.[68] Although the Commission acknowledged that at least some of the information sought by Sun could be protected by intellectual property rights, the Commission found that the refusal to provide the information was nonetheless abusive, because the case presented exceptional circumstances. Rejecting Microsoft's contention that Article 82 could not be applicable because this case did not satisfy the *Magill* criteria, the decision states that there is no exhaustive checklist of exceptional circumstances.[69] The Decision states that the Commission is entitled to take into account 'other circumstances of exceptional character when assessing a **10.50**

---

[67] Case T-201/04 R *Microsoft* v *Commission* (see order of the Preseident of the Court of First Instance, 26 July 2004).

[68] See Press Release Conclusion of Microsoft investigation, 24 March 2004, IP/04/382.

[69] The argument that the *Magill* criteria are not met is based on the suggestion that Sun and others did not seek to satisfy an unmet consumer need. Although the Commission does not address this argument directly, there is a suggestion that access to interface information might still be important to allow development of as yet undefined products that would meet new consumer needs in the future.

refusal to supply'.[70] The Decision concludes that it was necessary to analyse the 'entirety of the circumstances' and take a 'decision based on the results of a comprehensive investigation'.[71]

10.51 The reliance by the Commission on the 'entirety' or 'the totality of the circumstances' as a basis for identifying exceptional circumstances may be too open-ended to attract support from the CFI or ECJ. The whole purpose of the 'exceptional circumstances' test in the first place was to ensure that the category was narrow and confined to those cases where the owner of the intellectual property right was engaging in conduct that was so clearly anticompetitive under the Treaty that competition law had to limit the exercise of the intellectual property right. The decision does, however, disclose additional factors that might usefully qualify as creating exceptional circumstances in addition to the criteria identified in *Magill*.

10.52 The first such factor is that Sun was not a new entrant seeking to copy a Microsoft product. Instead Sun Microsystems was offering an innovative product for which there was substantial and demonstrable demand. The Sun Solaris workgroup server OS was not only *not* a clone of the Microsoft server, it actually preceded it in the market. The Commission could therefore legitimately impute to Microsoft the exclusionary motive of using its control over the PC OS market to evict an *innovating* competitor, i.e. conduct that amounted to an abuse of 'technological' leveraging of its dominance.[72]

10.53 The second potential factor that could support a finding of exceptional circumstances is the Commission finding that Microsoft's refusal to provide interface information represented a change of policy. According to the Commission, Microsoft had made such information available to Sun and to others previously. As already noted above, a feature of the general law on obligations to deal under Article 82 is that termination of relationships with existing customers is regulated more strictly than creation of new relationships. In a non-IP case, a dominant firm with a *de facto* monopoly on a necessary input cannot terminate an existing customer of a product where it competes itself on the downstream market, unless it can offer an objective justification. Thus the relevance of existing relationships as a criterion for establishing special duties to deal under Article 82 is already recognised. The argument for recognising this criterion in an intellectual property licensing context becomes more compelling where the dominant technology owner initially opted for an 'open' system as a strategy to grow and achieve dominance, creating legitimate expectations that such cooperation would continue.

[70] *Microsoft* para 555.
[71] Ibid. para 558.
[64A] [1980] ECR 119.

(3) *Refusals to license—unresolved issues*

As the discussion above indicates, the biggest unresolved issue in mandatory licensing under Article 82 involves the nature of 'exceptional circumstances'. The CFI and ultimately the ECJ can be expected to consider this question exhaustively in the years ahead. The better view must be that the *Magill* criteria are not exhaustive. It would be surprising, however, if the courts allow the Commission to find 'exceptional circumstances' based on the totality of circumstances in a given case. In the meantime, there are still unanswered questions about the *Magill* criteria, as applied in *IMS Health*. How different must a new product be from the one offered by the owner of the industrial standard? Must there be a showing that the new product was developed by the new entrant to the point that it simply requires the indispensable component supplied by the intellectual property right's owner? How strong must the potential unmet demand be to qualify a new entrant to a compulsory licence? How important is the extent of the investment by the intellectual property owner? What weight should be given to the fact that the incumbent intellectual property owner has its own product under development to meet the same need? These questions are likely to feature in litigation in the years ahead. **10.54**

Appendices

APPENDIX 1

The main features of the new Technology Transfer Regulation

1. The Scope of the Regulation (Articles 1 and 2)

The TTBER offers a wider scope for IPRs to be licensed than its predecessor because it extends the exemption to 'core' technologies based upon designs and software licensing as well as pure and mixed patents and know-how. The scope of the exemption has been delineated in two other major respects: First, 'technology transfer agreements' must consist of a contract made only '*between two undertakings*'.[1] Under Council Regulation 19/65 the Commission is not empowered to block exempt technology transfer agreements concluded between more than two undertakings. The agreement can stipulate conditions for more than one level of trade as long as it is concluded by two undertakings. For example, 'a licence agreement concerning not only the production stage but also the distribution stage, stipulating the obligations that the licensee must or may impose on resellers of the products produced under the license will be covered by the TTBER'.[2]

Secondly, the TTBER requires that the intellectual property licensing agreement must be '*for the production of contract products*', ie the licence must be primarily for the purpose of allowing the licensee to exploit the licensed technology in the form of the production of goods or services.[3] The term '*contract products*' includes goods manufactured or services provided with the licensed technology and applies to 'products incorporating the licensed technology'. The Guidelines clarify that this term is meant to cover two situations: (i) where the licensed technology is used in the production process, and (ii) where it is incorporated into the product itself. Furthermore, since the TTBER applies in all cases where technology is licensed for the purposes of producing goods and services, it can also apply where the licensor undertakes not to exercise his intellectual property rights against the licensee. Since the essence of a pure patent licence is the right to operate inside the scope of the exclusive right of the patent, 'it follows that the TTBER also covers so-called non-assertion agreements and settlement agreements whereby the licensor permits the licensee to produce within the scope of the patent'.[4]

Hence, the TTBER does not cover technology pools which tend to take the form either of agreements by two or more parties to pool their respective technologies and license them as a package or arrangements by two or more undertakings to license a third party and authorise it to license on the package of technologies.[5]

Nor does it apply to agreements that have sub-licensing as their primary object such as 'master licensing' agreements, authorising a licensee to grant licences to third parties (sub-licensees) for the

---

1 'Licence agreements concluded between more than two undertakings often give rise to the same issues as licence agreements of the same nature concluded between two undertakings. In its individual assessment of licence agreements which are of the same nature as those covered by the block exemption but which are concluded between more than two undertakings, the Commission will apply by analogy the principles set out in the TTBER.' Guidelines para 40.

2 See Recital 19 of the TTBER and Guidelines para 39.

3 See Recital 7 of the TTBER and Guidelines para 41.

4 Guidelines para 43.

5 See Guidelines para 41. Technology pools are dealt with in Chapter 9.

exploitation of the licensor's technology.[6] Agreements between the licensee and sub-licensees will be covered by the TTBER because they are bilateral agreements and involve the production of contract products. Moreover, the TTBER also covers agreements where the licensee is permitted to sub-license the licensed technology to third parties as long as the primary object of the agreement is the production of contract products.[7]

The TTBER can also cover 'subcontracting' where the licensor licenses technology to the licensee to manufacture certain contract products exclusively for the licensor.[8] It can apply as well to agreements for the licensee to carry out development work before obtaining a product or a process that is ready for commercial exploitation provided that a contract product has been identified. However, the Guidelines make it plain that the TTBER and the Guidelines will not apply to agreements whereby a technology is licensed purely for the purposes of enabling the licensee to carry out further research and development in various fields, such as the licensing of a technological research tool used in the process of further research activity.[9]

In cases of pure R&D agreements, however, it is often questionable whether Article 81 applies in the first place. There is rarely a restriction of competition unless some limit is placed on the marketing activity after a contract product has been identified. The Guidelines state this point more obliquely, but do not isolate the Article 81(1) issue from the Article 81(3) issue.

'The framework of the TTBER and the guidelines is based on the premise that there is a direct link between the licensed technology and an identified contract product. In cases where no such link exists the main object of the agreement is research and development as opposed to bringing a particular product to the market; in that case the analytical framework of the TTBER and the guidelines may not be appropriate. For the same reasons the TTBER and the guidelines do not cover research and development subcontracting whereby the licensee undertakes to carry out research and development in the field of the licensed technology and to hand back the improved technology package to the licensor. The main object of such agreements is the provision of research and development services aiming at improving the technology as opposed to the production of goods and services on the basis of the licensed technology.' [See Chapter 5]

The third condition is that the TTBER only applies to agreements in the form of a 'transfer'[10] with the primary object of transferring one of four defined 'core' technologies.

---

[6] In its individual assessment, the Commission will apply by analogy the principles set out in the TTBER and Guidelines to licensing agreements of the same nature as those covered by the block exemption but involving more than two undertakings. Guidelines para 40.

[7] Guidelines para 42.

[8] Where subcontracting involves the supply of equipment by the licensor to be used in the production of the goods and services covered by the agreement the TTBER will only apply where the licensed technology constitutes the primary object of the agreement. Guidelines para 44. Subcontracting is also covered by the Commission's Notice concerning the assessment of certain subcontracting agreements in relation to Article 81(1) of the Treaty (See OJ 1979 C 1, p 2) According to this notice, which remains applicable, subcontracting agreements whereby the subcontractor undertakes to manufacture certain products exclusively for the contractor generally fall outside Article 81(1). However, other restrictions imposed on the contractor such as the obligation not to conduct or exploit his own research and development may be caught by Article 81 (See point 3 of the subcontracting notice.)

[9] 'the TTBER and the guidelines do not cover research and development sub-contracting whereby the licensee undertakes to carry out research and development in the field of the licensed technology and to hand back the improved technology package to the licensor. The main object of such agreements is the provision of research and development services aiming at improving the technology as opposed to the production of goods and services on the basis of the licensed technology.' Guidelines para 45.

[10] The concept of 'transfer' implies that technology must flow from one undertaking to another normally taking the form of a licence for the use of the technology in exchange for royalty payments. It can take the form of an assignment provided that part of the risk associated with the exploitation of the technology remains with the assignor. Article 1(1)(b). It can also take the form of sub-licensing, whereby a licensee, with the licensor's consent, grants licences to sub-licensees for the exploitation of the technology.

The concept of 'transfer' implies that technology must flow from one undertaking to another normally taking the form of a licence for the use of the technology in exchange for royalty payments. It can take the form of an assignment providing that part of the risk associated with the exploitation of the technology remains with the assignor.[11] It can also take the form of sub-licensing, whereby a licensee, with the licensor's consent, grants licences to sub-licensees for the exploitation of the technology.

The transfer must have as its primary object the licensing one of four defined 'core' technologies. If that requirement is met, the licensed technology can include a package of other IPRs. The 'core' technologies as defined in TTBER[12] consist of patents, know how, copyright, software and design rights. The Commission has chosen to define 'patents' widely to include not only patents and patent applications, but also other IPRs such as utility models, applications for registration of utility models, designs, topographies of semiconductor products, supplementary protection certificates for medicinal products or other products for which such supplementary protection certificates may be obtained, and plant breeder's certificates. A major change in 2004 has been the inclusion of designs as a core technology by the device of defining it as a patent. The Commission has decided to retain its rather narrow definition of 'know-how' from the 1996 TTBER by continuing to define 'know-how' 'as a package of non-patented practical information, resulting from experience and testing,' which is 'secret,[13] substantial,[14] and identified'.[15]

A further new entrant to the category of core technology is computer programs. This offers belated recognition that transfers of technology frequently include software programs showing methods for production and processing. The computer software envisaged as core technology is included for the primary purpose of assisting in the production of the contract products and can be contrasted with the software licence that accompanies shrink-wrapped software sold purely to consumers.

The Regulation also allows a second tier of IPRs to be included in the licensing[16] package along with the core 'technology' as long as they are 'ancillary' provisions.[17] To meet the test of ancillarity, they must (i) not constitute the primary object of the agreement and (ii) be directly related to the manufacture or provision of the contract products. This formula is clearer than its predecessor. The test is whether the other IPR is included essentially to enable the licensee to better exploit the *core licensed technology*. If it appears that licensing the 'ancillary' IPR, say a trademark, is the real purpose of the agreement, it will not be exempted under TTBER.

---

[11] See Article 1(1)(b). [12] Article 1(1)(h).

[13] Article 1(1)(g) 'Secret' means that the know-how is not generally known or easily accessible.

[14] 'Substantial' means that the know-how includes information which is significant and useful for the manufacture or provision of the products covered by the licence agreement or the application of the process covered by the licence agreement. The Guidelines elaborate, 'the information must significantly contribute to or facilitate the manufacture or provision of the contract products. In cases where the licensed know-how relates to a product as opposed to a process, this condition implies that the know-how is necessary to produce the contract product. This condition is not satisfied where the contract product can be manufactured on the basis of freely available technology. However, the condition does not require that the contract product is of higher value than products produced with freely available technology. In the case of process technologies, this condition implies that the know-how is useful in the sense that it can reasonably be expected at the date of conclusion of the agreement to be capable of significantly improving the competitive position of the licensee, for instance by reducing his production costs.'

[15] 'Identified' means that it is possible to verify that the licensed know-how fulfils the criteria of secrecy and substantiality. This condition is satisfied where the licensed know-how is described in manuals or other written form. However, in some cases this may not be reasonably possible. The licensed know-how may consist of practical knowledge possessed by the licensor's employees. For instance, the licensor's employees may possess secret and substantial knowledge about a certain production process that is passed on to the licensee in the form of training of the licensee's employees. In such cases it is sufficient to describe the general nature of the know-how and to list the employees that will be or have been involved in passing it on to the licensee.

[16] Guidelines para 53. [17] Recital 9 Article 1(1)(b).

A good example is offered by the licensor authorising the licensee to use his trademark on the products incorporating the licensed technology. As the Guidelines state, 'the trademark licence may allow the licensee to better exploit the licensed technology by allowing consumers to make an immediate link between the product and the characteristics imputed to it by the licensed technology. An obligation on the licensee to use the licensor's trademark may also promote the dissemination of technology by allowing the licensor to identify himself as the source of the underlying technology.' The Guidelines go on to point out that 'where the value of the licensed technology to the licensee is limited because he already employs an identical or very similar technology and the main object of the agreement is the trademark, the TTBER does not apply'.[18]

The TTBER and Guidelines treat non-core IPRs such as trademark and copyright rather differently. On the one hand, for copyright, other than performance rights,[19] 'the Commission will as a general rule apply the principles set out in the TTBER and these guidelines when assessing under Article 81 such licensing of copyright'.[20] In contrast, the Commission will not extend the principles developed in the TTBER and Guidelines to trademark licensing where that constitutes the primary object of the agreement.[21]

II. The Duration of the Exemption Conferred on the Licensing Agreement

The exemption conferred on intellectual property licensing agreements by the new Regulation has a potentially longer duration than its predecessor. It can last 'as long as *the* intellectual property right in the licensed technology has not expired, lapsed or been declared invalid or, in the case of know-how for as long as the know how remains secret'.[22] If the know-how becomes publicly known as a result of action by the licensee, the exemption will continue to apply for the duration of the agreement. The Guidelines interpret this rather unclear statement of duration in Article 2 to mean that the block exemption will apply separately to each licensed property right covered by the agreement *but will continue in effect* until the date of expiry, invalidity or the coming into the public domain of the *last* intellectual property right which constitutes the core 'technology' as defined by Article 1 of TTBER.[23] This suggests that the royalty and other provisions relevant to the package of technologies continue in effect even if some of the core IPRs in the package have expired. The TTBER itself expires in 2014 so contracts cannot be expected to remain block exempted after that date. On the other hand licensing agreements which are self certified as

[18] See in this respect Commission Decision in *Moosehead/Whitbread*, OJ 1990 L 100, p 32.

[19] 'the licensing of rights in performances and other rights related to copyright is considered to raise particular issues and it may not be warranted to assess such licensing on the basis of the principles developed in these guidelines. In the case of the various rights related to performances value is created not by the reproduction and sale of copies of a product but by each individual performance of the protected work. Such exploitation can take various forms including the performance, showing or the renting of protected material such as movies, music or sporting events. In the application of Article 81 the specificities of the work and the way in which it is exploited must be taken into account. The Commission will therefore not apply by way of analogy the TTBER and the present guidelines to the licensing of these other rights.'

[20] 'The licensing of copyright for the purpose of reproduction and distribution of the protected work, i.e. the production of copies for resale, is considered to be similar to technology licensing. Since such licence agreements relate to the production and sale of products on the basis of an intellectual property right, they are considered to be of the same nature as technology transfer agreements and normally raise similar issues. See e.g. *Coditel (II)*, Case 262/81 [1982] ECR 3381.

[21] The Commission's view is that trademark licensing is more closely related to distribution agreements than technology licensing because it often occurs in the context of distribution and resale of goods. The Commission states in the Guidelines that where a trademark licence is directly related to the use, sale or resale of goods and services and does not constitute the primary object of the agreement, the licence agreement is covered by Commission Regulation (EC) 2790/1999 on the application of Article 81(3) of the Treaty to categories of vertical agreements and concerted practices [1999] OJ L 336, p 21.

[22] Article 2. [23] Guidelines para 55.

exempted can last longer than ten years, the limit imposed by TTBER 1996 on know-how, if the know-how remains secret or the patent remains valid

III. The Distinction Between 'Licensing Agreements Between Competitors' and 'Licensing Agreements Between Non-competitors (Article 1) [See Chapter 6]

The 1996 version of TTBER was based on the view that most intellectual property licensing agreements should be treated as potentially 'horizontal' agreements between competitors in part because the licensee often evolves into one as the result of the experience with manufacturing the new technology. Yet the overwhelming evidence was that at the time they are signed, most licensing agreements are actually 'vertical,' i.e., agreements between non-competitors. The economic realism of the new Regulation has resulted in a division of licensing agreements into two main categories—agreements between non-competitors or agreements between competitors and a recognition that the regulatory concerns are considerably greater in the case of horizontal agreements. This has resulted in a re-evaluation of prohibited restraints in the case of vertical licensing agreements and the creation of a deservedly more benign regulatory regime.

Moreover, the Regulation gives an economically enlightened definition of vertical 'agreements'. It applies not only to the paradigm vertical case of an agreement between an inventor and a manufacturer but also to an agreement between two manufacturers as long as they are not competitors in respect of the licensed product. In 'product markets' competitors are defined as 'actual' competitors, ie competing undertakings who in the absence of the technology transfer agreement would have been active on the relevant product and geographic markets on which the contract products are sold without infringing each other's intellectual property rights.[24] The one complication is that in product markets, 'competitors' also includes 'potential' competitors, i.e., those who are realistically in a position to undertake the necessary investments and accept the switching costs of entering the same market had the price of the product been raised.[25] In 'technology markets,' the definition of competitor is limited to 'actual' competitors.[26]

A further feature of the Regulation that is helpful to intellectual property licensing is that it defines the parties as 'competitors' or 'non-competitors' *at the time the contract is made* and will not allow the natural competition that may develop as the result of the licensing agreement as the manufacturing expertise of the licensee matures to affect the designation of the contract. Thus, if the parties are non-competitors at the time the agreement is made, they will not be re-designated for the purposes of the exemption during the duration of the agreement unless the agreement is materially amended.[27] The distinction between agreements between competitors and non-competitors has led to two specific legal effects within the TTBER: different market share limits to the safe harbour of the TTBER and different hard core restrictions in licensing agreements.

IV. The Market Share Thresholds (Article 3) [See Chapter 6]

If the new Regulation has been helpful to IP licensing in other respects, it has created complications for IP licensing by introducing a system of market share limits in order to harmonise it with the economic approach to Article 81 used in the design of the Vertical Agreements BER. By introducing a new legal regime whereby its 'safe harbour' is limited by market share limits,[28] the TTBER radically alters the nature of the block exemption and the overall legal framework for intellectual property licensing. Under the guise of giving greater recognition of the economic

[24] Article 1(j)(ii); Guidelines paras 24 and 31. [25] Article 1(j)(ii). [26] Article 1(j)(i).

[27] Article 4 (3); Guidelines para 31; see too paras 32–33.

[28] Recitals 10 and 11; Guidelines 26–33.

realities of intellectual property licensing, it creates legal uncertainty for the parties in volatile new technology markets. For licensing agreements between non-competitors, the block exemption will not apply where the licensed product *of each of the parties* exceeds 30% of the relevant market[29] because such agreements normally impose a lower risk to competition.[30] For the parties to agreements between competitors or 'horizontal' licensing agreements, the exemption will not apply where the licensed product *of both parties* exceeds a 20% market share.[31] In defining the market for the licensed *product*, both actual and potential competition are relevant.[32] If the agreement is to license *technology* only actual competition will be considered.[33]

If the product that is the subject of a technology transfer agreement exceeds the market share ceiling at any time during the course of the contract, it will lose the benefit of the block exemption after an uninterrupted[34] transition period of two years. As mentioned, the Regulation makes no concession to the volatility of relationships during the term of the licensing agreement as it did in respect of status of the parties as competitors or non-competitors whereby if the parties are non-competitors at the time the agreement is made, they will not be re-designated for the purposes of the exemption during the duration of the agreement unless the agreement is materially amended.[35]

V. The Hard Core Restrictions (Article 4) [See Chapters 7 and 8]

The new Regulation places considerable emphasis upon a narrow blacklist of prohibited 'hard core' restrictions whose presence in a licensing agreement make it unexemptible under the BER, but also almost always unenforceable under Article 81 generally. The hard core restrictions have been drafted on the supposition that they are 'almost always anti-competitive'.[36] Yet it is important to see that the effect of Article 4 is to take the licensing agreement outside the safe harbour of the BER, but not necessarily make the agreement irredeemably void as did the blacklist in the previous BER. Under the new regime a provision which is hard core restricted can still be rescued under Article 81(3) analysis outside the safe haven of the TTBER.

The hard core restrictions have been defined differently depending upon whether the licensing agreement that contains them is horizontal, ie, between competing undertakings or vertical ie, between non-competing undertakings.

1. The hard core restrictions in agreements between non-competitors (Article 4(2)) [See Chapter 7]

For agreements between non-competitors, the general hard core restrictions[37] are restricted to two main anticompetitive features of 'vertical' agreements. They include price-fixing (Article 4(2)(a) and territorial restrictions on sales Article 4 (2)(b): ['the restriction of the territory into which, or of the customers to whom, the licensee may passively sell the contract products'].

Moreover, even under Article 4(2)(b) there is a series of exceptions consisting of permitted territorial restrictions, including obligations on licensees not to sell **passively** into the exclusive territory of another licensee, Article 4(2)(b)(ii) not sell at all into the exclusive territory of the

---

29 Article 3(2). 30 Recital 11; Guidelines paras 68 and 69.

31 Article 3(1); Recital 10; Guidelines paras 66, 67, 69, 131.

32 Article 1(j)(ii); Guidelines paras 28, 29, 67.

33 Article 1(j)(i); Guidelines paras 30, 66.

34 Presumably, if the market share falls at any time during the two year period, the transitional period ends and if it rises again above the market share threshold, a new two year transitional period commences.

35 Article 3.1; Guidelines para 3. 36 Guidelines para 74.

37 There is a third hard core restriction that applies to territorial restrictions and to restrictions of active and passive sales to end users by a licensee who is part of a selective distribution system.

licensor, Article 4(2)(b)(i) requiring the licensee to manufacture or provide contract products only for its own use, etc.

The Commission has acknowledged the indispensability of protection against **passive** sales for licensees:

> 'it is unlikely that licensees would not enter into the licence without protection for a certain period of time against passive (and active) sales into the exclusive territory of a licensee by other licensees'.

This statement recognises the strategic importance of passive sales protection as an incentive in the technology transfer agreement and the Regulation provides that licensors can provide every licensee with protection for two years from the sale of the licensed product in its territory against passive sales by other licensees manufacturing the same licensed product in other territories. The theory is that the two years should be sufficient for each licensee to familiarise itself with the production process to achieve the efficiencies to allow it to catch and compete on equal terms with other licensees.[38]

2. The hard core restrictions in agreements between competitors (Article 4(1) [See Chapter 8]

Where the licensee competes with the licensor at the time the agreement is concluded, the Regulation contains four main hard core restrictions. The first three are basic anti-cartel competition rules such as bans on price fixing,[39] reciprocal output limitations,[40] and market allocation clauses.[41] The fourth is a prohibition on licensors restricting the licensee's ability to carry out their R&D and exploit their own technology.[42]

The Regulation also creates a special category of agreement between competitors which takes the form of a non-reciprocal licensing agreement. In such a case, the licensor is allowed under an exception to Article 4(1)(c) to offer an exclusive license, that is a license to produce and sell the contract products without the licensor himself producing goods in that territory or selling the contract goods from that territory. In such a case, the licensee will merely be doing what the licensor was entitled to do and hence that restriction, on its own, cannot be viewed as anticompetitive. Indeed, it may even be argued that Article 81(1) does not apply to a simple exclusive licence between licensor and licensee as long as the agreement involves no third parties such as other licensees.[43] A second analogous exception consists of field of use provisions. A field of use restriction limits the exploitation of the licensed technology by the licensee to one or more particular fields of use leaving untouched the licensor's ability to exploit the licensed technology in another field. A good example is offered by a maize seed variety that is licensed for animal food only with the licensor retaining exclusive rights to exploit the seed variety for human foodstuffs. Field of use restrictions may be 'exclusive' or 'sole' and are treated for competition purposes as analogous to exclusive or sole territorial licences. Again, as long as the field of use obligation is limited to the licensing agreement, it is little more than a subdivision of the licensor's own powers and may not even be caught by Article 81(1).

---

[38] See discussion in Guidelines paras 107–116.

[39] Article 4(1)(a) Guidelines paras 79–80, 156.

[40] Article 4((1)(b) Guidelines paras 82–83, 175.

[41] Article 4(1)(c); Guidelines para 84.

[42] Unless it is to prevent the licensed know-how being disclosed to third parties (in which case, the restriction must be proportionate and necessary to prevent the disclosure of the licensed technology. Article 4(1)(d). See too Guidelines para 94.

[43] Cf. Article 4(2) of Regulation 17/62 discussed in Anderman op cit pp 52–53.

VI. Excluded Restrictions (Article 5) [See Chapters 7 and 8]

The Commission has also created a short list of prima facie excluded restrictive conditions in Article 5 which, unlike hard core restrictions, are only void in themselves; they will not affect the remainder of the agreement. The Commission has in effect introduced a severability rule for such clauses. While they cannot be exempted as part of the block exemption process, they can be exempted individually if they meet the four conditions of Article 81(3). There are three main excluded restrictions: (a) any direct or indirect obligation by the licensee to assign or to grant an exclusive license in respect of its own severable improvements to the licensed technology; (b) any direct or indirect obligation by the licensee to assign or to grant an exclusive license in respect of its own severable improvements to the licensed technology; and (c) any direct or indirect obligation on the licensee not to challenge the validity of the IPRs held by the licensor.

Article 5(2) then adds that where the parties are not competing undertakings, any direct or indirect obligation limiting the licensee's ability to exploit its own technology or limiting the ability of any of the parties to the agreement to carry out research and development unless indispensable to prevent the disclosure of the licensed know-how to third parties. It is worth noting that this latter exclusion, as well as Article 4(1)(d) for agreements between competitors, may place some limits on the drafting of provisions in respect of the non-severable improvements of the licensee.

VII. Withdrawal and Non-application of TTBER

The Regulation provides a procedure for the competition authorities of the Member States[44] as well the Commission[45] itself to withdraw the benefit of the block exemption where they can show that they have effects incompatible with Article 81(3). This can include cases where access to third parties technologies to the market is restricted; where access of potential licensees is restricted; and where the parties, without any objectively valid reason, do not exploit the licensed technology. These are the main examples where the agreement reduces incentives to innovate or hinders access to markets.[46]

The TTBER also provides a procedure for the non-application of the BER in cases of parallel networks of similar agreements covering more than 50% of a relevant market.[47]

---

44 Article 6.2. 45 Article 6.1.

46 Article 6 and Guidelines paras 117–122; Recital 17.

47 Article 7 and Guidelines paras 123–129. This seems to overlap with Article6(1)(a).

APPENDIX 2

Commission Regulation (EC) No 772/2004 on the application of Article 81(3) of the Treaty to categories of technology transfer agreements

(Text with EEA relevance) ([2004] OJ L 123/11)

THE COMMISSION OF THE EUROPEAN COMMUNITIES,

Having regard to the Treaty establishing the European Community,

Having regard to Council Regulation No 19/65/EEC of 2 March 1965 on application of Article 85(3) of the Treaty to certain categories of agreements and concerted practices,[1] and in particular Article 1 thereof,

Having published a draft of this Regulation,[2]

After consulting the Advisory Committee on Restrictive Practices and Dominant Positions,

Whereas:

(1) Regulation No 19/65/EEC empowers the Commission to apply Article 81(3) of the Treaty by Regulation to certain categories of technology transfer agreements and corresponding concerted practices to which only two undertakings are party which fall within Article 81(1).

(2) Pursuant to Regulation No 19/65/EEC, the Commission has, in particular, adopted Regulation (EC) No 240/96 of 31 January 1996 on the application of Article 85(3) of the Treaty to certain categories of technology transfer agreements.[3]

(3) On 20 December 2001 the Commission published an evaluation report on the transfer of technology block exemption Regulation (EC) No 240/96.[4] This generated a public debate on the application of Regulation (EC) No 240/96 and on the application in general of Article 81(1) and (3) of the Treaty to technology transfer agreements. The response to the evaluation report from Member States and third parties has been generally in favour of reform of Community competition policy on technology transfer agreements. It is therefore appropriate to repeal Regulation (EC) No 240/96.

(4) This Regulation should meet the two requirements of ensuring effective competition and providing adequate legal security for undertakings. The pursuit of these objectives should take account of the need to simplify the regulatory framework and its application. It is appropriate to move away from the approach of listing exempted clauses and to place greater emphasis on defining the categories of agreements which are exempted up to a certain level of market power and on specifying the restrictions or clauses which are not to be contained in such agreements. This is consistent with an economics-based approach which assesses the impact of agreements on the relevant market. It is also consistent with such an approach to

[1] OJ 36, 6.3.1965, p 533/65. Regulation as last amended by Regulation (EC) No 1/2003 (OJ L 1, 4.1.2003, p 1).

[2] OJ C 235, 1.10.2003, p 10.

[3] OJ L 31, 9.2.1996, p 2. Regulation as amended by the 2003 Act of Accession.

[4] COM(2001) 786 final.

make a distinction between agreements between competitors and agreements between non-competitors.

(5) Technology transfer agreements concern the licensing of technology. Such agreements will usually improve economic efficiency and be pro-competitive as they can reduce duplication of research and development, strengthen the incentive for the initial research and development, spur incremental innovation, facilitate diffusion and generate product market competition.

(6) The likelihood that such efficiency-enhancing and pro-competitive effects will outweigh any anti-competitive effects due to restrictions contained in technology transfer agreements depends on the degree of market power of the undertakings concerned and, therefore, on the extent to which those undertakings face competition from undertakings owning substitute technologies or undertakings producing substitute products.

(7) This Regulation should only deal with agreements where the licensor permits the licensee to exploit the licensed technology, possibly after further research and development by the licensee, for the production of goods or services. It should not deal with licensing agreements for the purpose of subcontracting research and development. It should also not deal with licensing agreements to set up technology pools, that is to say, agreements for the pooling of technologies with the purpose of licensing the created package of intellectual property rights to third parties.

(8) For the application of Article 81(3) by regulation, it is not necessary to define those technology transfer agreements that are capable of falling within Article 81(1). In the individual assessment of agreements pursuant to Article 81(1), account has to be taken of several factors, and in particular the structure and the dynamics of the relevant technology and product markets.

(9) The benefit of the block exemption established by this Regulation should be limited to those agreements which can be assumed with sufficient certainty to satisfy the conditions of Article 81(3). In order to attain the benefits and objectives of technology transfer, the benefit of this Regulation should also apply to provisions contained in technology transfer agreements that do not constitute the primary object of such agreements, but are directly related to the application of the licensed technology.

(10) For technology transfer agreements between competitors it can be presumed that, where the combined share of the relevant markets accounted for by the parties does not exceed 20% and the agreements do not contain certain severely anti-competitive restraints, they generally lead to an improvement in production or distribution and allow consumers a fair share of the resulting benefits.

(11) For technology transfer agreements between non-competitors it can be presumed that, where the individual share of the relevant markets accounted for by each of the parties does not exceed 30% and the agreements do not contain certain severely anti-competitive restraints, they generally lead to an improvement in production or distribution and allow consumers a fair share of the resulting benefits.

(12) There can be no presumption that above these market-share thresholds technology transfer agreements do fall within the scope of Article 81(1). For instance, an exclusive licensing agreement between non-competing undertakings does often not fall within the scope of Article 81(1). There can also be no presumption that, above these market-share thresholds, technology transfer agreements falling within the scope of Article 81(1) will not satisfy the conditions for exemption. However, it can also not be presumed that they will usually give rise to objective advantages of such a character and size as to compensate for the disadvantages which they create for competition.

(13) This Regulation should not exempt technology transfer agreements containing restrictions which are not indispensable to the improvement of production or distribution. In particular, technology transfer agreements containing certain severely anti-competitive restraints

such as the fixing of prices charged to third parties should be excluded from the benefit of the block exemption established by this Regulation irrespective of the market shares of the undertakings concerned. In the case of such hardcore restrictions the whole agreement should be excluded from the benefit of the block exemption.

(14) In order to protect incentives to innovate and the appropriate application of intellectual property rights, certain restrictions should be excluded from the block exemption. In particular exclusive grant back obligations for severable improvements should be excluded. Where such a restriction is included in a licence agreement only the restriction in question should be excluded from the benefit of the block exemption.

(15) The market-share thresholds, the non-exemption of technology transfer agreements containing severely anti-competitive restraints and the excluded restrictions provided for in this Regulation will normally ensure that the agreements to which the block exemption applies do not enable the participating undertakings to eliminate competition in respect of a substantial part of the products in question.

(16) In particular cases in which the agreements falling under this Regulation nevertheless have effects incompatible with Article 81(3), the Commission should be able to withdraw the benefit of the block exemption. This may occur in particular where the incentives to innovate are reduced or where access to markets is hindered.

(17) Council Regulation (EC) No 1/2003 of 16 December 2002 on the implementation of the rules on competition laid down in Articles 81 and 82 of the Treaty[5] empowers the competent authorities of Member States to withdraw the benefit of the block exemption in respect of technology transfer agreements having effects incompatible with Article 81(3), where such effects are felt in their respective territory, or in a part thereof, and where such territory has the characteristics of a distinct geographic market. Member States must ensure that the exercise of this power of withdrawal does not prejudice the uniform application throughout the common market of the Community competition rules or the full effect of the measures adopted in implementation of those rules.

(18) In order to strengthen supervision of parallel networks of technology transfer agreements which have similar restrictive effects and which cover more than 50% of a given market, the Commission should be able to declare this Regulation inapplicable to technology transfer agreements containing specific restraints relating to the market concerned, thereby restoring the full application of Article 81 to such agreements.

(19) This Regulation should cover only technology transfer agreements between a licensor and a licensee. It should cover such agreements even if conditions are stipulated for more than one level of trade, by, for instance, requiring the licensee to set up a particular distribution system and specifying the obligations the licensee must or may impose on resellers of the products produced under the licence. However, such conditions and obligations should comply with the competition rules applicable to supply and distribution agreements. Supply and distribution agreements concluded between a licensee and its buyers should not be exempted by this Regulation.

(20) This Regulation is without prejudice to the application of Article 82 of the Treaty,

HAS ADOPTED THIS REGULATION:

Article 1

Definitions

1. For the purposes of this Regulation, the following definitions shall apply:

   (a) 'agreement' means an agreement, a decision of an association of undertakings or a concerted practice;

[5] OJ L 1, 4.1.2003, p 1. Regulation as amended by Regulation (EC) No 411/2004 (OJ L 68, 6.3.2004, p 1).

(b) 'technology transfer agreement' means a patent licensing agreement, a know-how licensing agreement, a software copyright licensing agreement or a mixed patent, know-how or software copyright licensing agreement, including any such agreement containing provisions which relate to the sale and purchase of products or which relate to the licensing of other intellectual property rights or the assignment of intellectual property rights, provided that those provisions do not constitute the primary object of the agreement and are directly related to the production of the contract products; assignments of patents, know-how, software copyright or a combination thereof where part of the risk associated with the exploitation of the technology remains with the assignor, in particular where the sum payable in consideration of the assignment is dependent on the turnover obtained by the assignee in respect of products produced with the assigned technology, the quantity of such products produced or the number of operations carried out employing the technology, shall also be deemed to be technology transfer agreements;

(c) 'reciprocal agreement' means a technology transfer agreement where two undertakings grant each other, in the same or separate contracts, a patent licence, a know-how licence, a software copyright licence or a mixed patent, know-how or software copyright licence and where these licences concern competing technologies or can be used for the production of competing products;

(d) 'non-reciprocal agreement' means a technology transfer agreement where one undertaking grants another undertaking a patent licence, a know-how licence, a software copyright licence or a mixed patent, know-how or software copyright licence, or where two undertakings grant each other such a licence but where these licences do not concern competing technologies and cannot be used for the production of competing products;

(e) 'product' means a good or a service, including both intermediary goods and services and final goods and services;

(f) 'contract products' means products produced with the licensed technology;

(g) 'intellectual property rights' includes industrial property rights, know-how, copyright and neighbouring rights;

(h) 'patents' means patents, patent applications, utility models, applications for registration of utility models, designs, topographies of semiconductor products, supplementary protection certificates for medicinal products or other products for which such supplementary protection certificates may be obtained and plant breeder's certificates;

(i) 'know-how' means a package of non-patented practical information, resulting from experience and testing, which is:
   (i) secret, that is to say, not generally known or easily accessible,
   (ii) substantial, that is to say, significant and useful for the production of the contract products, and
   (iii) identified, that is to say, described in a sufficiently comprehensive manner so as to make it possible to verify that it fulfils the criteria of secrecy and substantiality;

(j) 'competing undertakings' means undertakings which compete on the relevant technology market and/or the relevant product market, that is to say:
   (i) competing undertakings on the relevant technology market, being undertakings which license out competing technologies without infringing each others' intellectual property rights (actual competitors on the technology market); the relevant technology market includes technologies which are regarded by the licensees as interchangeable with or substitutable for the licensed technology, by reason of the technologies' characteristics, their royalties and their intended use,
   (ii) competing undertakings on the relevant product market, being undertakings which, in the absence of the technology transfer agreement, are both active on the relevant product and geographic market(s) on which the contract products are sold without infringing each others' intellectual property rights (actual competitors on the product

market) or would, on realistic grounds, undertake the necessary additional investments or other necessary switching costs so that they could timely enter, without infringing each others' intellectual property rights, the(se) relevant product and geographic market(s) in response to a small and permanent increase in relative prices (potential competitors on the product market); the relevant product market comprises products which are regarded by the buyers as interchangeable with or substitutable for the contract products, by reason of the products' characteristics, their prices and their intended use;

(k) 'selective distribution system' means a distribution system where the licensor undertakes to license the production of the contract products only to licensees selected on the basis of specified criteria and where these licensees undertake not to sell the contract products to unauthorised distributors;

(l) 'exclusive territory' means a territory in which only one undertaking is allowed to produce the contract products with the licensed technology, without prejudice to the possibility of allowing within that territory another licensee to produce the contract products only for a particular customer where this second licence was granted in order to create an alternative source of supply for that customer;

(m) 'exclusive customer group' means a group of customers to which only one undertaking is allowed actively to sell the contract products produced with the licensed technology;

(n) 'severable improvement' means an improvement that can be exploited without infringing the licensed technology.

2. The terms 'undertaking', 'licensor' and 'licensee' shall include their respective connected undertakings.

'Connected undertakings' means:

(a) undertakings in which a party to the agreement, directly or indirectly:
  (i) has the power to exercise more than half the voting rights, or
  (ii) has the power to appoint more than half the members of the supervisory board, board of management or bodies legally representing the undertaking, or
  (iii) has the right to manage the undertaking's affairs;

(b) undertakings which directly or indirectly have, over a party to the agreement, the rights or powers listed in (a);

(c) undertakings in which an undertaking referred to in (b) has, directly or indirectly, the rights or powers listed in (a);

(d) undertakings in which a party to the agreement together with one or more of the undertakings referred to in (a), (b) or (c), or in which two or more of the latter undertakings, jointly have the rights or powers listed in (a);

(e) undertakings in which the rights or the powers listed in (a) are jointly held by:
  (i) parties to the agreement or their respective connected undertakings referred to in (a) to (d), or
  (ii) one or more of the parties to the agreement or one or more of their connected undertakings referred to in (a) to (d) and one or more third parties.

Article 2

Exemption

Pursuant to Article 81(3) of the Treaty and subject to the provisions of this Regulation, it is hereby declared that Article 81(1) of the Treaty shall not apply to technology transfer agreements entered into between two undertakings permitting the production of contract products.

This exemption shall apply to the extent that such agreements contain restrictions of competition falling within the scope of Article 81(1). The exemption shall apply for as long as the intellectual property right in the licensed technology has not expired, lapsed or been declared invalid or, in the case of know-how, for as long as the know-how remains secret, except in the event where the know-how becomes publicly known as a result of action by the licensee, in which case the exemption shall apply for the duration of the agreement.

Article 3

Market-share thresholds

1. Where the undertakings party to the agreement are competing undertakings, the exemption provided for in Article 2 shall apply on condition that the combined market share of the parties does not exceed 20% on the affected relevant technology and product market.
2. Where the undertakings party to the agreement are not competing undertakings, the exemption provided for in Article 2 shall apply on condition that the market share of each of the parties does not exceed 30% on the affected relevant technology and product market.
3. For the purposes of paragraphs 1 and 2, the market share of a party on the relevant technology market(s) is defined in terms of the presence of the licensed technology on the relevant product market(s). A licensor's market share on the relevant technology market shall be the combined market share on the relevant product market of the contract products produced by the licensor and its licensees.

Article 4

Hardcore restrictions

1. Where the undertakings party to the agreement are competing undertakings, the exemption provided for in Article 2 shall not apply to agreements which, directly or indirectly, in isolation or in combination with other factors under the control of the parties, have as their object:
   (a) the restriction of a party's ability to determine its prices when selling products to third parties;
   (b) the limitation of output, except limitations on the output of contract products imposed on the licensee in a non-reciprocal agreement or imposed on only one of the licensees in a reciprocal agreement;
   (c) the allocation of markets or customers except:
      (i) the obligation on the licensee(s) to produce with the licensed technology only within one or more technical fields of use or one or more product markets,
      (ii) the obligation on the licensor and/or the licensee, in a non-reciprocal agreement, not to produce with the licensed technology within one or more technical fields of use or one or more product markets or one or more exclusive territories reserved for the other party,
      (iii) the obligation on the licensor not to license the technology to another licensee in a particular territory,
      (iv) the restriction, in a non-reciprocal agreement, of active and/or passive sales by the licensee and/or the licensor into the exclusive territory or to the exclusive customer group reserved for the other party,
      (v) the restriction, in a non-reciprocal agreement, of active sales by the licensee into the exclusive territory or to the exclusive customer group allocated by the licensor to another licensee provided the latter was not a competing undertaking of the licensor at the time of the conclusion of its own licence,
      (vi) the obligation on the licensee to produce the contract products only for its own use provided that the licensee is not restricted in selling the contract products actively and passively as spare parts for its own products,
      (vii) the obligation on the licensee, in a non-reciprocal agreement, to produce the contract products only for a particular customer, where the licence was granted in order to create an alternative source of supply for that customer;
   (d) the restriction of the licensee's ability to exploit its own technology or the restriction of the ability of any of the parties to the agreement to carry out research and development, unless

such latter restriction is indispensable to prevent the disclosure of the licensed know-how to third parties.

2. Where the undertakings party to the agreement are not competing undertakings, the exemption provided for in Article 2 shall not apply to agreements which, directly or indirectly, in isolation or in combination with other factors under the control of the parties, have as their object:

   (a) the restriction of a party's ability to determine its prices when selling products to third parties, without prejudice to the possibility of imposing a maximum sale price or recommending a sale price, provided that it does not amount to a fixed or minimum sale price as a result of pressure from, or incentives offered by, any of the parties;

   (b) the restriction of the territory into which, or of the customers to whom, the licensee may passively sell the contract products, except:

      (i) the restriction of passive sales into an exclusive territory or to an exclusive customer group reserved for the licensor,

      (ii) the restriction of passive sales into an exclusive territory or to an exclusive customer group allocated by the licensor to another licensee during the first two years that this other licensee is selling the contract products in that territory or to that customer group,

      (iii) the obligation to produce the contract products only for its own use provided that the licensee is not restricted in selling the contract products actively and passively as spare parts for its own products,

      (iv) the obligation to produce the contract products only for a particular customer, where the licence was granted in order to create an alternative source of supply for that customer,

      (v) the restriction of sales to end-users by a licensee operating at the wholesale level of trade,

      (vi) the restriction of sales to unauthorised distributors by the members of a selective distribution system;

   (c) the restriction of active or passive sales to end-users by a licensee which is a member of a selective distribution system and which operates at the retail level, without prejudice to the possibility of prohibiting a member of the system from operating out of an unauthorised place of establishment.

3. Where the undertakings party to the agreement are not competing undertakings at the time of the conclusion of the agreement but become competing undertakings afterwards, paragraph 2 and not paragraph 1 shall apply for the full life of the agreement unless the agreement is subsequently amended in any material respect.

Article 5

Excluded restrictions

1. The exemption provided for in Article 2 shall not apply to any of the following obligations contained in technology transfer agreements:

   (a) any direct or indirect obligation on the licensee to grant an exclusive licence to the licensor or to a third party designated by the licensor in respect of its own severable improvements to or its own new applications of the licensed technology;

   (b) any direct or indirect obligation on the licensee to assign, in whole or in part, to the licensor or to a third party designated by the licensor, rights to its own severable improvements to or its own new applications of the licensed technology;

   (c) any direct or indirect obligation on the licensee not to challenge the validity of intellectual property rights which the licensor holds in the common market, without prejudice to the possibility of providing for termination of the technology transfer agreement in the event that the licensee challenges the validity of one or more of the licensed intellectual property rights.

2. Where the undertakings party to the agreement are not competing undertakings, the exemption provided for in Article 2 shall not apply to any direct or indirect obligation limiting the licensee's ability to exploit its own technology or limiting the ability of any of the parties to the agreement to carry out research and development, unless such latter restriction is indispensable to prevent the disclosure of the licensed know-how to third parties.

Article 6

Withdrawal in individual cases

1. The Commission may withdraw the benefit of this Regulation, pursuant to Article 29(1) of Regulation (EC) No 1/2003, where it finds in any particular case that a technology transfer agreement to which the exemption provided for in Article 2 applies nevertheless has effects which are incompatible with Article 81(3) of the Treaty, and in particular where:
   (a) access of third parties' technologies to the market is restricted, for instance by the cumulative effect of parallel networks of similar restrictive agreements prohibiting licensees from using third parties' technologies;
   (b) access of potential licensees to the market is restricted, for instance by the cumulative effect of parallel networks of similar restrictive agreements prohibiting licensors from licensing to other licensees;
   (c) without any objectively valid reason, the parties do not exploit the licensed technology.
2. Where, in any particular case, a technology transfer agreement to which the exemption provided for in Article 2 applies has effects which are incompatible with Article 81(3) of the Treaty in the territory of a Member State, or in a part thereof, which has all the characteristics of a distinct geographic market, the competition authority of that Member State may withdraw the benefit of this Regulation, pursuant to Article 29(2) of Regulation (EC) No 1/2003, in respect of that territory, under the same circumstances as those set out in paragraph 1 of this Article.

Article 7

Non-application of this Regulation

1. Pursuant to Article 1a of Regulation No 19/65/EEC, the Commission may by regulation declare that, where parallel networks of similar technology transfer agreements cover more than 50% of a relevant market, this Regulation is not to apply to technology transfer agreements containing specific restraints relating to that market.
2. A regulation pursuant to paragraph 1 shall not become applicable earlier than six months following its adoption.

Article 8

Application of the market-share thresholds

1. For the purposes of applying the market-share thresholds provided for in Article 3 the rules set out in this paragraph shall apply.
   The market share shall be calculated on the basis of market sales value data. If market sales value data are not available, estimates based on other reliable market information, including market sales volumes, may be used to establish the market share of the undertaking concerned.

   The market share shall be calculated on the basis of data relating to the preceding calendar year. The market share held by the undertakings referred to in point (e) of the second subparagraph of Article 1(2) shall be apportioned equally to each undertaking having the rights or the powers listed in point (a) of the second subparagraph of Article 1(2).
2. If the market share referred to in Article 3(1) or (2) is initially not more than 20% respectively 30% but subsequently rises above those levels, the exemption provided for in Article 2 shall continue to apply for a period of two consecutive calendar years following the year in which the 20% threshold or 30% threshold was first exceeded.

Article 9

Repeal

Regulation (EC) No 240/96 is repealed.

References to the repealed Regulation shall be construed as references to this Regulation.

Article 10

Transitional period

The prohibition laid down in Article 81(1) of the Treaty shall not apply during the period from 1 May 2004 to 31 March 2006 in respect of agreements already in force on 30 April 2004 which do not satisfy the conditions for exemption provided for in this Regulation but which, on 30 April 2004, satisfied the conditions for exemption provided for in Regulation (EC) No 240/96.

Article 11

Period of validity

This Regulation shall enter into force on 1 May 2004.

It shall expire on 30 April 2014.

This Regulation shall be binding in its entirety and directly applicable in all Member States.

Done at Brussels, 27 April 2004.

For the Commission
Mario MONTI
Member of the Commission

APPENDIX 3

Commission Notice on guidelines on the application of Article 81 of the EC Treaty to technology transfer agreements

(Text with EEA relevance) ([2004] OJ C 101/2)

I. Introduction

1. These guidelines set out the principles for the assessment of technology transfer agreements under Article 81 of the Treaty. Technology transfer agreements concern the licensing of technology where the licensor permits the licensee to exploit the licensed technology for the production of goods or services, as defined in Article 1(1)(b) of Commission Regulation (EC) No 773/2004 on the application of Article 81(3) of the Treaty to categories of technology transfer agreements (the TTBER).[1]

2. The purpose of the guidelines is to provide guidance on the application of the TTBER as well as on the application of Article 81 to technology transfer agreements that fall outside the scope of the TTBER. The TTBER and the guidelines are without prejudice to the possible parallel application of Article 82 of the Treaty to licensing agreements.[2]

3. The standards set forth in these guidelines must be applied in light of the circumstances specific to each case. This excludes a mechanical application. Each case must be assessed on its own facts and the guidelines must be applied reasonably and flexibly. Examples given serve as illustrations only and are not intended to be exhaustive. The Commission will keep under review the functioning of the TTBER and the guidelines in the new enforcement system created by Regulation 1/2003[3] to consider whether changes need to be made.

4. The present guidelines are without prejudice to the interpretation of Article 81 and the TTBER that may be given by the Court of Justice and the Court of First Instance.

II. General Principles

1. Article 81 and intellectual property rights

5. The aim of Article 81 as a whole is to protect competition on the market with a view to promoting consumer welfare and an efficient allocation of resources. Article 81(1) prohibits all agreements and concerted practices between undertakings and decisions by associations of

---

[1] OJ L 123, 27.4.2004. The TTBER replaces Commission Regulation (EC) No 240/96 of 31 January 1996 on the application of Article 85(3) of the Treaty to certain categories of technology transfer agreements (OJ L 31, 9.2.1996, p 2).

[2] See Joined Cases C395/96 P and C396/96 P, *Compagnie Maritime Belge* [2000] ECR I-1365, para 130, and para 106 of the Commission Guidelines on the application of Article 81(3) of the Treaty, not yet published.

[3] Council Regulation (EC) No 1/2003 on the implementation of the rules on competition laid down in Articles 81 and 82 of the Treaty (OJ L 1, 4.1.2003, p 1).

undertakings[4] which may affect trade between Member States[5] and which have as their object or effect the prevention, restriction or distortion of competition.[6] As an exception to this rule Article 81(3) provides that the prohibition contained in Article 81(1) may be declared inapplicable in the case of agreements between undertakings which contribute to improving the production or distribution of products or to promoting technical or economic progress, while allowing consumers a fair share of the resulting benefits and which do not impose restrictions which are not indispensable to the attainment of these objectives and do not afford such undertakings the possibility of eliminating competition in respect of a substantial part of the products concerned.

6. Intellectual property laws confer exclusive rights on holders of patents, copyright, design rights, trademarks and other legally protected rights. The owner of intellectual property is entitled under intellectual property laws to prevent unauthorised use of his intellectual property and to exploit it, *inter alia*, by licensing it to third parties. Once a product incorporating an intellectual property right has been put on the market inside the EEA by the holder or with his consent, the intellectual property right is exhausted in the sense that the holder can no longer use it to control the sale of the product[7] (principle of Community exhaustion). The right holder has no right under intellectual property laws to prevent sales by licensees or buyers of such products incorporating the licensed technology.[8] The principle of Community exhaustion is in line with the essential function of intellectual property rights, which is to grant the holder the right to exclude others from exploiting his intellectual property without his consent.

7. The fact that intellectual property laws grant exclusive rights of exploitation does not imply that intellectual property rights are immune from competition law intervention. Articles 81 and 82 are in particular applicable to agreements whereby the holder licenses another undertaking to exploit his intellectual property rights.[9] Nor does it imply that there is an inherent conflict between intellectual property rights and the Community competition rules. Indeed, both bodies of law share the same basic objective of promoting consumer welfare and an efficient allocation of resources. Innovation constitutes an essential and dynamic component of an open and competitive market economy. Intellectual property rights promote dynamic competition by encouraging undertakings to invest in developing new or improved products and processes. So does competition by putting pressure on undertakings to innovate. Therefore, both intellectual property rights and competition are necessary to promote innovation and ensure a competitive exploitation thereof.

8. In the assessment of licence agreements under Article 81 it must be kept in mind that the creation of intellectual property rights often entails substantial investment and that it is often a risky endeavour. In order not to reduce dynamic competition and to maintain the incentive to innovate, the innovator must not be unduly restricted in the exploitation of intellectual property rights that turn out to be valuable. For these reasons the innovator should normally be free to seek compensation for successful projects that is sufficient to maintain investment incentives, taking failed projects into account. Technology licensing may also require the licensee to make

---

[4] In the following the term 'agreement' includes concerted practices and decisions of associations of undertakings.

[5] See Commission Notice on the concept of effect on trade between Member States contained in Articles 81 and 82 of the Treaty, not yet published.

[6] In the following the term 'restriction' includes the prevention and distortion of competition.

[7] This principle of Community exhaustion is for example enshrined in Article 7(1) of Directive 104/89/EEC to approximate the laws of the Member States relating to trade marks (OJ L 40, 11.2.1989, p 1), which provides that the trade mark shall not entitle the proprietor to prohibit its use in relation to goods which have been put on the market in the Community under that trade mark by the proprietor or with his consent.

[8] On the other hand, the sale of copies of a protected work does not lead to the exhaustion of performance rights, including rental rights, in the work, see in this respect Case 158/86, *Warner Brothers and Metronome Video*, [1988] ECR 2605, and Case C61/97, *Foreningen af danske videogramdistributører* [1998] ECR I-5171.

[9] See e.g. Joined Cases 56/64 and 58/64, *Consten and Grundig* [1966] ECR 429.

significant sunk investments in the licensed technology and production assets necessary to exploit it. Article 81 cannot be applied without considering such *ex ante* investments made by the parties and the risks relating thereto. The risk facing the parties and the sunk investment that must be committed may thus lead to the agreement falling outside Article 81(1) or fulfilling the conditions of Article 81(3), as the case may be, for the period of time required to recoup the investment.

9. In assessing licensing agreements under Article 81, the existing analytical framework is sufficiently flexible to take due account of the dynamic aspects of technology licensing. There is no presumption that intellectual property rights and licence agreements as such give rise to competition concerns. Most licence agreements do not restrict competition and create pro-competitive efficiencies. Indeed, licensing as such is pro-competitive as it leads to dissemination of technology and promotes innovation. In addition, even licence agreements that do restrict competition may often give rise to pro-competitive efficiencies, which must be considered under Article 81(3) and balanced against the negative effects on competition.[10] The great majority of licence agreements are therefore compatible with Article 81.

2. The general framework for applying Article 81

10. Article 81(1) prohibits agreements which have as their object or effect the restriction of competition. Article 81(1) applies both to restrictions of competition between the parties to an agreement and to restrictions of competition between any of the parties and third parties.

11. The assessment of whether a licence agreement restricts competition must be made within the actual context in which competition would occur in the absence of the agreement with its alleged restrictions.[11] In making this assessment it is necessary to take account of the likely impact of the agreement on inter-technology competition (i.e. competition between undertakings using competing technologies) and on intra-technology competition (i.e. competition between undertakings using the same technology).[12] Article 81(1) prohibits restrictions of both inter-technology competition and intra-technology competition. It is therefore necessary to assess to what extent the agreement affects or is likely to affect these two aspects of competition on the market.

12. The following two questions provide a useful framework for making this assessment. The first question relates to the impact of the agreement on inter-technology competition while the second question relates to the impact of the agreement on intra-technology competition. As restraints may be capable of affecting both inter-technology competition and intra-technology competition at the same time, it may be necessary to analyse a restraint in the light of both questions before it can be concluded whether or not competition within the meaning of Article 81(1) is restricted:

(a) Does the licence agreement restrict actual or potential competition that would have existed without the contemplated agreement? If so, the agreement may be caught by Article 81(1). In making this assessment it is necessary to take into account competition between the parties and competition from third parties. For instance, where two undertakings established in different Member States cross licence competing technologies and undertake not to sell products in each other's home markets, (potential) competition that existed prior to the agreement is restricted. Similarly, where a licensor imposes obligations on his licensees not to use competing technologies and these obligations foreclose third party technologies, actual or potential competition that would have existed in the absence of the agreement is restricted.

(b) Does the agreement restrict actual or potential competition that would have existed in the absence of the contractual restraint(s)? If so, the agreement may be caught by Article 81(1).

---

[10] The methodology for the application of Article 81(3) is set out in the Commission Guidelines on the application of Article 81(3) of the Treaty cited in note 2.

[11] See Case 56/65, *Société Technique Minière* [1966] ECR 337, and Case C7/95 P, *John Deere* [1998] ECR I-3111, paragraph 76.

[12] See in this respect e.g. judgment in *Consten and Grundig* cited in note 9.

For instance, where a licensor restricts its licensees from competing with each other, (potential) competition that could have existed between the licensees absent the restraints is restricted. Such restrictions include vertical price fixing and territorial or customer sales restrictions between licensees. However, certain restraints may in certain cases not be caught by Article 81(1) when the restraint is objectively necessary for the existence of an agreement of that type or that nature.[13] Such exclusion of the application of Article 81(1) can only be made on the basis of objective factors external to the parties themselves and not the subjective views and characteristics of the parties. The question is not whether the parties in their particular situation would not have accepted to conclude a less restrictive agreement, but whether, given the nature of the agreement and the characteristics of the market, a less restrictive agreement would not have been concluded by undertakings in a similar setting. For instance, territorial restraints in an agreement between non-competitors may fall outside Article 81(1) for a certain duration if the restraints are objectively necessary for a licensee to penetrate a new market. Similarly, a prohibition imposed on all licensees not to sell to certain categories of end users may not be restrictive of competition if such a restraint is objectively necessary for reasons of safety or health related to the dangerous nature of the product in question. Claims that in the absence of a restraint the supplier would have resorted to vertical integration are not sufficient. Decisions on whether or not to vertically integrate depend on a broad range of complex economic factors, a number of which are internal to the undertaking concerned.

13. In the application of the analytical framework set out in the previous paragraph it must be taken into account that Article 81(1) distinguishes between those agreements that have a restriction of competition as their object and those agreements that have a restriction of competition as their effect. An agreement or contractual restraint is only prohibited by Article 81(1) if its object or effect is to restrict inter-technology competition and/or intra-technology competition.

14. Restrictions of competition by object are those that by their very nature restrict competition. These are restrictions which in light of the objectives pursued by the Community competition rules have such a high potential for negative effects on competition that it is not necessary for the purposes of applying Article 81(1) to demonstrate any actual effects on the market.[14] Moreover, the conditions of Article 81(3) are unlikely to be fulfilled in the case of restrictions by object. The assessment of whether or not an agreement has as its object a restriction of competition is based on a number of factors. These factors include, in particular, the content of the agreement and the objective aims pursued by it. It may also be necessary to consider the context in which it is (to be) applied or the actual conduct and behaviour of the parties on the market.[15] In other words, an examination of the facts underlying the agreement and the specific circumstances in which it operates may be required before it can be concluded whether a particular restriction constitutes a hardcore restriction of competition. The way in which an agreement is actually implemented may reveal a restriction by object even where the formal agreement does not contain an express provision to that effect. Evidence of subjective intent on the part of the parties to restrict competition is a relevant factor but not a necessary condition. For licence agreements, the Commission considers that the restrictions covered by the list of hardcore restrictions of competition contained in Article 4 of the TTBER are restrictive by their very object.

15. If an agreement is not restrictive of competition by object it is necessary to examine whether it has restrictive effects on competition. Account must be taken of both actual and potential

[13] See in this respect the judgment in *Société Technique Minière* cited in note 11 and Case 258/78, *Nungesser* [1982] ECR 2015.

[14] See in this respect e.g. Case C49/92 P, *Anic Partecipazioni* [1999] ECR I-4125, para 99.

[15] See Joined Cases 29/83 and 30/83, *CRAM and Rheinzink* [1984] ECR 1679, para 26, and Joined Cases 96/82 and others, *ANSEAU-NAVEWA* [1983] ECR 3369, paras 23–25.

effects.[16] In other words the agreement must have likely anti-competitive effects. For licence agreements to be restrictive of competition by effect they must affect actual or potential competition to such an extent that on the relevant market negative effects on prices, output, innovation or the variety or quality of goods and services can be expected with a reasonable degree of probability. The likely negative effects on competition must be appreciable.[17] Appreciable anticompetitive effects are likely to occur when at least one of the parties has or obtains some degree of market power and the agreement contributes to the creation, maintenance or strengthening of that market power or allows the parties to exploit such market power. Market power is the ability to maintain prices above competitive levels or to maintain output in terms of product quantities, product quality and variety or innovation below competitive levels for a not insignificant period of time. The degree of market power normally required for a finding of an infringement under Article 81(1) is less than the degree of market power required for a finding of dominance under Article 82.

16. For the purposes of analysing restrictions of competition by effect it is normally necessary to define the relevant market and to examine and assess, *inter alia*, the nature of the products and technologies concerned, the market position of the parties, the market position of competitors, the market position of buyers, the existence of potential competitors and the level of entry barriers. In some cases, however, it may be possible to show anti-competitive effects directly by analysing the conduct of the parties to the agreement on the market. It may for example be possible to ascertain that an agreement has led to price increases.

17. Licence agreements, however, also have substantial pro-competitive potential. Indeed, the vast majority of licence agreements are pro-competitive. Licence agreements may promote innovation by allowing innovators to earn returns to cover at least part of their research and development costs. Licence agreements also lead to a dissemination of technologies, which may create value by reducing the production costs of the licensee or by enabling him to produce new or improved products. Efficiencies at the level of the licensee often stem from a combination of the licensor's technology with the assets and technologies of the licensee. Such integration of complementary assets and technologies may lead to a cost/output configuration that would not otherwise be possible. For instance, the combination of an improved technology of the licensor with more efficient production or distribution assets of the licensee may reduce production costs or lead to the production of a higher quality product. Licensing may also serve the pro-competitive purpose of removing obstacles to the development and exploitation of the licensee's own technology. In particular in sectors where large numbers of patents are prevalent licensing often occurs in order to create design freedom by removing the risk of infringement claims by the licensor. When the licensor agrees not to invoke his intellectual property rights to prevent the sale of the licensee's products, the agreement removes an obstacle to the sale of the licensee's product and thus generally promotes competition.

18. In cases where a licence agreement is caught by Article 81(1) the pro-competitive effects of the agreement must be balanced against its restrictive effects in the context of Article 81(3). When all four conditions of Article 81(3) are satisfied, the restrictive licence agreement in question is valid and enforceable, no prior decision to that effect being required.[18] Hardcore restrictions of competition only fulfil the conditions of Article 81(3) in exceptional circumstances. Such agreements generally fail (at least) one of the first two conditions of Article 81(3). They generally

---

[16] See the judgment in *John Deere* [1998] cited in note 11.

[17] Guidance on the issue of appreciability can be found in Commission notice on agreements of minor importance which do not appreciably restrict competition under Article 81(1) of the Treaty (OJ C 368, 22.12.2001, p 13). The notice defines appreciability in a negative way. Agreements, which fall outside the scope of the *de minimis* notice, do not necessarily have appreciable restrictive effects. An individual assessment is required.

[18] See Article 1(2) of Council Regulation No 1/2003 cited in note 3.

do not create objective economic benefits or benefits for consumers. Moreover, these types of agreements generally also fail the indispensability test under the third condition. For example, if the parties fix the price at which the products produced under the licence must be sold, this will generally lead to a lower output and a misallocation of resources and higher prices for consumers. The price restriction is also not indispensable to achieve the possible efficiencies resulting from the availability to both competitors of the two technologies.

3. Market definition

19. The Commission's approach to defining the relevant market is laid down in its market definition guidelines.[19] The present guidelines only address aspects of market definition that are of particular importance in the field of technology licensing.

20. Technology is an input, which is integrated either into a product or a production process. Technology licensing can therefore affect competition both in input markets and in output markets. For instance, an agreement between two parties which sell competing products and which cross license technologies relating to the production of these products may restrict competition on the product market concerned. It may also restrict competition on the market for technology and possibly also on other input markets. For the purposes of assessing the competitive effects of licence agreements it may therefore be necessary to define relevant goods and service markets (product markets) as well as technology markets.[20] The term 'product market' used in Article 3 of the TTBER refers to relevant goods and service markets in both their geographic and product dimension. As is clear from Article 1(1)(j) of the TTBER, the term is used merely to distinguish relevant goods and service markets from relevant technology markets.

21. The TTBER and these guidelines are concerned with effects both on product markets for final products and on product markets for intermediate products. The relevant product market includes products which are regarded by the buyers as interchangeable with or substitutable for the contract products incorporating the licensed technology, by reason of the products' characteristics, their prices and their intended use.

22. Technology markets consist of the licensed technology and its substitutes, i.e. other technologies which are regarded by the licensees as interchangeable with or substitutable for the licensed technology, by reason of the technologies' characteristics, their royalties and their intended use. The methodology for defining technology markets follows the same principles as the definition of product markets. Starting from the technology which is marketed by the licensor, one needs to identify those other technologies to which licensees could switch in response to a small but permanent increase in relative prices, i.e. the royalties. An alternative approach is to look at the market for products incorporating the licensed technology (cf. paragraph below).

23. Once relevant markets have been defined, market shares can be assigned to the various sources of competition in the market and used as an indication of the relative strength of market players. In the case of technology markets one way to proceed is to calculate market shares on the basis of each technology's share of total licensing income from royalties, representing a technology's share of the market where competing technologies are licensed. However, this may often be a mere theoretical and not a practical way to proceed because of lack of clear information on royalties etc. An alternative approach, which is the one used in Article 3(3) of the TTBER, is to calculate market shares on the technology market on the basis of sales of products incorporating the licensed technology on downstream product markets (see paragraph 70 below). Under this approach all sales on the relevant product market are taken into account, irrespective of whether

[19] Commission notice on the definition of the relevant market for the purposes of Community competition law (OJ C 372, 9.12.1997, p 5).

[20] As to these distinctions see also Commission Guidelines on the applicability of Article 81 of the EC Treaty to horizontal cooperation agreements (OJ C 3, 6.1.2001, p 2, paras 44 to 52).

the product incorporates a technology that is being licensed. In the case of technology markets the approach of Article 3(3) to take into account technologies that are (only) being used in-house, is justified. Indeed, this approach is in general a good indicator of the strength of the technology. First, it captures any potential competition from undertakings that are producing with their own technology and that are likely to start licensing in the event of a small but permanent increase in the price for licenses. Secondly, even where it is unlikely that other technology owners would start licensing, the licensor does not necessarily have market power on the technology market even if he has a high share of licensing income. If the downstream product market is competitive, competition at this level may effectively constrain the licensor. An increase in royalties upstream affects the costs of the licensee, making him less competitive, causing him to lose sales. A technology's market share on the product market also captures this element and is thus normally a good indicator of licensor market power. In individual cases outside the safe harbour of the TTBER it may be necessary, where practically possible, to apply both of the described approaches in order to assess more accurately the market strength of the licensor.

24. Moreover, outside the safe harbour of the TTBER it must also be taken into account that market share may not always be a good indication of the relative strength of available technologies. The Commission will therefore, *inter alia*, also have regard to the number of independently controlled technologies available in addition to the technologies controlled by the parties to the agreement that may be substitutable for the licensed technology at a comparable cost to the user (see paragraph 131 below).

25. Some licence agreements may affect innovation markets. In analysing such effects, however, the Commission will normally confine itself to examining the impact of the agreement on competition within existing product and technology markets.[21] Competition on such markets may be affected by agreements that delay the introduction of improved products or new products that over time will replace existing products. In such cases innovation is a source of potential competition which must be taken into account when assessing the impact of the agreement on product markets and technology markets. In a limited number of cases, however, it may be useful and necessary to also define innovation markets. This is particularly the case where the agreement affects innovation aiming at creating new products and where it is possible at an early stage to identify research and development poles.[22] In such cases it can be analysed whether after the agreement there will be a sufficient number of competing research and development poles left for effective competition in innovation to be maintained.

4. The distinction between competitors and non-competitors

26. In general, agreements between competitors pose a greater risk to competition than agreements between non-competitors. However, competition between undertakings that use the same technology (intra-technology competition between licensees) constitutes an important complement to competition between undertakings that use competing technologies (inter-technology competition). For instance, intra-technology competition may lead to lower prices for the products incorporating the technology in question, which may not only produce direct and immediate benefits for consumers of these products, but also spur further competition between undertakings that use competing technologies. In the context of licensing it must also be taken into account that licensees are selling their own product. They are not re-selling a product supplied by another undertaking. There may thus be greater scope for product differentiation and quality-based competition between licensees than in the case of vertical agreements for the resale of products.

27. In order to determine the competitive relationship between the parties it is necessary to

[21] See to that effect paras 50 to 52 of the Guidelines on horizontal cooperation agreements, cited in the previous note.

[22] Idem, para 51.

examine whether the parties would have been actual or potential competitors in the absence of the agreement. If without the agreement the parties would not have been actual or potential competitors in any relevant market affected by the agreement they are deemed to be non-competitors.

28. Where the licensor and the licensee are both active on the same product market or the same technology market without one or both parties infringing the intellectual property rights of the other party, they are actual competitors on the market concerned. The parties are deemed to be actual competitors on the technology market if the licensee is already licensing out his technology and the licensor enters the technology market by granting a license for a competing technology to the licensee.

29. The parties are considered to be potential competitors on the product market if in the absence of the agreement and without infringing the intellectual property rights of the other party it is likely that they would have undertaken the necessary additional investment to enter the relevant market in response to a small but permanent increase in product prices. In order to constitute a realistic competitive constraint entry has to be likely to occur within a short period. Normally a period of one to two years is appropriate. However, in individual cases longer periods can be taken into account. The period of time needed for undertakings already on the market to adjust their capacities can be used as a yardstick to determine this period. For instance, the parties are likely to be considered potential competitors on the product market where the licensee produces on the basis of its own technology in one geographic market and starts producing in another geographic market on the basis of a licensed competing technology. In such circumstances, it is likely that the licensee would have been able to enter the second geographic market on the basis of its own technology, unless such entry is precluded by objective factors, including the existence of blocking patents (see paragraph 32 below).

30. The parties are considered to be potential competitors on the technology market where they own substitutable technologies if in the specific case the licensee is not licensing his own technology, provided that he would be likely to do so in the event of a small but permanent increase in technology prices. However, for the application of the TTBER potential competition on the technology market is not taken into account (see paragraph 66 below).

31. In some cases the parties may become competitors subsequent to the conclusion of the agreement because the licensee develops and starts exploiting a competing technology. In such cases it must be taken into account that the parties were non-competitors at the time of conclusion of the agreement and that the agreement was concluded in that context. The Commission will therefore mainly focus on the impact of the agreement on the licensee's ability to exploit his own (competing) technology. In particular, the list of hardcore restrictions applying to agreements between competitors will not be applied to such agreements unless the agreement is subsequently amended in any material respect after the parties have become competitors (cf. Article 4(3) of the TTBER). The undertakings party to an agreement may also become competitors subsequent to the conclusion of the agreement where the licensee was already active on the product market prior to the licence and where the licensor subsequently enters the product market either on the basis of the licensed technology or a new technology. Also in this case the hardcore list relevant for agreements between non-competitors will continue to apply to the agreement unless the agreement is subsequently amended in any material respect (cf. article 4(3) of the TTBER).

32. If the parties own technologies that are in a one-way or two-way blocking position, the parties are considered to be non-competitors on the technology market. A one-way blocking position exists when a technology cannot be exploited without infringing upon another technology. This is for instance the case where one patent covers an improvement of a technology covered by another patent. In that case the exploitation of the improvement patent pre-supposes that the holder obtains a licence to the basic patent. A two-way blocking position exists where neither technology can be exploited without infringing upon the other technology and where the holders

thus need to obtain a licence or a waiver from each other. In assessing whether a blocking position exists the Commission will rely on objective factors as opposed to the subjective views of the parties. Particularly convincing evidence of the existence of a blocking position is required where the parties may have a common interest in claiming the existence of a blocking position in order to be qualified as non-competitors, for instance where the claimed two-way blocking position concerns technologies that are technological substitutes. Relevant evidence includes court decisions including injunctions and opinions of independent experts. In the latter case the Commission will, in particular, closely examine how the expert has been selected. However, also other convincing evidence, including expert evidence from the parties that they have or had good and valid reasons to believe that a blocking position exists or existed, can be relevant to substantiate the existence of a blocking position.

33. In some cases it may also be possible to conclude that while the licensor and the licensee produce competing products, they are non-competitors on the relevant product market and the relevant technology market because the licensed technology represents such a drastic innovation that the technology of the licensee has become obsolete or uncompetitive. In such cases the licensor's technology either creates a new market or excludes the licensee's technology from the market. Often, however, it is not possible to come to this conclusion at the time the agreement is concluded. It is usually only when the technology or the products incorporating it have been available to consumers for some time that it becomes apparent that the older technology has become obsolete or uncompetitive. For instance, when CD technology was developed and players and discs were put on the market, it was not obvious that this new technology would replace LP technology. This only became apparent some years later. The parties will therefore be considered to be competitors if at the time of the conclusion of the agreement it is not obvious that the licensee's technology is obsolete or uncompetitive. However, given that both Articles 81(1) and Article 81(3) must be applied in light of the actual context in which the agreement occurs, the assessment is sensitive to material changes in the facts. The classification of the relationship between the parties will therefore change into a relationship of non-competitors, if at a later point in time the licensee's technology becomes obsolete or uncompetitive on the market.

III. Application of the Block Exemption Regulation

1. The effects of the Block Exemption Regulation

34. Technology transfer agreements that fulfil the conditions set out in the TTBER are block exempted from the prohibition rule contained in Article 81(1). Block exempted agreements are legally valid and enforceable. Such agreements can only be prohibited for the future and only upon withdrawal of the block exemption by the Commission or a Member State competition authority. Block exempted agreements cannot be prohibited under Article 81 by national courts in the context of private litigation.

35. Block exemption of categories of technology transfer agreements is based on the presumption that such agreements—to the extent that they are caught by Article 81(1)—fulfil the four conditions laid down in Article 81(3). It is thus presumed that the agreements give rise to economic efficiencies, that the restrictions contained in the agreements are indispensable to the attainment of these efficiencies, that consumers within the affected markets receive a fair share of the efficiency gains and that the agreements do not afford the undertakings concerned the possibility of eliminating competition in respect of a substantial part of the products in question. The market share thresholds (Article 3), the hardcore list (Article 4) and the excluded restrictions (Article 5) set out in the TTBER aim at ensuring that only restrictive agreements that can reasonably be presumed to fulfil the four conditions of Article 81(3) are block exempted.

36. As set out in section IV below, many licence agreements fall outside Article 81(1), either because they do not restrict competition at all or because the restriction of competition is not

appreciable.[23] To the extent that such agreements would anyhow fall within the scope of the TTBER, there is no need to determine whether they are caught by Article 81(1).[24]

37. Outside the scope of the block exemption it is relevant to examine whether in the individual case the agreement is caught by Article 81(1) and if so whether the conditions of Article 81(3) are satisfied. There is no presumption that technology transfer agreements falling outside the block exemption are caught by Article 81(1) or fail to satisfy the conditions of Article 81(3). In particular, the mere fact that the market shares of the parties exceed the market share thresholds set out in Article 3 of the TTBER is not a sufficient basis for finding that the agreement is caught by Article 81(1). Individual assessment of the likely effects of the agreement is required. It is only when agreements contain hardcore restrictions of competition that it can normally be presumed that they are prohibited by Article 81.

2. Scope and duration of the Block Exemption Regulation

2.1. Agreements between two parties

38. According to Article 2(1) of the TTBER, the Regulation covers technology transfer agreements 'between two undertakings'. Technology transfer agreements between more than two undertakings are not covered by the TTBER.[25] The decisive factor in terms of distinguishing between agreements between two undertakings and multiparty agreements is whether the agreement in question is concluded between more than two undertakings.

39. Agreements concluded by two undertakings fall within the scope of the TTBER even if the agreement stipulates conditions for more than one level of trade. For instance, the TTBER applies to a licence agreement concerning not only the production stage but also the distribution stage, stipulating the obligations that the licensee must or may impose on resellers of the products produced under the licence.[26]

40. Licence agreements concluded between more than two undertakings often give rise to the same issues as licence agreements of the same nature concluded between two undertakings. In its individual assessment of licence agreements which are of the same nature as those covered by the block exemption but which are concluded between more than two undertakings, the Commission will apply by analogy the principles set out in the TTBER.

2.2. Agreements for the production of contract products

41. It follows from Article 2 that for licence agreements to be covered by the TTBER they must concern 'the production of contract products', i.e. products incorporating or produced with the licensed technology. In other words, to be covered by the TTBER the licence must permit the licensee to exploit the licensed technology for production of goods or services (see recital 7 of the TTBER). The TTBER does not cover technology pools. The notion of technology pools covers agreements whereby two or more parties agree to pool their respective technologies and license them as a package. The notion of technology pools also covers arrangements whereby two or more undertakings agree to license a third party and authorise him to license on the package of technologies. Technology pools are dealt with in section IV.4 below.

42. The TTBER applies to licence agreements for the production of contract products whereby the licensee is also permitted to sublicense the licensed technology to third parties provided, however, that the production of contract products constitutes the primary object of the agreement.

---

23 See in this respect the Notice on agreements of minor importance cited in note 17.

24 According to Article 3(2) of Regulation 1/2003, agreements which may affect trade between Member States but which are not prohibited by Article 81 cannot be prohibited by national competition law.

25 Under Council Regulation 19/65, OJ Special Edition Series I 1965–1966, p 35, the Commission is not empowered to block exempt technology transfer agreements concluded between more than two undertakings.

26 See recital 19 of the TTBER and further section 2.5 below.

Conversely, the TTBER does not apply to agreements that have sublicensing as their primary object. However, the Commission will apply by analogy the principles set out in the TTBER and these guidelines to such 'master licensing' agreements between licensor and licensee. Agreements between the licensee and sub-licensees are covered by the TTBER.

43. The term 'contract products' encompasses goods and services produced with the licensed technology. This is the case both where the licensed technology is used in the production process and where it is incorporated into the product itself. In these guidelines the term 'products incorporating the licensed technology' covers both situations. The TTBER applies in all cases where technology is licensed for the purposes of producing goods and services. It is sufficient in this respect that the licensor undertakes not to exercise his intellectual property rights against the licensee. Indeed, the essence of a pure patent licence is the right to operate inside the scope of the exclusive right of the patent. It follows that the TTBER also covers so-called non-assertion agreements and settlement agreements whereby the licensor permits the licensee to produce within the scope of the patent.

44. The TTBER covers 'subcontracting' whereby the licensor licenses technology to the licensee who undertakes to produce certain products on the basis thereof exclusively for the licensor. Subcontracting may also involve the supply of equipment by the licensor to be used in the production of the goods and services covered by the agreement. For the latter type of subcontracting to be covered by the TTBER, the licensed technology and not the supplied equipment must constitute the primary object of the agreement. Subcontracting is also covered by the Commission's Notice concerning the assessment of certain subcontracting agreements in relation to Article 81(1) of the Treaty.[27] According to this notice, which remains applicable, subcontracting agreements whereby the subcontractor undertakes to produce certain products exclusively for the contractor generally fall outside Article 81(1). However, other restrictions imposed on the subcontractor such as the obligation not to conduct or exploit his own research and development may be caught by Article 81.[28]

45. The TTBER also applies to agreements whereby the licensee must carry out development work before obtaining a product or a process that is ready for commercial exploitation, provided that a contract product has been identified. Even if such further work and investment is required, the object of the agreement is the production of an identified contract product. On the other hand, the TTBER and the guidelines do not cover agreements whereby a technology is licensed for the purpose of enabling the licensee to carry out further research and development in various fields. For instance, the TTBER and the guidelines do not cover the licensing of a technological research tool used in the process of further research activity. The framework of the TTBER and the guidelines is based on the premise that there is a direct link between the licensed technology and an identified contract product. In cases where no such link exists the main object of the agreement is research and development as opposed to bringing a particular product to the market; in that case the analytical framework of the TTBER and the guidelines may not be appropriate. For the same reasons the TTBER and the guidelines do not cover research and development subcontracting whereby the licensee undertakes to carry out research and development in the field of the licensed technology and to hand back the improved technology package to the licensor. The main object of such agreements is the provision of research and development services aimed at improving the technology as opposed to the production of goods and services on the basis of the licensed technology.

2.3. The concept of technology transfer agreements

46. The TTBER and these guidelines cover agreements for the transfer of technology. According to Article 1(1)(b) and (h) of the TTBER the concept of 'technology' covers patents and patent

[27] OJ C 1, 3.1.1979, p 2. [28] See para 3 of the subcontracting notice.

applications, utility models and applications for utility models, design rights, plant breeders rights, topographies of semiconductor products, supplementary protection certificates for medicinal products or other products for which such supplementary protection certificates may be obtained, software copyright, and know-how. The licensed technology should allow the licensee with or without other inputs to produce the contract products.

47. Know-how is defined in Article 1(1)(i) as a package of non-patented practical information, resulting from experience and testing, which is secret, substantial and identified. 'Secret' means that the know-how is not generally known or easily accessible. 'Substantial' means that the know-how includes information which is significant and useful for the production of the products covered by the licence agreement or the application of the process covered by the licence agreement. In other words, the information must significantly contribute to or facilitate the production of the contract products. In cases where the licensed know-how relates to a product as opposed to a process, this condition implies that the know-how is useful for the production the contract product. This condition is not satisfied where the contract product can be produced on the basis of freely available technology. However, the condition does not require that the contract product is of higher value than products produced with freely available technology. In the case of process technologies, this condition implies that the know-how is useful in the sense that it can reasonably be expected at the date of conclusion of the agreement to be capable of significantly improving the competitive position of the licensee, for instance by reducing his production costs. 'Identified' means that it is possible to verify that the licensed know-how fulfils the criteria of secrecy and substantiality. This condition is satisfied where the licensed know-how is described in manuals or other written form. However, in some cases this may not be reasonably possible. The licensed know-how may consist of practical knowledge possessed by the licensor's employees. For instance, the licensor's employees may possess secret and substantial knowledge about a certain production process which is passed on to the licensee in the form of training of the licensee's employees. In such cases it is sufficient to describe in the agreement the general nature of the know-how and to list the employees that will be or have been involved in passing it on to the licensee.

48. The concept of 'transfer' implies that technology must flow from one undertaking to another. Such transfers normally take the form of licensing whereby the licensor grants the licensee the right to use his technology against payment of royalties. It can also take the form of sub-licensing, whereby a licensee, having been authorised to do so by the licensor, grants licenses to third parties (sub-licensees) for the exploitation of the technology.

49. The TTBER only applies to agreements that have as their primary object the transfer of technology as defined in that Regulation as opposed to the purchase of goods and services or the licensing of other types of intellectual property. Agreements containing provisions relating to the purchase and sale of products are only covered by the TTBER to the extent that those provisions do not constitute the primary object of the agreement and are directly related to the application of the licensed technology. This is likely to be the case where the tied products take the form of equipment or process input which is specifically tailored to efficiently exploit the licensed technology. If, on the other hand, the product is simply another input into the final product, it must be carefully examined whether the licensed technology constitutes the primary object of the agreement. For instance, in cases where the licensee is already manufacturing a final product on the basis of another technology, the licence must lead to a significant improvement of the licensee's production process, exceeding the value of the product purchased from the licensor. The requirement that the tied products must be related to the licensing of technology implies that the TTBER does not cover the purchase of products that have no relation with the products incorporating the licensed technology. This is for example the case where the tied product is not intended to be used with the licensed product, but relates to an activity on a separate product market.

50. The TTBER only covers the licensing of other types of intellectual property such as trademarks and copyright, other than software copyright, to the extent that they are directly related to the exploitation of the licensed technology and do not constitute the primary object of the agreement. This condition ensures that agreements covering other types of intellectual property rights are only block exempted to the extent that these other intellectual property rights serve to enable the licensee to better exploit the licensed technology. The licensor may for instance authorise the licensee to use his trademark on the products incorporating the licensed technology. The trademark licence may allow the licensee to better exploit the licensed technology by allowing consumers to make an immediate link between the product and the characteristics imputed to it by the licensed technology. An obligation on the licensee to use the licensor's trademark may also promote the dissemination of technology by allowing the licensor to identify himself as the source of the underlying technology. However, where the value of the licensed technology to the licensee is limited because he already employs an identical or very similar technology and the main object of the agreement is the trademark, the TTBER does not apply.[29]

51. The licensing of copyright for the purpose of reproduction and distribution of the protected work, i.e. the production of copies for resale, is considered to be similar to technology licensing. Since such licence agreements relate to the production and sale of products on the basis of an intellectual property right, they are considered to be of a similar nature as technology transfer agreements and normally raise comparable issues. Although the TTBER does not cover copyright other than software copyright, the Commission will as a general rule apply the principles set out in the TTBER and these guidelines when assessing such licensing of copyright under Article 81.

52. On the other hand, the licensing of rights in performances and other rights related to copyright is considered to raise particular issues and it may not be warranted to assess such licensing on the basis of the principles developed in these guidelines. In the case of the various rights related to performances value is created not by the reproduction and sale of copies of a product but by each individual performance of the protected work. Such exploitation can take various forms including the performance, showing or the renting of protected material such as films, music or sporting events. In the application of Article 81 the specificities of the work and the way in which it is exploited must be taken into account.[30] For instance, resale restrictions may give rise to less competition concerns whereas particular concerns may arise where licensors impose on their licensees to extend to each of the licensors more favourable conditions obtained by one of them. The Commission will therefore not apply the TTBER and the present guidelines by way of analogy to the licensing of these other rights.

53. The Commission will also not extend the principles developed in the TTBER and these guidelines to trademark licensing. Trademark licensing often occurs in the context of distribution and resale of goods and services and is generally more akin to distribution agreements than technology licensing. Where a trademark licence is directly related to the use, sale or resale of goods and services and does not constitute the primary object of the agreement, the licence agreement is covered by Commission Regulation (EC) No 2790/1999 on the application of Article 81(3) of the Treaty to categories of vertical agreements and concerted practices.[31]

2.4. Duration

54. Subject to the duration of the TTBER, the block exemption applies for as long as the licensed property right has not lapsed, expired or been declared invalid. In the case of know-how the block exemption applies as long as the licensed know-how remains secret, except where the know-how becomes publicly known as a result of action by the licensee, in which case the exemption shall apply for the duration of the agreement (cf. Article 2 of the TTBER).

[29] See in this respect Commission Decision in *Moosehead/Whitbread* (OJ L 100, 20.4.1990, p 32).

[30] See in this respect Case 262/81, *Coditel (II)* [1982] ECR 3381. [31] OJ L 336, 29.12.1999, p 21.

55. The block exemption applies to each licensed property right covered by the agreement and ceases to apply on the date of expiry, invalidity or the coming into the public domain of the last intellectual property right which constitutes 'technology' within the meaning of the TTBER (cf. paragraph above).

2.5. Relationship with other block exemption regulations

56. The TTBER covers agreements between two undertakings concerning the licensing of technology for the purpose of the production of contract products. However, technology can also be an element of other types of agreements. In addition, the products incorporating the licensed technology are subsequently sold on the market. It is therefore necessary to address the interface between the TTBER and Commission Regulation (EC) No 2658/2000 on the application of Article 81(3) of the Treaty to categories of specialisation agreements,[32] Commission Regulation 2659/2000 on the application of Article 81(3) to categories of research and development agreements[33] and Commission Regulation (EC) No 2790/1999 on the application of Article 81(3) of the Treaty to categories of vertical agreements and concerted practices.[34]

2.5.1. The Block Exemption Regulations on specialisation and R&D agreements

57. According to Article 1(1)(c) of Regulation 2658/2000 on specialisation agreements, that Regulation covers, *inter alia*, joint production agreements by virtue of which two or more undertakings agree to produce certain products jointly. The Regulation extends to provisions concerning the assignment or use of intellectual property rights, provided that they do not constitute the primary object of the agreement, but are directly related to and necessary for its implementation.

58. Where undertakings establish a production joint venture and license the joint venture to exploit technology, which is used in the production of the products produced by the joint venture, such licensing is subject to Regulation 2658/2000 and not the TTBER. Accordingly, licensing in the context of a production joint venture normally falls to be considered under Regulation 2658/2000. However, where the joint venture engages in licensing of the technology to third parties, the activity is not linked to production by the joint venture and therefore not covered by that Regulation. Such licensing arrangements, which bring together the technologies of the parties, constitute technology pools, which are dealt with in section IV.4 below.

59. Regulation 2659/2000 on research and development agreements covers agreements whereby two or more undertakings agree to jointly carry out research and development and to jointly exploit the results thereof. According to Article 2(11), research and development and the exploitation of the results are carried out jointly where the work involved is carried out by a joint team, organisation or undertakings, jointly entrusted to a third party or allocated between the parties by way of specialisation in research, development, production and distribution, including licensing.

60. It follows that Regulation 2659/2000 covers licensing between the parties and by the parties to a joint entity in the context of a research and development agreement. In the context of such agreements the parties can also determine the conditions for licensing the fruits of the research and development agreement to third parties. However, since third party licensees are not party to the research and development agreement, the individual licence agreement concluded with third parties is not covered by Regulation 2659/2000. Such licence agreements are block exempted by the TTBER where they fulfil the conditions of that Regulation.

2.5.2. The Block Exemption Regulation on vertical agreements

61. Commission Regulation (EC) No 2790/1999 on vertical agreements covers agreements entered into between two or more undertakings each operating, for the purposes of the agreement, at different levels of the production or distribution chain, and relating to the conditions

[32] OJ L 304, 5.12.2000, p 3. [33] OJ L 304, 5.12.2000, p 7. [34] See note 31.

under which the parties may purchase, sell or resell certain goods or services. It thus covers supply and distribution agreements.[35]

62. Given that the TTBER only covers agreements between two parties and that a licensee, selling products incorporating the licensed technology, is a supplier for the purposes of Regulation 2790/1999, these two block exemption regulations are closely related. The agreement between licensor and licensee is subject to the TTBER whereas agreements concluded between a licensee and buyers are subject to Regulation 2790/1999 and the Guidelines on Vertical Restraints.[36]

63. The TTBER also block exempts agreements between the licensor and the licensee where the agreement imposes obligations on the licensee as to the way in which he must sell the products incorporating the licensed technology. In particular, the licensee can be obliged to establish a certain type of distribution system such as exclusive distribution or selective distribution. However, the distribution agreements concluded for the purposes of implementing such obligations must, in order to be block exempted, comply with Regulation 2790/1999. For instance, the licensor can oblige the licensee to establish a system based on exclusive distribution in accordance with specified rules. However, it follows from Article 4(b) of Regulation 2790/1999 that distributors must be free to make passive sales into the territories of other exclusive distributors.

64. Furthermore, distributors must in principle be free to sell both actively and passively into territories covered by the distribution systems of other licensees producing their own products on the basis of the licensed technology. This is because for the purposes of Regulation 2790/1999 each licensee is a separate supplier. However, the reasons underlying the block exemption contained in that Regulation may also apply where the products incorporating the licensed technology are sold by the licensees under a common brand belonging to the licensor. When the products incorporating the licensed technology are sold under a common brand identity there may be the same efficiency reasons for applying the same types of restraints between licensees' distribution systems as within a single vertical distribution system. In such cases the Commission would be unlikely to challenge restraints where by analogy the requirements of Regulation 2790/1999 are fulfilled. For a common brand identity to exist the products must be sold and marketed under a common brand, which is predominant in terms of conveying quality and other relevant information to the consumer. It does not suffice that in addition to the licensees' brands the product carries the licensor's brand, which identifies him as the source of the licensed technology.

3. The safe harbour established by the Block Exemption Regulation

65. According to Article 3 of the TTBER the block exemption of restrictive agreements is subject to market share thresholds, confining the scope of the block exemption to agreements that although they may be restrictive of competition can generally be presumed to fulfil the conditions of Article 81(3). Outside the safe harbour created by the market share thresholds individual assessment is required. The fact that market shares exceed the thresholds does not give rise to any presumption either that the agreement is caught by Article 81(1) or that the agreement does not fulfil the conditions of Article 81(3). In the absence of hardcore restrictions, market analysis is required.

66. The market share threshold to be applied for the purpose of the safe harbour of the TTBER depends on whether the agreement is concluded between competitors or non-competitors. For the purposes of the TTBER undertakings are competitors on the relevant technology market when they license competing technologies. Potential competition on the technology market is not taken into account for the application of the market share threshold or the hardcore list. Outside

[35] See the guide 'Competition policy in Europe—The competition rules for supply and distribution agreements', 2002.

[36] OJ C 291, 13.10.2000, p 1, and note 31.

the safe harbour of the TTBER potential competition on the technology market is taken into account but does not lead to the application of the hardcore list relating to agreements between competitors (see also paragraph 31 above).

67. Undertakings are competitors on the relevant product market where both undertakings are active on the same product and geographic market(s) on which the products incorporating the licensed technology are sold (actual competitors). They are also considered competitors where they would be likely, on realistic grounds, to undertake the necessary additional investments or other necessary switching costs to enter the relevant product and geographic market(s) within a reasonably short period of time[37] in response to a small and permanent increase in relative prices (potential competitors).

68. It follows from paragraphs 66 and 67 that two undertakings are not competitors for the purposes of the TTBER where the licensor is neither an actual nor a potential supplier of products on the relevant market and the licensee, already present on the product market, is not licensing out a competing technology even if he owns a competing technology and produces on the basis of that technology. However, the parties become competitors if at a later point in time the licensee starts licensing out his technology or the licensor becomes an actual or potential supplier of products on the relevant market. In that case the hardcore list relevant for agreements between non-competitors will continue to apply to the agreement unless the agreement is subsequently amended in any material respect, see Article 4(3) of the TTBER and paragraph 31 above.

69. In the case of agreements between competitors the market share threshold is 20 % and in the case of agreements between non-competitors it is 30 % (cf. Article 3(1) and (2) of the TTBER). Where the undertakings party to the licensing agreement are not competitors the agreement is covered if the market share of neither party exceeds 30 % on the affected relevant technology and product markets. Where the undertakings party to the licensing agreement are competitors the agreement is covered if the combined market shares of the parties do not exceed 20 % on the relevant technology and product markets. The market share thresholds apply both to technology markets and markets for products incorporating the licensed technology. If the applicable market share threshold is exceeded on an affected relevant market, the block exemption does not apply to the agreement for that relevant market. For instance, if the licence agreement concerns two separate product markets or two separate geographic markets, the block exemption may apply to one of the markets and not to the other.

70. In the case of technology markets, it follows from Article 3(3) of the TTBER that the licensor's market share is to be calculated on the basis of the sales of the licensor and all his licensees of products incorporating the licensed technology and this for each relevant market separately.[38] Where the parties are competitors on the technology market, sales of products incorporating the licensee's own technology must be combined with the sales of the products incorporating the licensed technology. In the case of new technologies that have not yet generated any sales, a zero market share is assigned. When sales commence the technology will start accumulating market share.

71. In the case of product markets, the licensee's market share is to be calculated on the basis of the licensee's sales of products incorporating the licensor's technology and competing products, i.e. the total sales of the licensee on the product market in question. Where the licensor is also a supplier of products on the relevant market, the licensor's sales on the product market in question must also be taken into account. In the calculation of market shares for product markets, however, sales made by other licensees are not taken into account when calculating the licensee's and/or licensor's market share.

[37] See para 29 above. [38] The reasons for this calculation rule are explained in para 23 above.

72. Market shares should be calculated on the basis of sales value data where such data are available. Such data normally provide a more accurate indication of the strength of a technology than volume data. However, where value based data are not available, estimates based on other reliable market information may be used, including market sales volume data.

73. The principles set out above can be illustrated by the following examples:

Licensing between non-competitors

Example 1

Company A is specialised in developing bio-technological products and techniques and has developed a new product Xeran. It is not active as a producer of Xeran, for which it has neither the production nor the distribution facilities. Company B is one of the producers of competing products, produced with freely available non-proprietary technologies. In year 1, B was selling EUR 25 million worth of products produced with the freely available technologies. In year 2, A gives a licence to B to produce Xeran. In that year B sells EUR 15 million produced with the help of the freely available technologies and EUR 15 million of Xeran. In year 3 and the following years B produces and sells only Xeran worth EUR 40 million annually. In addition in year 2, A is also licensing to C. C was not active on that product market before. C produces and sells only Xeran, EUR 10 million in year 2 and EUR 15 million in year 3 and thereafter. It is established that the total market of Xeran and its substitutes where B and C are active is worth EUR 200 million in each year.

In year 2, the year the licence agreement is concluded, A's market share on the technology market is 0 % as its market share has to be calculated on the basis of the total sales of Xeran in the preceding year. In year 3 A's market share on the technology market is 12,5 %, reflecting the value of Xeran produced by B and C in the preceding year 2. In year 4 and thereafter A's market share on the technology market is 27,5 %, reflecting the value of Xeran produced by B and C in the preceding year.

In year 2 B's market share on the product market is 12,5 %, reflecting B's EUR 25 million sales in year 1. In year 3 B's market share is 15 % because its sales have increased to EUR 30 million in year 2. In year 4 and thereafter B's market share is 20 % as its sales are EUR 40 million annually. C's market share on the product market is 0 % in year 1 and 2, 5 % in year 3 and 7,5 % thereafter.

As the licence agreements are between non-competitors and the individual market shares of A, B and C are below 30 % each year, the agreements fall within the safe harbour of the TTBER.

Example 2

The situation is the same as in example 1, however now B and C are operating in different geographic markets. It is established that the total market of Xeran and its substitutes is worth EUR 100 million annually in each geographic market.

In this case, A's market share on the technology market has to be calculated for each of the two geographic markets. In the market where B is active A's market share depends on the sale of Xeran by B. As in this example the total market is assumed to be EUR 100 million, i.e. half the size of the market in example 1, the market share of A is 0 % in year 2, 15 % in year 3 and 40 % thereafter. B's market share is 25 % in year 2, 30 % in year 3 and 40 % thereafter. In year 2 and 3 both A's and B's market share does not exceed the 30 %

threshold. The threshold is however exceeded from year 4 and this means that, in line with Article 8(2) of the TTBER, after year 6 the licence agreement between A and B can no longer benefit from the safe harbour but has to be assessed on an individual basis.

In the market where C is active A's market share depends on the sale of Xeran by C. A's market share on the technology market, based on C's sales in the previous year, is therefore 0 % in year 2, 10 % in year 3 and 15 % thereafter. The market share of C on the product market is the same: 0 % in year 2, 10 % in year 3 and 15 % thereafter. The licence agreement between A and C therefore falls within the safe harbour for the whole period.

Licensing between competitors

Example 3

Companies A and B are active on the same relevant product and geographic market for a certain chemical product. They also each own a patent on different technologies used to produce this product. In year 1 A and B sign a cross licence agreement licensing each other to use their respective technologies. In year 1 A and B produce only with their own technology and A sells EUR 15 million of the product and B sells EUR 20 million of the product. From year 2 they both use their own and the other's technology. From that year onward A sells EUR 10 million of the product produced with its own technology and EUR 10 million of the product produced with B's technology. B sells from year 2 EUR 15 million of the product produced with its own technology and EUR 10 million of the product produced with A's technology. It is established that the total market of the product and its substitutes is worth EUR 100 million in each year.

To assess the licence agreement under the TTBER, the market shares of A and B have to be calculated both on the technology market and the product market. The market share of A on the technology market depends on the amount of the product sold in the preceding year that was produced, by both A and B, with A's technology. In year 2 the market share of A on the technology market is therefore 15 %, reflecting its own production and sales of EUR 15 million in year 1. From year 3 A's market share on the technology market is 20 %, reflecting the EUR 20 million sale of the product produced with A's technology and produced and sold by A and B (EUR 10 million each). Similarly, in year 2 B's market share on the technology market is 20 % and thereafter 25 %.

The market shares of A and B on the product market depend on their respective sales of the product in the previous year, irrespective of the technology used. The market share of A on the product market is 15 % in year 2 and 20 % thereafter. The market share of B on the product market is 20 % in year 2 and 25 % thereafter.

As the agreement is between competitors, their combined market share, both on the technology and on the product market, has to be below the 20 % market share threshold in order to benefit from the safe harbour. It is clear that this is not the case here. The combined market share on the technology market and on the product market is 35 % in year 2 and 45 % thereafter. This agreement between competitors will therefore have to be assessed on an individual basis.

4. Hardcore restrictions of competition under the Block Exemption Regulation

4.1. General principles

74. Article 4 of the TTBER contains a list of hardcore restrictions of competition. The classification of a restraint as a hardcore restriction of competition is based on the nature of the restriction and experience showing that such restrictions are almost always anti-competitive. In line with the case law of the Community Courts[39] such a restriction may result from the clear objective of the agreement or from the circumstances of the individual case (cf. paragraph 14 above).

75. When a technology transfer agreement contains a hardcore restriction of competition, it follows from Article 4(1) and 4(2) of the TTBER that the agreement as a whole falls outside the scope of the block exemption. For the purposes of the TTBER hardcore restrictions cannot be severed from the rest of the agreement. Moreover, the Commission considers that in the context of individual assessment hardcore restrictions of competition will only in exceptional circumstances fulfil the four conditions of Article 81(3) (cf. paragraph 18 above).

76. Article 4 of the TTBER distinguishes between agreements between competitors and agreements between non-competitors.

4.2. Agreements between competitors

77. Article 4(1) lists the hardcore restrictions for licensing between competitors. According to Article 4(1), the TTBER does not cover agreements which, directly or indirectly, in isolation or in combination with other factors under the control of the parties, have as their object:

(a) The restriction of a party's ability to determine its prices when selling products to third parties;

(b) The limitation of output, except limitations on the output of contract products imposed on the licensee in a non-reciprocal agreement or imposed on only one of the licensees in a reciprocal agreement;

(c) The allocation of markets or customers except

 (i) the obligation on the licensee(s) to produce with the licensed technology only within one or more technical fields of use or one or more product markets;

 (ii) the obligation on the licensor and/or the licensee, in a non-reciprocal agreement, not to produce with the licensed technology within one or more technical fields of use or one or more product markets or one or more exclusive territories reserved for the other party;

 (iii) the obligation on the licensor not to license the technology to another licensee in a particular territory;

 (iv) the restriction, in a non-reciprocal agreement, of active and/or passive sales by the licensee and/or the licensor into the exclusive territory or to the exclusive customer group reserved for the other party;

 (v) the restriction, in a non-reciprocal agreement, of active sales by the licensee into the exclusive territory or to the exclusive customer group allocated by the licensor to another licensee provided that the latter was not a competing undertaking of the licensor at the time of the conclusion of its own licence;

 (vi) the obligation on the licensee to produce the contract products only for its own use provided that the licensee is not restricted in selling the contract products actively and passively as spare parts for its own products;

 (vii) the obligation on the licensee in a non-reciprocal agreement to produce the contract

[39] See e.g. the case law cited in note 15.

products only for a particular customer, where the licence was granted in order to create an alternative source of supply for that customer;

(d) The restriction of the licensee's ability to exploit its own technology or the restriction of the ability of any of the parties to the agreement to carry out research and development, unless such latter restriction is indispensable to prevent the disclosure of the licensed know-how to third parties.

78. For a number of hardcore restrictions the TTBER makes a distinction between reciprocal and non-reciprocal agreements. The hardcore list is stricter for reciprocal agreements than for non-reciprocal agreements between competitors. Reciprocal agreements are cross-licensing agreements where the licensed technologies are competing technologies or can be used for the production of competing products. A non-reciprocal agreement is an agreement where only one of the parties is licensing its technology to the other party or where in case of cross-licensing the licensed technologies are not competing technologies and cannot be used for the production of competing products. An agreement is not reciprocal merely because the agreement contains a grant back obligation or because the licensee licenses back own improvements of the licensed technology. In case at a later point in time a non-reciprocal agreement becomes a reciprocal agreement due to the conclusion of a second licence between the same parties, they may have to revise the first licence in order to avoid that the agreement contains a hardcore restriction. In the assessment of the individual case the Commission will take into account the time lapsed between the conclusion of the first and the second licence.

79. The hardcore restriction of competition contained in Article 4(1)(a) concerns agreements between competitors that have as their object the fixing of prices for products sold to third parties, including the products incorporating the licensed technology. Price fixing between competitors constitutes a restriction of competition by its very object. Price fixing can for instance take the form of a direct agreement on the exact price to be charged or on a price list with certain allowed maximum rebates. It is immaterial whether the agreement concerns fixed, minimum, maximum or recommended prices. Price fixing can also be implemented indirectly by applying disincentives to deviate from an agreed price level, for example, by providing that the royalty rate will increase if product prices are reduced below a certain level. However, an obligation on the licensee to pay a certain minimum royalty does not in itself amount to price fixing.

80. When royalties are calculated on the basis of individual product sales, the amount of the royalty has a direct impact on the marginal cost of the product and thus a direct impact on product prices.[40] Competitors can therefore use cross licensing with reciprocal running royalties as a means of co-ordinating prices on downstream product markets.[41] However, the Commission will only treat cross licences with reciprocal running royalties as price fixing where the agreement is devoid of any pro-competitive purpose and therefore does not constitute a bona fide licensing arrangement. In such cases where the agreement does not create any value and therefore has no valid business justification, the arrangement is a sham and amounts to a cartel.

81. The hardcore restriction contained in Article 4(1)(a) also covers agreements whereby royalties are calculated on the basis of all product sales irrespective of whether the licensed technology is being used. Such agreements are also caught by Article 4(1)(d) according to which the licensee must not be restricted in his ability to use his own technology (see paragraph 95 below). In general such agreements restrict competition since the agreement raises the cost of using the licensee's

[40] See in this respect para 98 of the Guidelines on the application of Article 81(3) of the Treaty cited in note 2.

[41] This is also the case where one party grants a licence to the other party and accepts to buy a physical input from the licensee. The purchase price can serve the same function as the royalty.

own competing technology and restricts competition that existed in the absence of the agreement.[42] This is so both in the case of reciprocal and non-reciprocal arrangements. Exceptionally, however, an agreement whereby royalties are calculated on the basis of all product sales may fulfil the conditions of Article 81(3) in an individual case where on the basis of objective factors it can be concluded that the restriction is indispensable for pro-competitive licensing to occur. This may be the case where in the absence of the restraint it would be impossible or unduly difficult to calculate and monitor the royalty payable by the licensee, for instance because the licensor's technology leaves no visible trace on the final product and practicable alternative monitoring methods are unavailable.

82. The hardcore restriction of competition set out in Article 4(1)(b) concerns reciprocal output restrictions on the parties. An output restriction is a limitation on how much a party may produce and sell. Article 4(1)(b) does not cover output limitations on the licensee in a non-reciprocal agreement or output limitations on one of the licensees in a reciprocal agreement provided that the output limitation only concerns products produced with the licensed technology. Article 4(1)(b) thus identifies as hardcore restrictions reciprocal output restrictions on the parties and output restrictions on the licensor in respect of his own technology. When competitors agree to impose reciprocal output limitations, the object and likely effect of the agreement is to reduce output in the market. The same is true of agreements that reduce the incentive of the parties to expand output, for example by obliging each other to make payments if a certain level of output is exceeded.

83. The more favourable treatment of non-reciprocal quantity limitations is based on the consideration that a one-way restriction does not necessarily lead to a lower output on the market while also the risk that the agreement is not a bona fide licensing arrangement is less when the restriction is non-reciprocal. When a licensee is willing to accept a one-way restriction, it is likely that the agreement leads to a real integration of complementary technologies or an efficiency enhancing integration of the licensor's superior technology with the licensee's productive assets. In a reciprocal agreement an output restriction on one of the licensees is likely to reflect the higher value of the technology licensed by one of the parties and may serve to promote pro-competitive licensing.

84. The hardcore restriction of competition set out in Article 4(1)(c) concerns the allocation of markets and customers. Agreements whereby competitors share markets and customers have as their object the restriction of competition. It is a hardcore restriction where competitors in a reciprocal agreement agree not to produce in certain territories or not to sell actively and/or passively into certain territories or to certain customers reserved for the other party.

85. Article 4(1)(c) applies irrespective of whether the licensee remains free to use his own technology. Once the licensee has tooled up to use the licensor's technology to produce a given product, it may be costly to maintain a separate production line using another technology in order to serve customers covered by the restrictions. Moreover, given the anti-competitive potential of the restraint the licensee may have little incentive to produce under his own technology. Such restrictions are also highly unlikely to be indispensable for pro-competitive licensing to occur.

86. Under Article 4(1)(c)(ii) it is not a hardcore restriction for the licensor in a non-reciprocal agreement to grant the licensee an exclusive licence to produce on the basis of the licensed technology in a particular territory and thus agree not to produce himself the contract products in or provide the contract products from that territory. Such exclusive licences are block exempted irrespective of the scope of the territory. If the licence is world-wide, the exclusivity implies that

[42] See in this respect Case 193/83, *Windsurfing International* [1986] ECR 611, para 67.

the licensor abstains from entering or remaining on the market. The block exemption also applies where the licence is limited to one or more technical fields of use or one or more product markets. The purpose of agreements covered by Article 4(1)(c)(ii) may be to give the licensee an incentive to invest in and develop the licensed technology. The object of the agreement is therefore not necessarily to share markets.

87. According to Article 4(1)(c)(iv) and for the same reason, the block exemption also applies to non-reciprocal agreements whereby the parties agree not to sell actively or passively[43] into an exclusive territory or to an exclusive customer group reserved for the other party.

88. According to Article 4(1)(c)(iii) it is also not a hardcore restriction if the licensor appoints the licensee as his sole licensee in a particular territory, implying that third parties will not be licensed to produce on the basis of the licensor's technology in the territory in question. In the case of such sole licences the block exemption applies irrespective of whether the agreement is reciprocal or not given that the agreement does not affect the ability of the parties to fully exploit their own technology in the respective territories.

89. Article 4(1)(c)(v) excludes from the hardcore list and thus block exempts up to the market share threshold restrictions in a non-reciprocal agreement on active sales by a licensee into the territory or to the customer group allocated by the licensor to another licensee. It is a condition, however, that the protected licensee was not a competitor of the licensor when the agreement was concluded. It is not warranted to hardcore such restrictions. By allowing the licensor to grant a licensee, who was not already on the market, protection against active sales by licensees which are competitors of the licensor and which for that reason are already established on the market, such restrictions are likely to induce the licensee to exploit the licensed technology more efficiently. On the other hand, if the licensees agree between themselves not to sell actively or passively into certain territories or to certain customer groups, the agreement amounts to a cartel amongst the licensees. Given that such agreements do not involve any transfer of technology they fall outside the scope of the TTBER.

90. According to Article 4(1)(c)(i) restrictions in agreements between competitors that limit the licence to one or more product markets or technical fields of use[44] are not hardcore restrictions. Such restrictions are block exempted up to the market share threshold of 20 % irrespective of whether the agreement is reciprocal or not. It is a condition for the application of the block exemption, however, that the field of use restrictions do not go beyond the scope of the licensed technologies. It is also a condition that licensees are not limited in the use of their own technology (see Article 4(1)(d)). Where licensees are limited in the use of their own technology the agreement amounts to market sharing.

91. The block exemption applies irrespective of whether the field of use restriction is symmetrical or asymmetrical. An asymmetrical field of use restriction in a reciprocal licence agreement implies that both parties are allowed to use the respective technologies that they license in only within different fields of use. As long as the parties are unrestricted in the use of their own technologies, it is not assumed that the agreement leads the parties to abandon or refrain from entering the field(s) covered by the licence to the other party. Even if the licensees tool up to use the licensed technology within the licensed field of use, there may be no impact on assets used to produce outside the scope of the licence. It is important in this regard that the restriction relates to distinct product markets or fields of use and not to customers, allocated by territory or by group, who purchase products falling within the same product market or technical field of use. The risk of market sharing is considered substantially greater in the latter case (see paragraph 85 above).

---

[43] For a general definition of active and passive sales, reference is made to para 50 of the Guidelines on vertical restraints cited in note 36.

[44] Field of use restrictions are further dealt with in section IV.2.4 below.

In addition, field of use restrictions may be necessary to promote pro-competitive licensing (see paragraph 182 below).

92. Article 4(1)(c)(vi) contains a further exception, namely captive use restrictions, i.e. a requirement whereby the licensee may produce the products incorporating the licensed technology only for his own use. Where the contract product is a component the licensee can thus be obliged to produce that component only for incorporation into his own products and can be obliged not to sell the components to other producers. The licensee must be able, however, to sell the components as spare parts for his own products and must thus be able to supply third parties that perform after sale services on these products. Captive use restrictions as defined may be necessary to encourage the dissemination of technology, particularly between competitors, and are covered by the block exemption. Such restrictions are also dealt with in section IV.2.5 below.

93. Finally, Article 4(1)(c)(vii) excludes from the hardcore list an obligation on the licensee in a non-reciprocal agreement to produce the contract products only for a particular customer with a view to creating an alternative source of supply for that customer. It is thus a condition for the application of Article 4(1)(c)(vii) that the licence is limited to creating an alternative source of supply for that particular customer. It is not a condition, however, that only one such licence is granted. Article 4(1)(c)(vii) also covers situations where more than one undertaking is licensed to supply the same specified customer. The potential of such agreements to share markets is limited where the licence is granted only for the purpose of supplying a particular customer. In particular, in such circumstances it cannot be assumed that the agreement will cause the licensee to cease exploiting his own technology.

94. The hardcore restriction of competition set out in Article 4(1)(d) covers firstly restrictions on any of the parties' ability to carry out research and development. Both parties must be free to carry out independent research and development. This rule applies irrespective of whether the restriction applies to a field covered by the licence or to other fields. However, the mere fact that the parties agree to provide each other with future improvements of their respective technologies does not amount to a restriction on independent research and development. The effect on competition of such agreements must be assessed in light of the circumstances of the individual case. Article 4(1)(d) also does not extend to restrictions on a party to carry out research and development with third parties, where such restriction is necessary to protect the licensor's know-how against disclosure. In order to be covered by the exception, the restrictions imposed to protect the licensor's know-how against disclosure must be necessary and proportionate to ensure such protection. For instance, where the agreement designates particular employees of the licensee to be trained in and responsible for the use of the licensed know-how, it may be sufficient to oblige the licensee not to allow those employees to be involved in research and development with third parties. Other safeguards may be equally appropriate.

95. According to Article 4(1)(d) the licensee must also be unrestricted in the use of his own competing technology provided that in so doing he does not make use of the technology licensed from the licensor. In relation to his own technology the licensee must not be subject to limitations in terms of where he produces or sells, how much he produces or sells and at what price he sells. He must also not be obliged to pay royalties on products produced on the basis of his own technology (cf. paragraph 81 above). Moreover, the licensee must not be restricted in licensing his own technology to third parties. When restrictions are imposed on the licensee's use of his own technology or to carry out research and development, the competitiveness of the licensee's technology is reduced. The effect of this is to reduce competition on existing product and technology markets and to reduce the licensee's incentive to invest in the development and improvement of his technology.

4.3. Agreements between non-competitors

96. Article 4(2) lists the hardcore restrictions for licensing between non-competitors. According to this provision, the TTBER does not cover agreements which, directly or indirectly, in isolation or in combination with other factors under the control of the parties, have as their object:

(a) the restriction of a party's ability to determine its prices when selling products to third parties, without prejudice to the possibility to impose a maximum sale price or recommend a sale price, provided that it does not amount to a fixed or minimum sale price as a result of pressure from, or incentives offered by, any of the parties;

(b) the restriction of the territory into which, or of the customers to whom, the licensee may passively sell the contract products, except:

   (i) the restriction of passive sales into an exclusive territory or to an exclusive customer group reserved for the licensor;

   (ii) the restriction of passive sales into an exclusive territory or to an exclusive customer group allocated by the licensor to another licensee during the first two years that this other licensee is selling the contract products in that territory or to that customer group;

   (iii) the obligation to produce the contract products only for its own use provided that the licensee is not restricted in selling the contract products actively and passively as spare parts for its own products;

   (iv) the obligation to produce the contract products only for a particular customer, where the licence was granted in order to create an alternative source of supply for that customer;

   (v) the restriction of sales to end users by a licensee operating at the wholesale level of trade;

   (vi) the restriction of sales to unauthorised distributors by the members of a selective distribution system;

(c) the restriction of active or passive sales to end users by a licensee which is a member of a selective distribution system and which operates at the retail level, without prejudice to the possibility of prohibiting a member of the system from operating out of an unauthorised place of establishment.

97. The hardcore restriction of competition set out in Article 4(2)(a) concerns the fixing of prices charged when selling products to third parties. More specifically, this provision covers restrictions which have as their direct or indirect object the establishment of a fixed or a minimum selling price or a fixed or minimum price level to be observed by the licensor or the licensee when selling products to third parties. In the case of agreements that directly establish the selling price, the restriction is clear-cut. However, the fixing of selling prices can also be achieved through indirect means. Examples of the latter are agreements fixing the margin, fixing the maximum level of discounts, linking the sales price to the sales prices of competitors, threats, intimidation, warnings, penalties, or contract terminations in relation to observance of a given price level. Direct or indirect means of achieving price fixing can be made more effective when combined with measures to identify price-cutting, such as the implementation of a price monitoring system, or the obligation on licensees to report price deviations. Similarly, direct or indirect price fixing can be made more effective when combined with measures that reduce the licensee's incentive to lower his selling price, such as the licensor obliging the licensee to apply a most-favoured-customer clause, i.e. an obligation to grant to a customer any more favourable terms granted to any other customer. The same means can be used to make maximum or recommended prices work as fixed or minimum selling prices. However, the provision of a list of recommended prices to or the imposition of a maximum price on the licensee by the licensor is not considered in itself as leading to fixed or minimum selling prices.

98. Article 4(2)(b) identifies as hardcore restrictions of competition agreements or concerted

practices that have as their direct or indirect object the restriction of passive sales by licensees of products incorporating the licensed technology.[45] Passive sales restrictions on the licensee may be the result of direct obligations, such as the obligation not to sell to certain customers or to customers in certain territories or the obligation to refer orders from these customers to other licensees. It may also result from indirect measures aimed at inducing the licensee to refrain from making such sales, such as financial incentives and the implementation of a monitoring system aimed at verifying the effective destination of the licensed products. Quantity limitations may be an indirect means to restrict passive sales. The Commission will not assume that quantity limitations as such serve this purpose. However, it will be otherwise where quantity limitations are used to implement an underlying market partitioning agreement. Indications thereof include the adjustment of quantities over time to cover only local demand, the combination of quantity limitations and an obligation to sell minimum quantities in the territory, minimum royalty obligations linked to sales in the territory, differentiated royalty rates depending on the destination of the products and the monitoring of the destination of products sold by individual licensees. The general hardcore restriction covering passive sales by licensees is subject to a number of exceptions, which are dealt with below.

99. Article 4(2)(b) does not cover sales restrictions on the licensor. All sales restrictions on the licensor are block exempted up to the market share threshold of 30 %. The same applies to all restrictions on active sales by the licensee, with the exception of what is said on active selling in paragraphs 105 and 106 below. The block exemption of restrictions on active selling is based on the assumption that such restrictions promote investments, non-price competition and improvements in the quality of services provided by the licensees by solving free rider problems and hold-up problems. In the case of restrictions of active sales between licensees' territories or customer groups, it is not a condition that the protected licensee has been granted an exclusive territory or an exclusive customer group. The block exemption also applies to active sales restrictions where more than one licensee has been appointed for a particular territory or customer group. Efficiency enhancing investment is likely to be promoted where a licensee can be ensured that he will only face active sales competition from a limited number of licensees inside the territory and not also from licensees outside the territory.

100. Restrictions on active and passive sales by licensees into an exclusive territory or to an exclusive customer group reserved for the licensor do not constitute hardcore restrictions of competition (cf. Article 4(2)(b)(i)). Indeed, they are block exempted. It is presumed that up to the market share threshold such restraints, where restrictive of competition, promote pro-competitive dissemination of technology and integration of such technology into the production assets of the licensee. For a territory or customer group to be reserved for the licensor, it is not required that the licensor is actually producing with the licensed technology in the territory or for the customer group in question. A territory or customer group can also be reserved by the licensor for later exploitation.

101. Restrictions on passive sales by licensees into an exclusive territory or customer group allocated to another licensee are block exempted for two years calculated from the date on which the protected licensee first markets the products incorporating the licensed technology inside his exclusive territory or to his exclusive customer group (cf. Article 4(2)(b)(ii)). Licensees often have to commit substantial investments in production assets and promotional activities in order to start up and develop a new territory. The risks facing the new licensee are therefore likely to be substantial, in particular since promotional expenses and investment in assets required to produce on the basis of a particular technology are often sunk, i.e. they cannot be recovered if the licensee

[45] This hardcore restriction applies to licence agreements concerning trade within the Community. As regards agreements concerning exports outside the Community or imports/re-imports from outside the Community see Case C-306/96, *Javico* [1998] ECR I-1983.

exits the market. In such circumstances, it is often the case that licensees would not enter into the licence agreement without protection for a certain period of time against (active and) passive sales into their territory by other licensees. Restrictions on passive sales into the exclusive territory of a licensee by other licensees therefore often fall outside Article 81(1) for a period of up to two years from the date on which the product incorporating the licensed technology was first put on the market in the exclusive territory by the licensee in question. However, to the extent that in individual cases such restrictions are caught by Article 81(1) they are block exempted. After the expiry of this two-year period restrictions on passive sales between licensees constitute hardcore restrictions. Such restrictions are generally caught by Article 81(1) and are unlikely to fulfil the conditions of Article 81(3). In particular, passive sales restrictions are unlikely to be indispensable for the attainment of efficiencies.[46]

102. Article 4(2)(b)(iii) brings under the block exemption a restriction whereby the licensee is obliged to produce products incorporating the licensed technology only for his own (captive) use. Where the contract product is a component the licensee can thus be obliged to use that product only for incorporation into his own products and can be obliged not to sell the product to other producers. The licensee must however be able to actively and passively sell the products as spare parts for his own products and must thus be able to supply third parties that perform after sale services on these products. Captive use restrictions are also dealt with in section IV.2.5 below.

103. As in the case of agreements between competitors (cf. paragraph 93 above) the block exemption also applies to agreements whereby the licensee is obliged to produce the contract products only for a particular customer in order to provide that customer with an alternative source of supply (cf. Article 4(2)(b)(iv)). In the case of agreements between non-competitors, such restrictions are unlikely to be caught by Article 81(1).

104. Article 4(2)(b)(v) brings under the block exemption an obligation on the licensee not to sell to end users and thus only to sell to retailers. Such an obligation allows the licensor to assign the wholesale distribution function to the licensee and normally falls outside Article 81(1).[47]

105. Finally Article 4(2)(b)(vi) brings under the block exemption a restriction on the licensee not to sell to unauthorised distributors. This exception allows the licensor to impose on the licensees an obligation to form part of a selective distribution system. In that case, however, the licensees must according to Article 4(2)(c) be permitted to sell both actively and passively to end users, without prejudice to the possibility to restrict the licensee to a wholesale function as foreseen in Article 4(2)(b)(v) (cf. the previous paragraph).

106. It is recalled (cf. paragraph 39 above) that the block exemption covers licence agreements whereby the licensor imposes obligations which the licensee must or may impose on his buyers, including distributors. However, these obligations must comply with the competition rules applicable to supply and distribution agreements. Since the TTBER is limited to agreements between two parties the agreements concluded between the licensee and his buyers implementing such obligations are not covered by the TTBER. Such agreements are only block exempted when they comply with Regulation 2790/1999 (cf. section 2.5.2 above).

5. Excluded restrictions

107. Article 5 of the TTBER lists four types of restrictions that are not block exempted and which thus require individual assessment of their anti-competitive and pro-competitive effects. It follows from Article 5 that the inclusion in a licence agreement of any of the restrictions contained in these provisions does not prevent the application of the block exemption to the rest

[46] See in this respect para 77 of the judgment in *Nungesser* cited in note 13.

[47] See in this respect Case 26/76, *Metro (I)* [1977] ECR 1875.

of the agreement. It is only the individual restriction in question that is not block exempted, implying that individual assessment is required. Accordingly, the rule of severability applies to the restrictions set out in Article 5.

108. Article 5(1) provides that the block exemption shall not apply to the following three obligations:

(a) Any direct or indirect obligation on the licensee to grant an exclusive licence to the licensor or to a third party designated by the licensor in respect of its own severable improvements to or its new applications of the licensed technology.
(b) Any direct or indirect obligation on the licensee to assign to the licensor or to a third party designated by the licensor rights to severable improvements to or new applications of the licensed technology.
(c) Any direct or indirect obligation on the licensee not to challenge the validity of intellectual property rights held by the licensor in the common market. However, the TTBER does cover the possibility for the licensor to terminate the licence agreement in the event that the licensee challenges the validity of the licensed technology.

The purpose of Article 5(1)(a), (b) and (c) is to avoid block exemption of agreements that may reduce the incentive of licensees to innovate.

109. Article 5(1)(a) and 5(1)(b) concerns exclusive grant backs or assignments to the licensor of severable improvements of the licensed technology. An improvement is severable if it can be exploited without infringing upon the licensed technology. An obligation to grant the licensor an exclusive licence to severable improvements of the licensed technology or to assign such improvements to the licensor is likely to reduce the licensee's incentive to innovate since it hinders the licensee in exploiting his improvements, including by way of licensing to third parties. This is the case both where the severable improvement concerns the same application as the licensed technology and where the licensee develops new applications of the licensed technology. According to Article 5(1)(a) and (b) such obligations are not block exempted. However, the block exemption does cover non-exclusive grant back obligations in respect of severable improvements. This is so even where the grant back obligation is non-reciprocal, i.e. only imposed on the licensee, and where under the agreement the licensor is entitled to feed-on the severable improvements to other licensees. A non-reciprocal grant back obligation may promote innovation and the dissemination of new technology by permitting the licensor to freely determine whether and to what extent to pass on his own improvements to his licensees. A feed-on clause may also promote the dissemination of technology because each licensee knows at the time of contracting that he will be on an equal footing with other licensees in terms of the technology on the basis of which he is producing. Exclusive grant backs and obligations to assign non-severable improvements are not restrictive of competition within the meaning of Article 81(1) since non-severable improvements cannot be exploited by the licensee without the licensor's permission.

110. The application of Article 5(1)(a) and (b) does not depend on whether or not the licensor pays consideration in return for acquiring the improvement or for obtaining an exclusive licence. However, the existence and level of such consideration may be a relevant factor in the context of an individual assessment under Article 81. When grant backs are made against consideration it is less likely that the obligation creates a disincentive for the licensee to innovate. In the assessment of exclusive grant backs outside the scope of the block exemption the market position of the licensor on the technology market is also a relevant factor. The stronger the position of the licensor, the more likely it is that exclusive grant back obligations will have restrictive effects on competition in innovation. The stronger the position of the licensor's technology the more likely it is that the licensee will be an important source of innovation and future competition. The negative impact of grant back obligations can also be increased in case of parallel networks of licence agreements containing such obligations. When available technologies are controlled by a

limited number of licensors that impose exclusive grant back obligations on licensees, the risk of anti-competitive effects is greater than where there are a number of technologies only some of which are licensed on exclusive grant back terms.

111. The risk of negative effects on innovation is higher in the case of cross licensing between competitors where a grant back obligation on both parties is combined with an obligation on both parties to share with the other party improvements of his own technology. The sharing of all improvements between competitors may prevent each competitor from gaining a competitive lead over the other (see also paragraph 208 below). However, the parties are unlikely to be prevented from gaining a competitive lead over each other where the purpose of the licence is to permit them to develop their respective technologies and where the licence does not lead them to use the same technological base in the design of their products. This is the case where the purpose of the licence is to create design freedom rather than to improve the technological base of the licensee.

112. The excluded restriction set out in Article 5(1)(c) concerns non-challenge clauses, i.e. obligations not to challenge the validity of the licensor's intellectual property. The reason for excluding non-challenge clauses from the scope of the block exemption is the fact that licensees are normally in the best position to determine whether or not an intellectual property right is invalid. In the interest of undistorted competition and in conformity with the principles underlying the protection of intellectual property, invalid intellectual property rights should be eliminated. Invalid intellectual property stifles innovation rather than promoting it. Article 81(1) is likely to apply to non-challenge clauses where the licensed technology is valuable and therefore creates a competitive disadvantage for undertakings that are prevented from using it or are only able to use it against payment of royalties.[48] In such cases the conditions of Article 81(3) are unlikely to be fulfilled.[49] However, the Commission takes a favourable view of non-challenge clauses relating to know-how where once disclosed it is likely to be impossible or very difficult to recover the licensed know-how. In such cases, an obligation on the licensee not to challenge the licensed know-how promotes dissemination of new technology, in particular by allowing weaker licensors to license stronger licensees without fear of a challenge once the know-how has been absorbed by the licensee.

113. The TTBER covers the possibility for the licensor to terminate the licence agreement in the event of a challenge of the licensed technology. Accordingly, the licensor is not forced to continue dealing with a licensee that challenges the very subject matter of the licence agreement, implying that upon termination any further use by the licensee of the challenged technology is at the challenger's own risk. Article 5(1)(c) ensures, however, that the TTBER does not cover contractual obligations obliging the licensee not to challenge the licensed technology, which would permit the licensor to sue the licensee for breach of contract and thereby create a further disincentive for the licensee to challenge the validity of the licensor's technology. The provision thereby ensures that the licensee is in the same position as third parties.

114. Article 5(2) excludes from the scope of the block exemption, in the case of agreements between non-competitors, any direct or indirect obligation limiting the licensee's ability to exploit his own technology or limiting the ability of the parties to the agreement to carry out research and development, unless such latter restriction is indispensable to prevent the disclosure of licensed know-how to third parties. The content of this condition is the same as that of Article 4(1)(d) of the hardcore list concerning agreements between competitors, which is dealt with in paragraphs 94 and 95 above. However, in the case of agreements between non-competitors it cannot be

[48] If the licensed technology is outdated no restriction of competition arises, see in this respect Case 65/86, *Bayer v Süllhofer* [1988] ECR 5249.

[49] As to non-challenge clauses in the context of settlement agreements see point 209 below.

considered that such restrictions generally have negative effects on competition or that the conditions of Article 81(3) are generally not satisfied.[50] Individual assessment is required.

115. In the case of agreements between non-competitors, the licensee normally does not own a competing technology. However, there may be cases where for the purposes of the block exemption the parties are considered non-competitors in spite of the fact that the licensee does own a competing technology. This is the case where the licensee owns a technology but does not license it and the licensor is not an actual or potential supplier on the product market. For the purposes of the block exemption the parties are in such circumstances neither competitors on the technology market nor competitors on the product market.[51] In such cases it is important to ensure that the licensee is not restricted in his ability to exploit his own technology and further develop it. This technology constitutes a competitive constraint in the market, which should be preserved. In such a situation restrictions on the licensee's use of his own technology or on research and development are normally considered to be restrictive of competition and not to satisfy the conditions of Article 81(3). For instance, an obligation on the licensee to pay royalties not only on the basis of products it produces with the licensed technology but also on the basis of products it produces with its own technology will generally limit the ability of the licensee to exploit its own technology and thus be excluded from the scope of the block exemption.

116. In cases where the licensee does not own a competing technology or is not already developing such a technology, a restriction on the ability of the parties to carry out independent research and development may be restrictive of competition where only a few technologies are available. In that case the parties may be an important (potential) source of innovation in the market. This is particularly so where the parties possess the necessary assets and skills to carry out further research and development. In that case the conditions of Article 81(3) are unlikely to be fulfilled. In other cases where several technologies are available and where the parties do not possess special assets or skills, the restriction on research and development is likely to either fall outside Article 81(1) for lack of an appreciable restrictive effect or satisfy the conditions of Article 81(3). The restraint may promote the dissemination of new technology by assuring the licensor that the licence does not create a new competitor and by inducing the licensee to focus on the exploitation and development of the licensed technology. Moreover, Article 81(1) only applies where the agreement reduces the licensee's incentive to improve and exploit his own technology. This is for instance not likely to be the case where the licensor is entitled to terminate the licence agreement once the licensee commences to produce on the basis of his own competing technology. Such a right does not reduce the licensee's incentive to innovate, since the agreement can only be terminated when a commercially viable technology has been developed and products produced on the basis thereof are ready to be put on the market.

6. Withdrawal and disapplication of the Block Exemption Regulation

6.1. Withdrawal procedure

117. According to Article 6 of the TTBER, the Commission and the competition authorities of the Member States may withdraw the benefit of the block exemption in respect of individual agreements that do not fulfil the conditions of Article 81(3). The power of the competition authorities of the Member States to withdraw the benefit of the block exemption is limited to cases where the relevant geographic market is no wider than the territory of the Member State in question.

118. The four conditions of Article 81(3) are cumulative and must all be fulfilled for the exception rule to be applicable.[52] The block exemption can therefore be withdrawn where a particular agreement fails one or more of the four conditions.

---

[50] See para 14 above. [51] See paras 66 and 67 above.

[52] See in this respect para 42 of the Guidelines on the application of Article 81(3) of the Treaty, cited in note 2.

119. Where the withdrawal procedure is applied, the withdrawing authority bears the burden of proving that the agreement falls within the scope of Article 81(1) and that the agreement does not satisfy all four conditions of Article 81(3). Given that withdrawal implies that the agreement in question restricts competition within the meaning of Article 81(1) and does not fulfil the conditions of Article 81(3), withdrawal is necessarily accompanied by a negative decision based on Articles 5, 7 or 9 of Regulation 1/2003.

120. According to Article 6, withdrawal may in particular be warranted in the following circumstances:

1. access of third parties' technologies to the market is restricted, for instance by the cumulative effect of parallel networks of similar restrictive agreements prohibiting licensees from using third party technology;
2. access of potential licensees to the market is restricted, for instance by the cumulative effect of parallel networks of similar restrictive agreements preventing licensors from licensing to other licensees;
3. without any objectively valid reason the parties refrain from exploiting the licensed technology.

121. Articles 4 and 5 of the TTBER, containing the list of hardcore restrictions of competition and excluded restrictions, aim at ensuring that block exempted agreements do not reduce the incentive to innovate, do not delay the dissemination of technology, and do not unduly restrict competition between the licensor and licensee or between licensees. However, the list of hardcore restrictions and the list of excluded restrictions do not take into account all the possible impacts of licence agreements. In particular, the block exemption does not take account of any cumulative effect of similar restrictions contained in networks of licence agreements. Licence agreements may lead to foreclosure of third parties both at the level of the licensor and at the level of the licensee. Foreclosure of other licensors may stem from the cumulative effect of networks of licence agreements prohibiting the licensees from exploiting competing technologies, leading to the exclusion of other (potential) licensors. Foreclosure of licensors is likely to arise in cases where most of the undertakings on the market that could (efficiently) take a competing licence are prevented from doing so as a consequence of restrictive agreements and where potential licensees face relatively high barriers to entry. Foreclosure of other licensees may stem from the cumulative effect of licence agreements prohibiting licensors from licensing other licensees and thereby preventing potential licensees from gaining access to the necessary technology. The issue of foreclosure is examined in more detail in section IV.2.7 below. In addition, the Commission is likely to withdraw the benefit of the block exemption where a significant number of licensors of competing technologies in individual agreements impose on their licensees to extend to them more favourable conditions agreed with other licensors.

122. The Commission is also likely to withdraw the benefit of the block exemption where the parties refrain from exploiting the licensed technology, unless they have an objective justification for doing so. Indeed, when the parties do not exploit the licensed technology, no efficiency enhancing activity takes place, in which case the very rationale of the block exemption disappears. However, exploitation does not need to take the form of an integration of assets. Exploitation also occurs where the licence creates design freedom for the licensee by allowing him to exploit his own technology without facing the risk of infringement claims by the licensor. In the case of licensing between competitors, the fact that the parties do not exploit the licensed technology may be an indication that the arrangement is a disguised cartel. For these reasons the Commission will examine very closely cases of non-exploitation.

6.2. Disapplication of the Block Exemption Regulation

123. Article 7 of the TTBER enables the Commission to exclude from the scope of the TTBER, by means of regulation, parallel networks of similar agreements where these cover more than 50%

of a relevant market. Such a measure is not addressed to individual undertakings but concerns all undertakings whose agreements are defined in the regulation disapplying the TTBER.

124. Whereas withdrawal of the benefit of the TTBER by the Commission under Article 6 implies the adoption of a decision under Articles 7 or 9 of Regulation 1/2003, the effect of a Commission disapplication regulation under Article 7 of the TTBER is merely to remove, in respect of the restraints and the markets concerned, the benefit of the TTBER and to restore the full application of Article 81(1) and (3). Following the adoption of a regulation declaring the TTBER inapplicable for a particular market in respect of agreements containing certain restraints, the criteria developed by the relevant case law of the Community Courts and by notices and previous decisions adopted by the Commission will give guidance on the application of Article 81 to individual agreements. Where appropriate, the Commission will take a decision in an individual case, which can provide guidance to all the undertakings operating on the market concerned.

125. For the purpose of calculating the 50 % market coverage ratio, account must be taken of each individual network of licence agreements containing restraints, or combinations of restraints, producing similar effects on the market.

126. Article 7 does not entail an obligation on the part of the Commission to act where the 50 % market-coverage ratio is exceeded. In general, disapplication is appropriate when it is likely that access to the relevant market or competition therein is appreciably restricted. In assessing the need to apply Article 7, the Commission will consider whether individual withdrawal would be a more appropriate remedy. This may depend, in particular, on the number of competing undertakings contributing to a cumulative effect on a market or the number of affected geographic markets within the Community.

127. Any regulation adopted under Article 7 must clearly set out its scope. This means, first, that the Commission must define the relevant product and geographic market(s) and, secondly, that it must identify the type of licensing restraint in respect of which the TTBER will no longer apply. As regards the latter aspect, the Commission may modulate the scope of its regulation according to the competition concern which it intends to address. For instance, while all parallel networks of non-compete arrangements will be taken into account for the purpose of establishing the 50 % market coverage ratio, the Commission may nevertheless restrict the scope of the disapplication regulation only to non-compete obligations exceeding a certain duration. Thus, agreements of a shorter duration or of a less restrictive nature might be left unaffected, due to the lesser degree of foreclosure attributable to such restraints. Where appropriate, the Commission may also provide guidance by specifying the market share level which, in the specific market context, may be regarded as insufficient to bring about a significant contribution by an individual undertaking to the cumulative effect. In general, when the market share of the products incorporating a technology licensed by an individual licensor does not exceed 5 %, the agreement or network of agreements covering that technology is not considered to contribute significantly to a cumulative foreclosure effect.[53]

128. The transitional period of not less than six months that the Commission will have to set under Article 7(2) should allow the undertakings concerned to adapt their agreements to take account of the regulation disapplying the TTBER.

129. A regulation disapplying the TTBER will not affect the block exempted status of the agreements concerned for the period preceding its entry into force.

[53] See in this respect para 8 of the Commission Notice on agreements of minor importance, cited in note 17.

IV. Application of Article 81(1) and 81(3) Outside the Scope of the Block Exemption Regulation

1. The general framework for analysis

130. Agreements that fall outside the block exemption, for example because the market share thresholds are exceeded or the agreement involves more than two parties, are subject to individual assessment. Agreements that either do not restrict competition within the meaning of Article 81(1) or which fulfil the conditions of Article 81(3) are valid and enforceable. It is recalled that there is no presumption of illegality of agreements that fall outside the scope of the block exemption provided that they do not contain hardcore restrictions of competition. In particular, there is no presumption that Article 81(1) applies merely because the market share thresholds are exceeded. Individual assessment based on the principles described in these guidelines is required.

131. In order to promote predictability beyond the application of the TTBER and to confine detailed analysis to cases that are likely to present real competition concerns, the Commission takes the view that outside the area of hardcore restrictions Article 81 is unlikely to be infringed where there are four or more independently controlled technologies in addition to the technologies controlled by the parties to the agreement that may be substitutable for the licensed technology at a comparable cost to the user. In assessing whether the technologies are sufficiently substitutable the relative commercial strength of the technologies in question must be taken into account. The competitive constraint imposed by a technology is limited if it does not constitute a commercially viable alternative to the licensed technology. For instance, if due to network effects in the market consumers have a strong preference for products incorporating the licensed technology, other technologies already on the market or likely to come to market within a reasonable period of time may not constitute a real alternative and may therefore impose only a limited competitive constraint. The fact that an agreement falls outside the safe harbour described in this paragraph does not imply that the agreement is caught by Article 81(1) and, if so, that the conditions of Article 81(3) are not satisfied. As for the market share safe harbour of the TTBER, this additional safe harbour merely creates a negative presumption that the agreement is not prohibited by Article 81. Outside the safe harbour individual assessment of the agreement based on the principles developed in these guidelines is required.

1.1. The relevant factors

132. In the application of Article 81 to individual cases it is necessary to take due account of the way in which competition operates on the market in question. The following factors are particularly relevant in this respect:

(a) the nature of the agreement;
(b) the market position of the parties;
(c) the market position of competitors;
(d) the market position of buyers of the licensed products;
(e) entry barriers;
(f) maturity of the market; and
(g) other factors.

The importance of individual factors may vary from case to case and depends on all other factors. For instance, a high market share of the parties is usually a good indicator of market power, but in the case of low entry barriers it may not be indicative of market power. It is therefore not possible to provide firm rules on the importance of the individual factors.

133. Technology transfer agreements can take many shapes and forms. It is therefore important to analyse the nature of the agreement in terms of the competitive relationship between the parties and the restraints that it contains. In the latter regard it is necessary to go beyond the express terms

of the agreement. The existence of implicit restraints may be derived from the way in which the agreement has been implemented by the parties and the incentives that they face.

134. The market position of the parties provides an indication of the degree of market power, if any, possessed by the licensor, the licensee or both. The higher their market share the greater their market power is likely to be. This is particularly so where the market share reflects cost advantages or other competitive advantages vis-à-vis competitors. These competitive advantages may for instance result from being a first mover in the market, from holding essential patents or from having superior technology.

135. In analysing the competitive relationship between the parties it is sometimes necessary to go beyond the analysis set out in the above sections II.3 on market definition and II.4 on the distinction between competitors and non-competitors. Even where the licensor is not an actual or potential supplier on the product market and the licensee is not an actual or potential competitor on the technology market, it is relevant to the analysis whether the licensee owns a competing technology, which is not being licensed. If the licensee has a strong position on the product market, an agreement granting him an exclusive licence to a competing technology can restrict competition significantly compared to the situation where the licensor does not grant an exclusive licence or licences other undertakings.

136. Market shares and possible competitive advantages and disadvantages are also used to assess the market position of competitors. The stronger the actual competitors and the greater their number the less risk there is that the parties will be able to individually exercise market power. However, if the number of competitors is rather small and their market position (size, costs, R&D potential, etc.) is rather similar, this market structure may increase the risk of collusion.

137. The market position of buyers provides an indication of whether or not one or more buyers possess buyer power. The first indicator of buying power is the market share of the buyer on the purchase market. This share reflects the importance of his demand for possible suppliers. Other indicators focus on the position of the buyer on his resale market, including characteristics such as a wide geographic spread of his outlets, and his brand image amongst final consumers. In some circumstances buyer power may prevent the licensor and/or the licensee from exercising market power on the market and thereby solve a competition problem that would otherwise have existed. This is particularly so when strong buyers have the capacity and the incentive to bring new sources of supply on to the market in the case of a small but permanent increase in relative prices. Where the strong buyers merely extract favourable terms from the supplier or simply pass on any price increase to their customers, the position of the buyers is not such as to prevent the exercise of market power by the licensee on the product market and therefore not such as to solve the competition problem on that market.[54]

138. Entry barriers are measured by the extent to which incumbent companies can increase their price above the competitive level without attracting new entry. In the absence of entry barriers, easy and quick entry would render price increases unprofitable. When effective entry, preventing or eroding the exercise of market power, is likely to occur within one or two years, entry barriers can, as a general rule, be said to be low. Entry barriers may result from a wide variety of factors such as economies of scale and scope, government regulations, especially where they establish exclusive rights, state aid, import tariffs, intellectual property rights, ownership of resources where the supply is limited due to for instance natural limitations, essential facilities, a first mover advantage or brand loyalty of consumers created by strong advertising over a period of time. Restrictive agreements entered into by undertakings may also work as an entry barrier by making access more difficult and foreclosing (potential) competitors. Entry barriers may be present at all stages of the research and development, production and distribution process. The question

[54] See in this respect Case T-228/97, *Irish Sugar* [1999] ECR II-2969, para 101.

whether certain of these factors should be described as entry barriers depends particularly on whether they entail sunk costs. Sunk costs are those costs which have to be incurred to enter or be active on a market but which are lost when the market is exited. The more costs are sunk, the more potential entrants have to weigh the risks of entering the market and the more credibly incumbents can threaten that they will match new competition, as sunk costs make it costly for incumbents to leave the market. In general, entry requires sunk costs, sometimes minor and sometimes major. Therefore, actual competition is in general more effective and will weigh more heavily in the assessment of a case than potential competition.

139. A mature market is a market that has existed for some time, where the technology used is well known and widespread and not changing very much and in which demand is relatively stable or declining. In such a market restrictions of competition are more likely to have negative effects than in more dynamic markets.

140. In the assessment of particular restraints other factors may have to be taken into account. Such factors include cumulative effects, i.e. the coverage of the market by similar agreements, the duration of the agreements, the regulatory environment and behaviour that may indicate or facilitate collusion like price leadership, pre-announced price changes and discussions on the 'right' price, price rigidity in response to excess capacity, price discrimination and past collusive behaviour.

1.2. Negative effects of restrictive licence agreements

141. The negative effects on competition on the market that may result from restrictive technology transfer agreements include the following:

1. reduction of inter-technology competition between the companies operating on a technology market or on a market for products incorporating the technologies in question, including facilitation of collusion, both explicit and tacit;
2. foreclosure of competitors by raising their costs, restricting their access to essential inputs or otherwise raising barriers to entry; and
3. reduction of intra-technology competition between undertakings that produce products on the basis of the same technology.

142. Technology transfer agreements may reduce inter-technology competition, i.e. competition between undertakings that license or produce on the basis of substitutable technologies. This is particularly so where reciprocal obligations are imposed. For instance, where competitors transfer competing technologies to each other and impose a reciprocal obligation to provide each other with future improvements of their respective technologies and where this agreement prevents either competitor from gaining a technological lead over the other, competition in innovation between the parties is restricted (see also paragraph 208 below).

143. Licensing between competitors may also facilitate collusion. The risk of collusion is particularly high in concentrated markets. Collusion requires that the undertakings concerned have similar views on what is in their common interest and on how the co-ordination mechanisms function. For collusion to work the undertakings must also be able to monitor each other's market behaviour and there must be adequate deterrents to ensure that there is an incentive not to depart from the common policy on the market, while entry barriers must be high enough to limit entry or expansion by outsiders. Agreements can facilitate collusion by increasing transparency in the market, by controlling certain behaviour and by raising barriers to entry. Collusion can also exceptionally be facilitated by licensing agreements that lead to a high degree of commonality of costs, because undertakings that have similar costs are more likely to have similar views on the terms of coordination.[55]

[55] See in this respect para 23 of the Guidelines on horizontal cooperation agreements, cited in note 20.

144. Licence agreements may also affect inter-technology competition by creating barriers to entry for and expansion by competitors. Such foreclosure effects may stem from restraints that prevent licensees from licensing from third parties or create disincentives for them to do so. For instance, third parties may be foreclosed where incumbent licensors impose non-compete obligations on licensees to such an extent that an insufficient number of licensees are available to third parties and where entry at the level of licensees is difficult. Suppliers of substitutable technologies may also be foreclosed where a licensor with a sufficient degree of market power ties together various parts of a technology and licenses them together as a package while only part of the package is essential to produce a certain product.

145. Licence agreements may also reduce intra-technology competition, i.e. competition between undertakings that produce on the basis of the same technology. An agreement imposing territorial restraints on licensees, preventing them from selling into each other's territory reduces competition between them. Licence agreements may also reduce intra-technology competition by facilitating collusion between licensees. Moreover, licence agreements that reduce intra-technology competition may facilitate collusion between owners of competing technologies or reduce inter-technology competition by raising barriers to entry.

1.3. Positive effects of restrictive licence agreements and the framework for analysing such effects

146. Even restrictive licence agreements mostly also produce pro-competitive effects in the form of efficiencies, which may outweigh their anti-competitive effects. This assessment takes place within the framework of Article 81(3), which contains an exception from the prohibition rule of Article 81(1). For this exception to be applicable the licence agreement must produce objective economic benefits, the restrictions on competition must be indispensable to attain the efficiencies, consumers must receive a fair share of the efficiency gains, and the agreement must not afford the parties the possibility of eliminating competition in respect of a substantial part of the products concerned.

147. The assessment of restrictive agreements under Article 81(3) is made within the actual context in which they occur[56] and on the basis of the facts existing at any given point in time. The assessment is sensitive to material changes in the facts. The exception rule of Article 81(3) applies as long as the four conditions are fulfilled and ceases to apply when that is no longer the case.[57] However, when applying Article 81(3) in accordance with these principles it is necessary to take into account the initial sunk investments made by any of the parties and the time needed and the restraints required to commit and recoup an efficiency enhancing investment. Article 81 cannot be applied without considering the *ex ante* investment and the risks relating thereto. The risk facing the parties and the sunk investment that must be committed to implement the agreement can thus lead to the agreement falling outside Article 81(1) or fulfilling the conditions of Article 81(3), as the case may be, for the period of time required to recoup the investment.

148. The first condition of Article 81(3) requires an assessment of what are the objective benefits in terms of efficiencies produced by the agreement. In this respect, licence agreements have the potential of bringing together complementary technologies and other assets allowing new or improved products to be put on the market or existing products to be produced at lower cost. Outside the context of hardcore cartels, licensing often occurs because it is more efficient for the licensor to licence the technology than to exploit it himself. This may particularly be the case where the licensee already has access to the necessary production assets. The agreement allows the

[56] See Joined Cases 25/84 and 26/84, *Ford* [1985] ECR 2725.

[57] See in this respect for example Commission Decision in *TPS* (OJ L 90, 2.4.1999, p 6). Similarly, the prohibition of Article 81(1) also only applies as long as the agreement has a restrictive object or restrictive effects.

licensee to gain access to a technology that can be combined with these assets, allowing him to exploit new or improved technologies. Another example of potentially efficiency enhancing licensing is where the licensee already has a technology and where the combination of this technology and the licensor's technology gives rise to synergies. When the two technologies are combined the licensee may be able to attain a cost/output configuration that would not otherwise be possible. Licence agreements may also give rise to efficiencies at the distribution stage in the same way as vertical distribution agreements. Such efficiencies can take the form of cost savings or the provision of valuable services to consumers. The positive effects of vertical agreements are described in the Guidelines on Vertical Restraints.[58] A further example of possible efficiency gains is agreements whereby technology owners assemble a technology package for licensing to third parties. Such pooling arrangements may in particular reduce transaction costs, as licensees do not have to conclude separate licence agreements with each licensor. Pro-competitive licensing may also occur to ensure design freedom. In sectors where large numbers of intellectual property rights exist and where individual products may infringe upon a number of existing and future property rights, licence agreements whereby the parties agree not to assert their property rights against each other are often pro-competitive because they allow the parties to develop their respective technologies without the risk of subsequent infringement claims.

149. In the application of the indispensability test contained in Article 81(3) the Commission will in particular examine whether individual restrictions make it possible to perform the activity in question more efficiently than would have been the case in the absence of the restriction concerned. In making this assessment the market conditions and the realities facing the parties must be taken into account. Undertakings invoking the benefit of Article 81(3) are not required to consider hypothetical and theoretical alternatives. They must, however, explain and demonstrate why seemingly realistic and significantly less restrictive alternatives would be significantly less efficient. If the application of what appears to be a commercially realistic and less restrictive alternative would lead to a significant loss of efficiencies, the restriction in question is treated as indispensable. In some cases, it may also be necessary to examine whether the agreement as such is indispensable to achieve the efficiencies. This may for example be so in the case of technology pools that include complementary but non-essential technologies,[59] in which case it must be examined to what extent such inclusion gives rise to particular efficiencies or whether, without a significant loss of efficiencies, the pool could be limited to technologies for which there are no substitutes. In the case of simple licensing between two parties it is generally not necessary to go beyond an examination of the indispensability of individual restraints. Normally there is no less restrictive alternative to the licence agreement as such.

150. The condition that consumers must receive a fair share of the benefits implies that consumers of the products produced under the licence must at least be compensated for the negative effects of the agreement.[60] This means that the efficiency gains must fully off-set the likely negative impact on prices, output and other relevant factors caused by the agreement. They may do so by changing the cost structure of the undertakings concerned, giving them an incentive to reduce price, or by allowing consumers to gain access to new or improved products, compensating for any likely price increase.[61]

151. The last condition of Article 81(3), according to which the agreement must not afford the parties the possibility of eliminating competition in respect of a substantial part of the products concerned, presupposes an analysis of remaining competitive pressures on the market and the

[58] Cited in note 36. See in particular paras 115 *et seq.*

[59] As to these concepts see section IV.4.1 below.

[60] See para 85 of the Guidelines on the application of Article 81(3) of the Treaty, cited in note 2.

[61] Idem, paras 98 and 102.

impact of the agreement on such sources of competition. In the application of the last condition of Article 81(3) the relationship between Article 81(3) and Article 82 must be taken into account. According to settled case law, the application of Article 81(3) cannot prevent the application of Article 82 of the Treaty.[62] Moreover, since Articles 81 and 82 both pursue the aim of maintaining effective competition on the market, consistency requires that Article 81(3) be interpreted as precluding any application of the exception rule to restrictive agreements that constitute an abuse of a dominant position.[63]

152. The fact that the agreement substantially reduces one dimension of competition does not necessarily mean that competition is eliminated within the meaning of Article 81(3). A technology pool, for instance, can result in an industry standard, leading to a situation in which there is little competition in terms of the technological format. Once the main players in the market adopt a certain format, network effects may make it very difficult for alternative formats to survive. This does not imply, however, that the creation of a de facto industry standard always eliminates competition within the meaning of the last condition of Article 81(3). Within the standard, suppliers may compete on price, quality and product features. However, in order for the agreement to comply with Article 81(3), it must be ensured that the agreement does not unduly restrict competition and does not unduly restrict future innovation.

2. The application of Article 81 to various types of licensing restraints

153. This section deals with various types of restraints that are commonly included in licence agreements. Given their prevalence it is useful to provide guidance as to how they are assessed outside the safe harbour of the TTBER. Restraints that have already been dealt with in the preceding parts of these guidelines, in particular sections III.4 and III.5, are only dealt with briefly in the present section.

154. This section covers both agreements between non-competitors and agreements between competitors. In respect of the latter a distinction is made—where appropriate—between reciprocal and non-reciprocal agreements. No such distinction is required in the case of agreements between non-competitors. When undertakings are neither actual nor potential competitors on a relevant technology market or on a market for products incorporating the licensed technology, a reciprocal licence is for all practical purposes no different from two separate licences. Arrangements whereby the parties assemble a technology package, which is then licensed to third parties, are technology pools, which are dealt with in section 4 below.

155. This section does not deal with obligations in licence agreements that are generally not restrictive of competition within the meaning of Article 81(1). These obligations include but are not limited to:

(a) confidentiality obligations;
(b) obligations on licensees not to sub-license;
(c) obligations not to use the licensed technology after the expiry of the agreement, provided that the licensed technology remains valid and in force;
(d) obligations to assist the licensor in enforcing the licensed intellectual property rights;
(e) obligations to pay minimum royalties or to produce a minimum quantity of products incorporating the licensed technology; and
(f) obligations to use the licensor's trade mark or indicate the name of the licensor on the product.

---

[62] See para 130 of the judgment cited in note 2. Similarly, the application of Article 81(3) does not prevent the application of the Treaty rules on the free movement of goods, services, persons and capital. These provisions are in certain circumstances applicable to agreements, decisions and concerted practices within the meaning of Article 81(1), see to that effect Case C-309/99, *Wouters*, [2002] ECR I-1577, para 120.

[63] See in this respect Case T-51/89, *Tetra Pak (I)* [1990] ECR II-309. See also para 106 of the Guidelines on the application of Article 81(3) of the Treaty cited in note 2 above.

2.1. Royalty obligations

156. The parties to a licence agreement are normally free to determine the royalty payable by the licensee and its mode of payment without being caught by Article 81(1). This principle applies both to agreements between competitors and agreements between non-competitors. Royalty obligations may for instance take the form of lump sum payments, a percentage of the selling price or a fixed amount for each product incorporating the licensed technology. In cases where the licensed technology relates to an input which is incorporated into a final product it is as a general rule not restrictive of competition that royalties are calculated on the basis of the price of the final product, provided that it incorporates the licensed technology. In the case of software licensing royalties based on the number of users and royalties calculated on a per machine basis are generally compatible with Article 81(1).

157. In the case of licence agreements between competitors it is recalled, see paragraphs and above, that in a limited number of circumstances royalty obligations may amount to price fixing, which is a hardcore restriction (cf. Article 4(1)(a)). It is a hardcore restriction under Article 4(1)(a) if competitors provide for reciprocal running royalties in circumstances where the licence is a sham, in that its purpose is not to allow an integration of complementary technologies or to achieve another pro-competitive aim. It is also a hardcore restriction under Article 4(1)(a) and 4(1)(d) if royalties extend to products produced solely with the licensee's own technology.

158. Other types of royalty arrangements between competitors are block exempted up to the market share threshold of 20 % even if they restrict competition. Outside the safe harbour of the block exemption Article 81(1) may be applicable where competitors cross license and impose running royalties that are clearly disproportionate compared to the market value of the licence and where such royalties have a significant impact on market prices. In assessing whether the royalties are disproportionate it is relevant to have regard to the royalties paid by other licensees on the product market for the same or substitute technologies. In such cases it is unlikely that the conditions of Article 81(3) are satisfied. Article 81(1) may also apply where reciprocal running royalties per unit increase as output increases. If the parties have a significant degree of market power, such royalties may have the effect of limiting output.

159. Notwithstanding the fact that the block exemption only applies as long as the technology is valid and in force, the parties can normally agree to extend royalty obligations beyond the period of validity of the licensed intellectual property rights without falling foul of Article 81(1). Once these rights expire, third parties can legally exploit the technology in question and compete with the parties to the agreement. Such actual and potential competition will normally suffice to ensure that the obligation in question does not have appreciable anti-competitive effects.

160. In the case of agreements between non-competitors the block exemption covers agreements whereby royalties are calculated on the basis of both products produced with the licensed technology and products produced with technologies licensed from third parties. Such arrangements may facilitate the metering of royalties. However, they may also lead to foreclosure by increasing the cost of using third party inputs and may thus have similar effects as a non-compete obligation. If royalties are paid not just on products produced with the licensed technology but also on products produced with third party technology, then the royalties will increase the cost of the latter products and reduce demand for third party technology. Outside the scope of the block exemption it must therefore be examined whether the restriction has foreclosure effects. For that purpose it is appropriate to use the analytical framework set out in section 2.7 below. In the case of appreciable foreclosure effects such agreements are caught by Article 81(1) and unlikely to fulfil the conditions of Article 81(3), unless there is no other practical way of calculating and monitoring royalty payments.

2.2. Exclusive licensing and sales restrictions

161. For the present purposes it is useful to distinguish between restrictions as to production within a given territory (exclusive or sole licences) and restrictions on the sale of products incorporating the licensed technology into a given territory and to a given customer group (sales restrictions).

2.2.1. Exclusive and sole licences

162. A licence is deemed to be exclusive if the licensee is the only one who is permitted to produce on the basis of the licensed technology within a given territory. The licensor thus undertakes not to produce itself or license others to produce within a given territory. This territory may cover the whole world. Where the licensor undertakes only not to licence third parties to produce within a given territory, the licence is a sole licence. Often exclusive or sole licensing is accompanied by sales restrictions that limit the parties in where they may sell products incorporating the licensed technology.

163. Reciprocal exclusive licensing between competitors falls under Article 4(1)(c), which identifies market sharing between competitors as a hardcore restriction. Reciprocal sole licensing between competitors is block exempted up to the market share threshold of 20 %. Under such an agreement the parties mutually commit not to license their competing technologies to third parties. In cases where the parties have a significant degree of market power such agreements may facilitate collusion by ensuring that the parties are the only sources of output in the market based on the licensed technologies.

164. Non-reciprocal exclusive licensing between competitors is block exempted up to the market share threshold of 20 %. Above the market share threshold it is necessary to analyse what are the likely anti-competitive effects of such exclusive licensing. Where the exclusive licence is worldwide it implies that the licensor leaves the market. In cases where exclusivity is limited to a particular territory such as a Member State the agreement implies that the licensor abstains from producing goods and services inside the territory in question. In the context of Article 81(1) it must in particular be assessed what is the competitive significance of the licensor. If the licensor has a limited market position on the product market or lacks the capacity to effectively exploit the technology in the licensee's territory, the agreement is unlikely to be caught by Article 81(1). A special case is where the licensor and the licensee only compete on the technology market and the licensor, for instance being a research institute or a small research based undertaking, lacks the production and distribution assets to effectively bring to market products incorporating the licensed technology. In such cases Article 81(1) is unlikely to be infringed.

165. Exclusive licensing between non-competitors—to the extent that it is caught by Article 81(1)[64]—is likely to fulfil the conditions of Article 81(3). The right to grant an exclusive licence is generally necessary in order to induce the licensee to invest in the licensed technology and to bring the products to market in a timely manner. This is in particular the case where the licensee must make large investments in further developing the licensed technology. To intervene against the exclusivity once the licensee has made a commercial success of the licensed technology would deprive the licensee of the fruits of his success and would be detrimental to competition, the dissemination of technology and innovation. The Commission will therefore only exceptionally intervene against exclusive licensing in agreements between non-competitors, irrespective of the territorial scope of the licence.

166. The main situation in which intervention may be warranted is where a dominant licensee obtains an exclusive licence to one or more competing technologies. Such agreements are likely to be caught by Article 81(1) and unlikely to fulfil the conditions of Article 81(3). It is a condition

---

[64] See the judgment in *Nungesser* cited in note 13.

however that entry into the technology market is difficult and the licensed technology constitutes a real source of competition on the market. In such circumstances an exclusive licence may foreclose third party licensees and allow the licensee to preserve his market power.

167. Arrangements whereby two or more parties cross licence each other and undertake not to licence third parties give rise to particular concerns when the package of technologies resulting from the cross licences creates a de facto industry standard to which third parties must have access in order to compete effectively on the market. In such cases the agreement creates a closed standard reserved for the parties. The Commission will assess such arrangements according to the same principles as those applied to technology pools (see section 4 below). It will normally be required that the technologies which support such a standard be licensed to third parties on fair, reasonable and non-discriminatory terms.[65] Where the parties to the arrangement compete with third parties on an existing product market and the arrangement relates to that product market a closed standard is likely to have substantial exclusionary effects. This negative impact on competition can only be avoided by licensing also to third parties.

2.2.2. Sales restrictions

168. Also as regards sales restrictions there is an important distinction to be made between licensing between competitors and between non-competitors.

169. Restrictions on active and passive sales by one or both parties in a reciprocal agreement between competitors are hardcore restrictions of competition under Article 4(1)(c). Sales restrictions on either party in a reciprocal agreement between competitors are caught by Article 81(1) and are unlikely to fulfil the conditions of Article 81(3). Such restrictions are generally considered market sharing, since they prevent the affected party from selling actively and passively into territories and to customer groups which he actually served or could realistically have served in the absence of the agreement.

170. In the case of non-reciprocal agreements between competitors the block exemption applies to restrictions on active and passive sales by the licensee or the licensor into the exclusive territory or to the exclusive customer group reserved for the other party (cf. Article 4(1)(c)(iv). Above the market share threshold of 20 % sales restrictions between licensor and licensee are caught by Article 81(1) when one or both of the parties have a significant degree of market power. Such restrictions, however, may be indispensable for the dissemination of valuable technologies and therefore fulfil the conditions of Article 81(3). This may be the case where the licensor has a relatively weak market position in the territory where he exploits himself the technology. In such circumstances restrictions on active sales in particular may be indispensable to induce the licensor to grant the licence. In the absence thereof the licensor would risk facing active competition in his main area of activity. Similarly, restrictions on active sales by the licensor may be indispensable, in particular, where the licensee has a relatively weak market position in the territory allocated to him and has to make significant investments in order to efficiently exploit the licensed technology.

171. The block exemption also covers restrictions on active sales into the territory or to the customer group allocated to another licensee, who was not a competitor of the licensor at the time when he concluded the licence agreement with the licensor. It is a condition, however, that the agreement between the parties in question is non-reciprocal. Above the market share threshold such active sales restrictions are likely to be caught by Article 81(1) when the parties have a significant degree of market power. However, the restraint is likely to be indispensable within the meaning of Article 81(3) for the period of time required for the protected licensee to penetrate a new market and establish a market presence in the allocated territory or vis-à-vis the allocated customer group. This protection against active sales allows the licensee to overcome the asymmetry,

[65] See in this respect the Commission's Notice in the *Canon/Kodak* Case (OJ C 330, 1.11.1997, p 10) and the *IGR Stereo Television* Case mentioned in the XI Report on Competition Policy, para 94.

which he faces due to the fact that some of the licensees are competing undertakings of the licensor and thus already established on the market. Restrictions on passive sales by licensees into a territory or to a customer group allocated to another licensee are hardcore restrictions under Article 4(1)(c) of the TTBER.

172. In the case of agreements between non-competitors sales restrictions between the licensor and a licensee are block exempted up to the market share threshold of 30 %. Above the market share threshold restrictions on active and passive sales by licensees to territories or customer groups reserved for the licensor may fall outside Article 81(1) where on the basis of objective factors it can be concluded that in the absence of the sales restrictions licensing would not occur. A technology owner cannot normally be expected to create direct competition with himself on the basis of his own technology. In other cases sales restrictions on the licensee may be caught by Article 81(1) both where the licensor individually has a significant degree of market power and in the case of a cumulative effect of similar agreements concluded by licensors which together hold a strong position on the market.

173. Sales restrictions on the licensor, when caught by Article 81(1), are likely to fulfil the conditions of Article 81(3) unless there are no real alternatives to the licensor's technology on the market or such alternatives are licensed by the licensee from third parties. Such restrictions and in particular restrictions on active sales are likely to be indispensable within the meaning of Article 81(3) in order to induce the licensee to invest in the production, marketing and sale of the products incorporating the licensed technology. It is likely that the licensee's incentive to invest would be significantly reduced if he would face direct competition from the licensor whose production costs are not burdened by royalty payments, possibly leading to sub-optimal levels of investment.

174. As regards restrictions on sales between licensees in agreements between non-competitors, the TTBER block exempts restrictions on active selling between territories or customer groups. Above the market share threshold restrictions on active sales between licensees' territories and customer groups limit intra-technology competition and are likely to be caught by Article 81(1) when the individual licensee has a significant degree of market power. Such restrictions, however, may fulfil the conditions of Article 81(3) where they are necessary to prevent free riding and to induce the licensee to make the investment necessary for efficient exploitation of the licensed technology inside his territory and to promote sales of the licensed product. Restrictions on passive sales are covered by the hardcore list of Article 4(2)(b), cf. paragraph 101 above, when they exceed two years from the date on which the licensee benefiting from the restrictions first put the product incorporating the licensed technology on the market inside his exclusive territory. Passive sales restrictions exceeding this two-year period are unlikely to fulfil the conditions of Article 81(3).

2.3. Output restrictions

175. Reciprocal output restrictions in licence agreements between competitors constitute a hardcore restriction covered by Article 4(1)(b) of the TTBER (cf. point 82 above). Article 4(1)(b) does not cover output restrictions imposed on the licensee in a non-reciprocal agreement or on one of the licensees in an reciprocal agreement. Such restrictions are block exempted up to the market share threshold of 20 %. Above the market share threshold, output restrictions on the licensee may restrict competition where the parties have a significant degree of market power. However, Article 81(3) is likely to apply in cases where the licensor's technology is substantially better than the licensee's technology and the output limitation substantially exceeds the output of the licensee prior to the conclusion of the agreement. In that case the effect of the output limitation is limited even in markets where demand is growing. In the application of Article 81(3) it must also be taken into account that such restrictions may be necessary in order to induce the licensor to disseminate his technology as widely as possible. For instance, a licensor may be reluctant to license his

competitors if he cannot limit the licence to a particular production site with a specific capacity (a site licence). Where the licence agreement leads to a real integration of complementary assets, output restrictions on the licensee may therefore fulfil the conditions of Article 81(3). However, this is unlikely to be the case where the parties have substantial market power.

176. Output restrictions in licence agreements between non-competitors are block exempted up to the market share threshold of 30 %. The main anti-competitive risk flowing from output restrictions on licensees in agreements between non-competitors is reduced intratechnology competition between licensees. The significance of such anti-competitive effects depends on the market position of the licensor and the licensees and the extent to which the output limitation prevents the licensee from satisfying demand for the products incorporating the licensed technology.

177. When output restrictions are combined with exclusive territories or exclusive customer groups, the restrictive effects are increased. The combination of the two types of restraints makes it more likely that the agreement serves to partition markets.

178. Output limitations imposed on the licensee in agreements between non-competitors may also have pro-competitive effects by promoting the dissemination of technology. As a supplier of technology, the licensor should normally be free to determine the output produced with the licensed technology by the licensee. If the licensor were not free to determine the output of the licensee, a number of licence agreements might not come into existence in the first place, which would have a negative impact on the dissemination of new technology. This is particularly likely to be the case where the licensor is also a producer, since in that case the output of the licensees may find their way back into the licensor's main area of operation and thus have a direct impact on these activities. On the other hand, it is less likely that output restrictions are necessary in order to ensure dissemination of the licensor's technology when combined with sales restrictions on the licensee prohibiting him from selling into a territory or customer group reserved for the licensor.

2.4. Field of use restrictions

179. Under a field of use restriction the licence is either limited to one or more technical fields of application or one or more product markets. There are many cases in which the same technology can be used to make different products or can be incorporated into products belonging to different product markets. A new moulding technology may for instance be used to make plastic bottles and plastic glasses, each product belonging to separate product markets. However, a single product market may encompass several technical fields of use. For instance a new engine technology may be employed in four cylinder engines and six cylinder engines. Similarly, a technology to make chipsets may be used to produce chipsets with up to four CPUs and more than four CPUs. A licence limiting the use of the licensed technology to produce say four cylinder engines and chipsets with up to four CPUs constitutes a technical field of use restriction.

180. Given that field of use restrictions are block exempted and that certain customer restrictions are hardcore restrictions under Articles 4(1)(c) and 4(2)(b) of the TTBER, it is important to distinguish the two categories of restraints. A customer restriction presupposes that specific customer groups are identified and that the parties are restricted in selling to such identified groups. The fact that a technical field of use restriction may correspond to certain groups of customers within a product market does not imply that the restraint is to be classified as a customer restriction. For instance, the fact that certain customers buy predominantly or exclusively chipsets with more than four CPUs does not imply that a licence which is limited to chipsets with up to four CPUs constitutes a customer restriction. However, the field of use must be defined objectively by reference to identified and meaningful technical characteristics of the licensed product.

181. A field of use restriction limits the exploitation of the licensed technology by the licensee to

one or more particular fields of use without limiting the licensor's ability to exploit the licensed technology. In addition, as with territories, these fields of use can be allocated to the licensee under an exclusive or sole licence. Field of use restrictions combined with an exclusive or sole licence also restrict the licensor's ability to exploit his own technology, by preventing him from exploiting it himself, including by way of licensing to others. In the case of a sole license only licensing to third parties is restricted. Field of use restrictions combined with exclusive and sole licences are treated in the same way as the exclusive and sole licenses dealt with in section 2.2.1 above. In particular, for licensing between competitors, this means that reciprocal exclusive licensing is hardcore under Article 4(1)(c).

182. Field of use restrictions may have pro-competitive effects by encouraging the licensor to license his technology for applications that fall outside his main area of focus. If the licensor could not prevent licensees from operating in fields where he exploits the technology himself or in fields where the value of the technology is not yet well established, it would be likely to create a disincentive for the licensor to license or would lead him to charge a higher royalty. It must also be taken into account that in certain sectors licensing often occurs to ensure design freedom by preventing infringement claims. Within the scope of the licence the licensee is able to develop his own technology without fearing infringement claims by the licensor.

183. Field of use restrictions on licensees in agreements between actual or potential competitors are block exempted up to the market share threshold of 20 %. The main competitive concern in the case of such restrictions is the risk that the licensee ceases to be a competitive force outside the licensed field of use. This risk is greater in the case of cross licensing between competitors where the agreement provides for asymmetrical field of use restrictions. A field of use restriction is asymmetrical where one party is permitted to use the licensed technology within one product market or technical field of use and the other party is permitted to use the other licensed technology within another product market or technical field of use. Competition concerns may in particular arise where the licensee's production facility, which is tooled up to use the licensed technology, is also used to produce with his own technology products outside the licensed field of use. If the agreement is likely to lead the licensee to reduce output outside the licensed field of use, the agreement is likely to be caught by Article 81(1). Symmetrical field of use restrictions, i.e. agreements whereby the parties are licensed to use each other's technologies within the same field(s) of use, are unlikely to be caught by Article 81(1). Such agreements are unlikely to restrict competition that existed in the absence of the agreement. Article 81(1) is also unlikely to apply in the case of agreements that merely enable the licensee to develop and exploit his own technology within the scope of the licence without fearing infringement claims by the licensor. In such circumstances field of use restrictions do not in themselves restrict competition that existed in the absence of the agreement. In the absence of the agreement the licensee also risked infringement claims outside the scope of the licensed field of use. However, if the licensee without business justification terminates or scales back his activities in the area outside the licensed field of use this may be an indication of an underlying market sharing arrangement amounting to a hardcore restriction under Article 4(1)(c) of the TTBER.

184. Field of use restrictions on licensee and licensor in agreements between non-competitors are block exempted up to the market share threshold of 30 %. Field of use restrictions in agreements between non-competitors whereby the licensor reserves one or more product markets or technical fields of use for himself are generally either non-restrictive of competition or efficiency enhancing. They promote dissemination of new technology by giving the licensor an incentive to license for exploitation in fields in which he does not want to exploit the technology himself. If the licensor could not prevent licensees from operating in fields where the licensor exploits the technology himself, it would be likely to create a disincentive for the licensor to licence.

185. In agreements between non-competitors the licensor is normally also entitled to grant sole or exclusive licences to different licensees limited to one or more fields of use. Such restrictions

limit intra-technology competition between licensees in the same way as exclusive licensing and are analysed in the same way (cf. section 2.2.1 above).

2.5. Captive use restrictions

186. A captive use restriction can be defined as an obligation on the licensee to limit his production of the licensed product to the quantities required for the production of his own products and for the maintenance and repair of his own products. In other words, this type of use restriction takes the form of an obligation on the licensee to use the products incorporating the licensed technology only as an input for incorporation into his own production; it does not cover the sale of the licensed product for incorporation into the products of other producers. Captive use restrictions are block exempted up to the respective market share thresholds of 20 % and 30 %. Outside the scope of the block exemption it is necessary to examine what are the pro-competitive and anti-competitive effects of the restraint. In this respect it is necessary to distinguish agreements between competitors from agreements between non-competitors.

187. In the case of licence agreements between competitors a restriction that imposes on the licensee to produce under the licence only for incorporation into his own products prevents him from being a supplier of components to third party producers. If prior to the conclusion of the agreement, the licensee was not an actual or likely potential supplier of components to other producers, the captive use restriction does not change anything compared to the pre-existing situation. In those circumstances the restriction is assessed in the same way as in the case of agreements between non-competitors. If, on the other hand, the licensee is an actual or likely component supplier, it is necessary to examine what is the impact of the agreement on this activity. If by tooling up to use the licensor's technology the licensee ceases to use his own technology on a stand alone basis and thus to be a component supplier, the agreement restricts competition that existed prior to the agreement. It may result in serious negative market effects when the licensor has a significant degree of market power on the component market.

188. In the case of licence agreements between non-competitors there are two main competitive risks stemming from captive use restrictions: (a) a restriction of intra-technology competition on the market for the supply of inputs and (b) an exclusion of arbitrage between licensees enhancing the possibility for the licensor to impose discriminatory royalties on licensees.

189. Captive use restrictions, however, may also promote pro-competitive licensing. If the licensor is a supplier of components, the restraint may be necessary in order for the dissemination of technology between non-competitors to occur. In the absence of the restraint the licensor may not grant the licence or may do so only against higher royalties, because otherwise he would create direct competition to himself on the component market. In such cases a captive use restriction is normally either not restrictive of competition or covered by Article 81(3). It is a condition, however, that the licensee is not restricted in selling the licensed product as replacement parts for his own products. The licensee must be able to serve the after market for his own products, including independent service organisations that service and repair the products produced by him.

190. Where the licensor is not a component supplier on the relevant market, the above reason for imposing captive use restrictions does not apply. In such cases a captive use restriction may in principle promote the dissemination of technology by ensuring that licensees do not sell to producers that compete with the licensor on other markets. However, a restriction on the licensee not to sell into certain customer groups reserved for the licensor normally constitutes a less restrictive alternative. Consequently, in such cases a captive use restriction is normally not necessary for the dissemination of technology to take place.

2.6. Tying and bundling

191. In the context of technology licensing tying occurs when the licensor makes the licensing of one technology (the tying product) conditional upon the licensee taking a licence for another technology or purchasing a product from the licensor or someone designated by him (the tied

product). Bundling occurs where two technologies or a technology and a product are only sold together as a bundle. In both cases, however, it is a condition that the products and technologies involved are distinct in the sense that there is distinct demand for each of the products and technologies forming part of the tie or the bundle. This is normally not the case where the technologies or products are by necessity linked in such a way that the licensed technology cannot be exploited without the tied product or both parts of the bundle cannot be exploited without the other. In the following the term 'tying' refers to both tying and bundling.

192. Article 3 of the TTBER, which limits the application of the block exemption by market share thresholds, ensures that tying and bundling are not block exempted above the market share thresholds of 20 % in the case of agreements between competitors and 30 % in the case of agreements between non-competitors. The market share thresholds apply to any relevant technology or product market affected by the licence agreement, including the market for the tied product. Above the market share thresholds it is necessary to balance the anti-competitive and pro-competitive effects of tying.

193. The main restrictive effect of tying is foreclosure of competing suppliers of the tied product. Tying may also allow the licensor to maintain market power in the market for the tying product by raising barriers to entry since it may force new entrants to enter several markets at the same time. Moreover, tying may allow the licensor to increase royalties, in particular when the tying product and the tied product are partly substitutable and the two products are not used in fixed proportion. Tying prevents the licensee from switching to substitute inputs in the face of increased royalties for the tying product. These competition concerns are independent of whether the parties to the agreement are competitors or not. For tying to produce likely anti-competitive effects the licensor must have a significant degree of market power in the tying product so as to restrict competition in the tied product. In the absence of market power in the tying product the licensor cannot use his technology for the anticompetitive purpose of foreclosing suppliers of the tied product. Furthermore, as in the case of non-compete obligations, the tie must cover a certain proportion of the market for the tied product for appreciable foreclosure effects to occur. In cases where the licensor has market power on the market for the tied product rather than on the market for the tying product, the restraint is analysed as non-compete or quantity forcing, reflecting the fact that any competition problem has its origin on the market for the 'tied' product and not on the market for the 'tying' product.[66]

194. Tying can also give rise to efficiency gains. This is for instance the case where the tied product is necessary for a technically satisfactory exploitation of the licensed technology or for ensuring that production under the licence conforms to quality standards respected by the licensor and other licensees. In such cases tying is normally either not restrictive of competition or covered by Article 81(3). Where the licensees use the licensor's trademark or brand name or where it is otherwise obvious to consumers that there is a link between the product incorporating the licensed technology and the licensor, the licensor has a legitimate interest in ensuring that the quality of the products are such that it does not undermine the value of his technology or his reputation as an economic operator. Moreover, where it is known to consumers that the licensees (and the licensor) produce on the basis of the same technology it is unlikely that licensees would be willing to take a licence unless the technology is exploited by all in a technically satisfactory way.

195. Tying is also likely to be pro-competitive where the tied product allows the licensee to exploit the licensed technology significantly more efficiently. For instance, where the licensor licenses a particular process technology the parties can also agree that the licensee buys a catalyst

[66] For the applicable analytical framework see section 2.7 below and paras 138 *et seq.* of the Guidelines on Vertical Restraints cited in note 36.

from the licensor which is developed for use with the licensed technology and which allows the technology to be exploited more efficiently than in the case of other catalysts. Where in such cases the restriction is caught by Article 81(1), the conditions of Article 81(3) are likely to be fulfilled even above the market share thresholds.

2.7. Non-compete obligations

196. Non-compete obligations in the context of technology licensing take the form of an obligation on the licensee not to use third party technologies which compete with the licensed technology. To the extent that a non-compete obligation covers a product or additional technology supplied by the licensor the obligation is dealt with in the preceding section on tying.

197. The TTBER exempts non-compete obligations both in the case of agreements between competitors and in the case of agreements between non-competitors up to the market share thresholds of 20 % and 30 % respectively.

198. The main competitive risk presented by non-compete obligations is foreclosure of third party technologies. Non-compete obligations may also facilitate collusion between licensors in the case of cumulative use. Foreclosure of competing technologies reduces competitive pressure on royalties charged by the licensor and reduces competition between the incumbent technologies by limiting the possibilities for licensees to substitute between competing technologies. As in both cases the main problem is foreclosure, the analysis can in general be the same in the case of agreements between competitors and agreements between non-competitors. However, in the case of cross licensing between competitors where both agree not to use third party technologies the agreement may facilitate collusion between them on the product market, thereby justifying the lower market share threshold of 20 %.

199. Foreclosure may arise where a substantial part of potential licensees are already tied to one or, in the case of cumulative effects, more sources of technology and are prevented from exploiting competing technologies. Foreclosure effects may result from agreements concluded by a single licensor with a significant degree of market power or by a cumulative effect of agreements concluded by several licensors, even where each individual agreement or network of agreements is covered by the TTBER. In the latter case, however, a serious cumulative effect is unlikely to arise as long as less than 50 % of the market is tied. Above this threshold significant foreclosure is likely to occur when there are relatively high barriers to entry for new licensees. If barriers to entry are low, new licensees are able to enter the market and exploit commercially attractive technologies held by third parties and thus represent a real alternative to incumbent licensees. In order to determine the real possibility for entry and expansion by third parties it is also necessary to take account of the extent to which distributors are tied to licensees by non-compete obligations. Third party technologies only have a real possibility of entry if they have access to the necessary production and distribution assets. In other words, the ease of entry depends not only on the availability of licensees but also the extent to which they have access to distribution. In assessing foreclosure effects at the distribution level the Commission will apply the analytical framework set out in section IV.2.1 of the Guidelines on Vertical Restraints.[67]

200. When the licensor has a significant degree of market power, obligations on licensees to obtain the technology only from the licensor can lead to significant foreclosure effects. The stronger the market position of the licensor the higher the risk of foreclosing competing technologies. For appreciable foreclosure effects to occur the non-compete obligations do not necessarily have to cover a substantial part of the market. Even in the absence thereof, appreciable foreclosure effects may occur where non-compete obligations are targeted at undertakings that are the most likely to license competing technologies. The risk of foreclosure is particularly high where there is

---

[67] See note 36.

only a limited number of potential licensees and the licence agreement concerns a technology which is used by the licensees to make an input for their own use. In such cases the entry barriers for a new licensor are likely to be high. Foreclosure may be less likely in cases where the technology is used to make a product that is sold to third parties; although in this case the restriction also ties production capacity for the input in question, it does not tie demand for the product incorporating the input produced with the licensed technology. To enter the market in the latter case licensors only need access to one or more licensee(s) that have suitable production capacity and unless only few undertakings possess or are able to obtain the assets required to take a licence, it is unlikely that by imposing non-compete obligations on its licensees the licensor is able to deny competitors access to efficient licensees.

201. Non-compete obligations may also produce pro-competitive effects. First, such obligations may promote dissemination of technology by reducing the risk of misappropriation of the licensed technology, in particular know-how. If a licensee is entitled to license competing technologies from third parties, there is a risk that particularly licensed know-how would be used in the exploitation of competing technologies and thus benefit competitors. When a licensee also exploits competing technologies, it normally also makes monitoring of royalty payments more difficult, which may act as a disincentive to licensing.

202. Second, non-compete obligations possibly in combination with an exclusive territory may be necessary to ensure that the licensee has an incentive to invest in and exploit the licensed technology effectively. In cases where the agreement is caught by Article 81(1) because of an appreciable foreclosure effect, it may be necessary in order to benefit from Article 81(3) to choose a less restrictive alternative, for instance to impose minimum output or royalty obligations, which normally have less potential to foreclose competing technologies.

203. Third, in cases where the licensor undertakes to make significant client specific investments for instance in training and tailoring of the licensed technology to the licensee's needs, non-compete obligations or alternatively minimum output or minimum royalty obligations may be necessary to induce the licensor to make the investment and to avoid hold-up problems. However, normally the licensor will be able to charge directly for such investments by way of a lump sum payment, implying that less restrictive alternatives are available.

3. Settlement and non-assertion agreements

204. Licensing may serve as a means of settling disputes or avoiding that one party exercises his intellectual property rights to prevent the other party from exploiting his own technology. Licensing including cross licensing in the context of settlement agreements and non-assertion agreements is not as such restrictive of competition since it allows the parties to exploit their technologies post agreement. However, the individual terms and conditions of such agreements may be caught by Article 81(1). Licensing in the context of settlement agreements is treated like other licence agreements. In the case of technologies that from a technical point of view are substitutes, it is therefore necessary to assess to what extent it is likely that the technologies in question are in a one-way or two-way blocking position (cf. paragraph 32 above). If so, the parties are not deemed to be competitors.

205. The block exemption applies provided that the agreement does not contain any hardcore restrictions of competition as set out in Article 4 of the TTBER. The hardcore list of Article 4(1) may in particular apply where it was clear to the parties that no blocking position exists and that consequently they are competitors. In such cases the settlement is merely a means to restrict competition that existed in the absence of the agreement.

206. In cases where it is likely that in the absence of the licence the licensee could be excluded from the market, the agreement is generally pro-competitive. Restrictions that limit intra-technology competition between the licensor and the licensee are often compatible with Article 81, see section 2 above.

207. Agreements whereby the parties cross license each other and impose restrictions on the use of their technologies, including restrictions on the licensing to third parties, may be caught by Article 81(1). Where the parties have a significant degree of market power and the agreement imposes restrictions that clearly go beyond what is required in order to unblock, the agreement is likely to be caught by Article 81(1) even if it is likely that a mutual blocking position exists. Article 81(1) is particularly likely to apply where the parties share markets or fix reciprocal running royalties that have a significant impact on market prices.

208. Where under the agreement the parties are entitled to use each other's technology and the agreement extends to future developments, it is necessary to assess what is the impact of the agreement on the parties' incentive to innovate. In cases where the parties have a significant degree of market power the agreement is likely to be caught by Article 81(1) where the agreement prevents the parties from gaining a competitive lead over each other. Agreements that eliminate or substantially reduce the possibilities of one party to gain a competitive lead over the other reduce the incentive to innovate and thus adversely affect an essential part of the competitive process. Such agreements are also unlikely to satisfy the conditions of Article 81(3). It is particularly unlikely that the restriction can be considered indispensable within the meaning of the third condition of Article 81(3). The achievement of the objective of the agreement, namely to ensure that the parties can continue to exploit their own technology without being blocked by the other party, does not require that the parties agree to share future innovations. However, the parties are unlikely to be prevented from gaining a competitive lead over each other where the purpose of the licence is to allow the parties to develop their respective technologies and where the licence does not lead them to use the same technological solutions. Such agreements merely create design freedom by preventing future infringement claims by the other party.

209. In the context of a settlement and non-assertion agreement, non-challenge clauses are generally considered to fall outside Article 81(1). It is inherent in such agreements that the parties agree not to challenge *ex post* the intellectual property rights covered by the agreement. Indeed, the very purpose of the agreement is to settle existing disputes and/or to avoid future disputes.

4. Technology pools

210. Technology pools are defined as arrangements whereby two or more parties assemble a package of technology which is licensed not only to contributors to the pool but also to third parties. In terms of their structure technology pools can take the form of simple arrangements between a limited number of parties or elaborate organisational arrangements whereby the organisation of the licensing of the pooled technologies is entrusted to a separate entity. In both cases the pool may allow licensees to operate on the market on the basis of a single licence.

211. There is no inherent link between technology pools and standards, but in some cases the technologies in the pool support (wholly or partly) a *de facto* or *de jure* industry standard. When technology pools do support an industry standard they do not necessarily support a single standard. Different technology pools may support competing standards.[68]

212. Agreements establishing technology pools and setting out the terms and conditions for their operation are not—irrespective of the number of parties—covered by the block exemption (cf. section III.2.2 above). Such agreements are addressed only by these guidelines. Pooling arrangements give rise to a number of particular issues regarding the selection of the included technologies and the operation of the pool, which do not arise in the context of other types of licensing. The individual licences granted by the pool to third party licensees, however, are treated

[68] See in this respect the Commission's press release IP/02/1651 concerning the licensing of patents for third generation (3G) mobile services. This case involved five technology pools creating five different technologies, each of which could be used to produce 3G equipment.

like other licence agreements, which are block exempted when the conditions set out in the TTBER are fulfilled, including the requirements of Article 4 of the TTBER containing the list of hardcore restrictions.

213. Technology pools may be restrictive of competition. The creation of a technology pool necessarily implies joint selling of the pooled technologies, which in the case of pools composed solely or predominantly of substitute technologies amounts to a price fixing cartel. Moreover, in addition to reducing competition between the parties, technology pools may also, in particular when they support an industry standard or establish a *de facto* industry standard, result in a reduction of innovation by foreclosing alternative technologies. The existence of the standard and the related technology pool may make it more difficult for new and improved technologies to enter the market.

214. Technology pools can also produce pro-competitive effects, in particular by reducing transaction costs and by setting a limit on cumulative royalties to avoid double marginalisation. The creation of a pool allows for one-stop licensing of the technologies covered by the pool. This is particularly important in sectors where intellectual property rights are prevalent and where in order to operate on the market licences need to be obtained from a significant number of licensors. In cases where licensees receive on-going services concerning the application of the licensed technology, joint licensing and servicing can lead to further cost reductions.

4.1. The nature of the pooled technologies

215. The competitive risks and the efficiency enhancing potential of technology pools depend to a large extent on the relationship between the pooled technologies and their relationship with technologies outside the pool. Two basic distinctions must be made, namely (a) between technological complements and technological substitutes and (b) between essential and non-essential technologies.

216. Two technologies[69] are complements as opposed to substitutes when they are both required to produce the product or carry out the process to which the technologies relate. Conversely, two technologies are substitutes when either technology allows the holder to produce the product or carry out the process to which the technologies relate. A technology is essential as opposed to non-essential if there are no substitutes for that technology inside or outside the pool and the technology in question constitutes a necessary part of the package of technologies for the purposes of producing the product(s) or carrying out the process(es) to which the pool relates. A technology for which there are no substitutes, remains essential as long as the technology is covered by at least one valid intellectual property right. Technologies that are essential are by necessity also complements.

217. When technologies in a pool are substitutes, royalties are likely to be higher than they would otherwise be, because licensees do not benefit from rivalry between the technologies in question. When the technologies in the pool are complements the arrangement reduces transaction costs and may lead to lower overall royalties because the parties are in a position to fix a common royalty for the package as opposed to each fixing a royalty which does not take account of the royalty fixed by others.

218. The distinction between complementary and substitute technologies is not clear-cut in all cases, since technologies may be substitutes in part and complements in part. When due to efficiencies stemming from the integration of two technologies licensees are likely to demand both technologies the technologies are treated as complements even if they are partly substitutable. In such cases it is likely that in the absence of the pool licensees would want to licence both technologies due to the additional economic benefit of employing both technologies as opposed to employing only one of them.

[69] The term 'technology' is not limited to patents. It covers also patent applications and intellectual property rights other than patents.

219. The inclusion in the pool of substitute technologies restricts inter-technology competition and amounts to collective bundling. Moreover, where the pool is substantially composed of substitute technologies, the arrangement amounts to price fixing between competitors. As a general rule the Commission considers that the inclusion of substitute technologies in the pool constitutes a violation of Article 81(1). The Commission also considers that it is unlikely that the conditions of Article 81(3) will be fulfilled in the case of pools comprising to a significant extent substitute technologies. Given that the technologies in question are alternatives, no transaction cost savings accrue from including both technologies in the pool. In the absence of the pool licensees would not have demanded both technologies. It is not sufficient that the parties remain free to license independently. In order not to undermine the pool, which allows them to jointly exercise market power, the parties are likely to have little incentive to do so.

220. When a pool is composed only of technologies that are essential and therefore by necessity also complements, the creation of the pool as such generally falls outside Article 81(1) irrespective of the market position of the parties. However, the conditions on which licences are granted may be caught by Article 81(1).

221. Where non-essential but complementary patents are included in the pool there is a risk of foreclosure of third party technologies. Once a technology is included in the pool and is licensed as part of the package, licensees are likely to have little incentive to license a competing technology when the royalty paid for the package already covers a substitute technology. Moreover, the inclusion of technologies which are not necessary for the purposes of producing the product(s) or carrying out the process(es) to which the technology pool relates also forces licensees to pay for technology that they may not need. The inclusion of complementary patents thus amounts to collective bundling. When a pool encompasses non-essential technologies, the agreement is likely to be caught by Article 81(1) where the pool has a significant position on any relevant market.

222. Given that substitute and complementary technologies may be developed after the creation of the pool, the assessment of essentiality is an on-going process. A technology may therefore become non-essential after the creation of the pool due to the emergence of new third party technologies. One way to ensure that such third party technologies are not foreclosed is to exclude from the pool technologies that have become non-essential. However, there may be other ways to ensure that third party technologies are not foreclosed. In the assessment of technology pools comprising non-essential technologies, i.e. technologies for which substitutes exist outside the pool or which are not necessary in order to produce one or more products to which the pool relates, the Commission will in its overall assessment, *inter alia*, take account of the following factors:

(a) whether there are any pro-competitive reasons for including the non-essential technologies in the pool;
(b) whether the licensors remain free to license their respective technologies independently. Where the pool is composed of a limited number of technologies and there are substitute technologies outside the pool, licensees may want to put together their own technological package composed partly of technology forming part of the pool and partly of technology owned by third parties;
(c) whether, in cases where the pooled technologies have different applications some of which do not require use of all of the pooled technologies, the pool offers the technologies only as a single package or whether it offers separate packages for distinct applications. In the latter case it is avoided that technologies which are not essential to a particular product or process are tied to essential technologies;
(d) whether the pooled technologies are available only as a single package or whether licensees have the possibility of obtaining a licence for only part of the package with a corresponding reduction of royalties. The possibility to obtain a licence for only part of the package may reduce the risk of foreclosure of third party technologies outside the pool, in particular where

the licensee obtains a corresponding reduction in royalties. This requires that a share of the overall royalty has been assigned to each technology in the pool. Where the licence agreements concluded between the pool and individual licensees are of relatively long duration and the pooled technology supports a de facto industry standard, it must also be taken into account that the pool may foreclose access to the market of new substitute technologies. In assessing the risk of foreclosure in such cases it is relevant to take into account whether or not licensees can terminate at reasonable notice part of the licence and obtain a corresponding reduction of royalties.

4.2. Assessment of individual restraints

223. The purpose of this section is to address a certain number of restraints that in one form or another are commonly found in technology pools and which need to be assessed in the overall context of the pool. It is recalled, cf. paragraph 212 above, that the TTBER applies to licence agreements concluded between the pool and third party licensees. This section is therefore limited to addressing the creation of the pool and licensing issues that are particular to licensing in the context of technology pools.

224. In making its assessment the Commission will be guided by the following main principles:

1. The stronger the market position of the pool the greater the risk of anti-competitive effects.
2. Pools that hold a strong position on the market should be open and non-discriminatory.
3. Pools should not unduly foreclose third party technologies or limit the creation of alternative pools.

225. Undertakings setting up a technology pool that is compatible with Article 81, and any industry standard that it may support, are normally free to negotiate and fix royalties for the technology package and each technology's share of the royalties either before or after the standard is set. Such agreement is inherent in the establishment of the standard or pool and cannot in itself be considered restrictive of competition and may in certain circumstances lead to more efficient outcomes. In certain circumstances it may be more efficient if the royalties are agreed before the standard is chosen and not after the standard is decided upon, to avoid that the choice of the standard confers a significant degree of market power on one or more essential technologies. On the other hand, licensees must remain free to determine the price of products produced under the licence. Where the selection of technologies to be included in the pool is carried out by an independent expert this may further competition between available technological solutions.

226. Where the pool has a dominant position on the market, royalties and other licensing terms should be fair and non-discriminatory and licences should be non-exclusive. These requirements are necessary to ensure that the pool is open and does not lead to foreclosure and other anticompetitive effects on down stream markets. These requirements, however, do not preclude different royalties for different uses. It is in general not considered restrictive of competition to apply different royalty rates to different product markets, whereas there should be no discrimination within product markets. In particular, the treatment of licensees should not depend on whether they are licensors or not. The Commission will therefore take into account whether licensors are also subject to royalty obligations.

227. Licensors and licensees must be free to develop competing products and standards and must also be free to grant and obtain licences outside the pool. These requirements are necessary in order to limit the risk of foreclosure of third party technologies and ensure that the pool does not limit innovation and preclude the creation of competing technological solutions. Where a pool supports a *(de facto)* industry standard and where the parties are subject to non-compete obligations, the pool creates a particular risk of preventing the development of new and improved technologies and standards.

228. Grant back obligations should be non-exclusive and be limited to developments that are

essential or important to the use of the pooled technology. This allows the pool to feed on and benefit from improvements to the pooled technology. It is legitimate for the parties to ensure that the exploitation of the pooled technology cannot be held up by licensees that hold or obtain essential patents.

229. One of the problems identified with regard to patent pools is the risk that they shield invalid patents. Pooling raises the costs/risks for a successful challenge, because the challenge fails if only one patent in the pool is valid. The shielding of invalid patents in the pool may oblige licensees to pay higher royalties and may also prevent innovation in the field covered by an invalid patent. In order to limit this risk any right to terminate a licence in the case of a challenge must be limited to the technologies owned by the licensor who is the addressee of the challenge and must not extend to the technologies owned by the other licensors in the pool.

4.3. The institutional framework governing the pool

230. The way in which a technology pool is created, organised and operated can reduce the risk of it having the object or effect of restricting competition and provide assurances to the effect that the arrangement is pro-competitive.

231. When participation in a standard and pool creation process is open to all interested parties representing different interests it is more likely that technologies for inclusion into the pool are selected on the basis of price/ quality considerations than when the pool is set up by a limited group of technology owners. Similarly, when the relevant bodies of the pool are composed of persons representing different interests, it is more likely that licensing terms and conditions, including royalties, will be open and non-discriminatory and reflect the value of the licensed technology than when the pool is controlled by licensor representatives.

232. Another relevant factor is the extent to which independent experts are involved in the creation and operation of the pool. For instance, the assessment of whether or not a technology is essential to a standard supported by a pool is often a complex matter that requires special expertise. The involvement in the selection process of independent experts can go a long way in ensuring that a commitment to include only essential technologies is implemented in practice.

233. The Commission will take into account how experts are selected and what are the exact functions that they are to perform. Experts should be independent from the undertakings that have formed the pool. If experts are connected to the licensors or otherwise depend on them, the involvement of the expert will be given less weight. Experts must also have the necessary technical expertise to perform the various functions with which they have been entrusted. The functions of independent experts may include, in particular, an assessment of whether or not technologies put forward for inclusion into the pool are valid and whether or not they are essential.

234. It is also relevant to consider the arrangements for exchanging sensitive information among the parties. In oligopolistic markets exchanges of sensitive information such as pricing and output data may facilitate collusion.[70] In such cases the Commission will take into account to what extent safeguards have been put in place, which ensure that sensitive information is not exchanged. An independent expert or licensing body may play an important role in this respect by ensuring that output and sales data, which may be necessary for the purposes of calculating and verifying royalties is not disclosed to undertakings that compete on affected markets.

235. Finally, it is relevant to take account of the dispute resolution mechanism foreseen in the instruments setting up the pool. The more dispute resolution is entrusted to bodies or persons that are independent of the pool and the members thereof, the more likely it is that the dispute resolution will operate in a neutral way.

---

[70] See in this respect the judgment in *John Deere* cited in note 11.

INDEX